The Alibi of Capital

The Alibi of Capital

How We Broke the Earth to Steal the Future on the Promise of a Better Tomorrow

Timothy Mitchell

VERSO
London • New York

First published by Verso 2026

The manufacturer's authorized representative in the EU for product safety (GPSR) is LOGOS EUROPE, 9 rue Nicolas Poussin, 17000, La Rochelle, France
Contact@logoseurope.eu

1 3 5 7 9 10 8 6 4 2

Verso
UK: 6 Meard Street, London W1F 0EG
US: 207 East 32nd Street, New York, NY 10016
versobooks.com

Verso is the imprint of New Left Books

ISBN-13: 978-1-83674-227-2
ISBN-13: 978-1-83674-230-2 (US EBK)
ISBN-13: 978-1-83674-229-6 (UK EBK)

British Library Cataloguing in Publication Data
A catalogue record for this book is available from the British Library

Library of Congress Cataloging-in-Publication Data
A catalog record for this book is available from the Library of Congress

To John, Chris, Nick, Ant, and Mary

Contents

Acknowledgements

Over more than a decade, I have had the opportunity to develop the ideas for this book in public lectures and seminars. I am grateful to all those who invited and hosted me and to the many commentators and participants who contributed their thoughts. The occasions include the AIME Workshop, Sciences Po, Paris, 2014; the Knowledge Culture Economy conference, University of Western Sydney, 2014; the Environmental Change Institute, Oxford, 2014; the Graduate School of Humanities and Social Sciences, University of Lucerne, 2014; the Ester Boserup Prize Lecture, University of Copenhagen, 2015; the Log Luncheon, Williams College, 2015; the Center for Global Peace and Conflict Studies, UC Irvine, 2015; the Oxford Centre for Global History, 2016; the Center for the Humanities, Wesleyan University, 2017; the Reading Lecture, Colgate University, 2017; the Cultures of Energy conference, Carnegie Mellon University, 2017; the Harvard Global History Seminar, 2017 and 2024; the Dialogue of San Giorgio, Venice, 2017; the Robert L. Heilbroner Center for Capitalism Studies, the New School, 2017; the Max Planck Institute for the History of Science, Berlin, 2017; the Mellon Sawyer Seminar on Energy Justice in Global Perspective, UC Santa Barbara, 2018; the Doha Institute for Graduate Studies, Qatar, 2018; the Klopsteg Lecture, Northwestern University, 2018; the Ibrahim Abu-Lughod Institute of International Studies, Birzeit University, 2018; the Multiple Carbons workshop, Harvard University, 2019; the Institute for Global Politics, University of Bath, 2019 and 2020; the Makerere Institute of Social Research, Kampala, 2020; the Centre for Gulf Studies, University of Exeter, 2022; the Walter

Rodney Collective for Historical Research, SOAS University of London, 2022; the Human Ecology Division, Lund University, 2023; the Eilert Sundt Lecture, University of Oslo, 2024; and the Archipelago Lecture, KTH Stockholm, 2024.

Two of the chapters of this book are revised versions of previous publications. An earlier version of chapter 5 appeared in 'Around 1948: Interdisciplinary Approaches to Global Transformation', edited by Leela Gandhi and Deborah Nelson, *Critical Inquiry*, 40: 4, 2014. Chapter 6 was originally published in *Do Economists Make Markets? On the Performativity of Economics*, edited by Donald MacKenzie, Fabian Muniesa, and Lucia Siu, Princeton University Press, 2008. Excerpts from three other chapters have also previously appeared in print or online: parts of the introduction were published in a catalogue for a 2020 exhibition at the ZKM Center for Art and Media, Karlsruhe, *Critical Zones: The Science and Politics of Landing on Earth*, edited by Bruno Latour and Peter Weibel, MIT Press, 2020; sections of chapter 7 appeared previously online in New Silk Roads, *e-flux architecture*, January 2020; and parts of chapter 8 as 'The Great EconoCon: Climate Crisis and the Misconception of Growth', *Critical Times*, 9: 1–2 (2026). I thank the editors and publishers for allowing me to reuse this material.

At Verso Books, the editorial director Sebastian Budgen has encouraged this project for many years. I am grateful to him, the production editor, Nick Walther, and all the staff at the press for their work. I was also fortunate to have the help of a series of research assistants: Rosa Klein-Baer, Dalia Ghaith, Rebecca Glade, Naye Idriss, and Noam Chen-Zion. Their input to the writing of the book was invaluable.

In 2022–3, I was invited to spend the year as Visiting Professor in the School of Social Science at the Institute for Advanced Study, Princeton. Much of this book was written or revised there. I am grateful to Wendy Brown for the invitation to codirect the School's theme seminar that year on Climate Crisis Politics, to her colleagues at the School, Didier Fassin, Alondra Nelson, and Joan Scott, to the staff, especially Miriam Harris and Munirah Bishop, and to the members who collaborated on the theme seminar, in many cases offering detailed comments on my draft chapters.

At Columbia University I have benefited over the years from discussions with colleagues in the Department of Middle Eastern, South Asian and African Studies, in particular Wael Hallaq, Gil Hochberg, Sudipta Kaviraj, Mahmood Mamdani, Debashree Mukherjee, and

Jennifer Wenzel, and from the support of the administrative staff, led by Jessica Rechtschaffer. My coeditors at the journal *Comparative Studies of South Asia, Africa and the Middle East*, in particular Anupama Rao, have been the source of numerous ideas. Many other friends and colleagues in and around New York have shared ideas and conversations, in some cases over decades, including Gil Anidjar, Talal Asad, Koray Çalışkan, Julia Elyachar, Michael Gilsenan, Robert Vitalis, and the late Brinkley Messick. I am also grateful to colleagues elsewhere who read and commented on drafts of various chapters, in particular Brett Christophers, Jennifer Derr, Jonathan Levy, Zachary Lockman, and Keith Tribe.

During the years of writing this book, numerous graduate students at Columbia have participated in my seminars on colonialism and technopolitics. Most of them contributed, directly or indirectly, to the thinking that has shaped the book. Several of them went on to write PhD dissertations from which I have directly drawn ideas, including Casey Primel, Shehab Ismail, Timothy Shenk, Onur Özgöde, Alaa El-Shafei, and Nadeem Mansour. Three of them were around throughout the writing of the book and shared many conversations: Rana Baker, Ibrahim Elhoudaiby, and Hengameh Ziai. To them I owe special thanks.

As I worked on this writing over the years, Adie and JJ Mitchell charted their own lives, along with Chloe and Jesse, while also sharing the journey of this book. Our numerous adventures together, on waters charted and uncharted, epitomise all the challenges and joys of that shared life. Lila Abu-Lughod shared every moment of these years. Her counsel, her intellect, and her love have been indispensable to the writing of the book and to every part our life together. Finally, my five siblings have always provided for every undertaking the proper mix of scepticism, humour, and support. The book is dedicated to them.

Introduction

Uber Eats – How Capitalism Consumes the Future

We live in an age in which extraordinary wealth seems to arrive from unfathomable sources. When the US firm Uber went public in 2019, the stock market set its value at $82 billion, an immense figure for a ten-year-old car service company that owned no cars and had never made a profit. To explain such events, the news media often turn to metaphors from meteorology, describing the investors' gains as 'stratospheric'. What other way to explain, for example, how the $5 million that Goldman Sachs had invested in Uber in 2011 was now worth over half a billion dollars, a return in eight years of more than 1,000 per cent? More critical commentators called the firm's value something conjured out of 'thin air'.[1]

The source of such windfalls is nothing meteorological. To understand this way of making money we need to come down to earth. While Uber is an extreme case, its mode of acquiring unearned wealth is commonplace. The mode is a defining feature of our contemporary form of life, a key to understanding how and why the methods of enrichment and impoverishment we call capitalism came into being, and a clue to grasping why we now face the catastrophe of climate collapse. The company created its value by constructing a practical means of consuming the future.

The aim of this book is to ask how we came to organise collective life on the principle of the impoverishment of the future; explore the development of this principle in the destruction of rivers, the colonising of territories, the expansion of cities, the building of infrastructure, and the burning of carbon; see how lives today are encumbered by the repayment of earlier extractions; and connect this mode of impoverishing the many and enriching the few to the climate crisis. Its aim is also to examine the political and economic language that developed to describe and order these forms of life – terms like capitalism, technology, finance, the economy and its growth – and ask whether they are adequate for making sense of how the impoverishment operates and of our inability to act in the face of climate breakdown.

Methods of extracting income from the future have been around for a long time. The device that Uber used, the joint-stock company, has existed in its current form for more than 150 years. Modern investor-owned firms first proliferated in the West in the nineteenth century to construct railways and other extended, terraforming, carbon-intensive, often imperial structures whose scale and durability, usually built at great ecological and human cost, promised their shareholders access to unearned wealth from the future. The history of joint-stock companies goes back even further, as we will see, to the armed trading corporations that European merchants began creating some three centuries earlier to conquer world trade, enabling them to colonise lands and subjugate or eliminate peoples across the Earth.

In the past, such methods of enrichment were usually the exception. Joint-stock companies could be established only by royal charter or act of parliament, the charter typically expiring after a limited number of years. The early colonising corporations led to speculative bubbles and the burdens of imperial war, and were liable to political opposition, closure, public rescue, or collapse. The older merchant networks of Asia and Africa that European colonisation sought to usurp, in particular those of the Indo-Islamic world, had placed limits on the use of business contracts for acquiring unspecified, unseen future goods, on the grounds that such speculative schemes allowed one party to extract unearned profit from another. Even so, long-distance trade across that world was routinely conducted on credit, the distance creating a delay in payment that justified a higher price – distance and delay thus operating as the very source of profit. Islamic legal practice also recognised arrangements that allowed merchants to profit from the speculative purchase of future assets, such as the widespread use of forward contracts to acquire agricultural crops cheaply in advance, from those compelled to find funds to pay taxes.[2] But the futures from which surplus was extracted were limited by the length of the crop cycle or the extent of a trans-regional trade route. With the colonial expansion of the West, and especially from the imperial age of the late nineteenth century, the investor-owned corporation became a means of controlling futures across continents and reorganising livelihoods and landscapes on an immense scale.

Today, a firm like Uber works through somewhat different methods but has a similarly expansive ambition. Following its stock market launch in 2019, the company borrowed $3 billion to purchase the Dubai-based firm Careem, in an effort to monopolise the car service business across the Middle East and North Africa – part of a plan to dominate transportation services on every continent. The geographical expansion of such firms, both past and present, serves as, and at the same time conceals, the source of their investors' wealth. What they seek to control is the acquisition of revenue from the future.

We have an everyday language for describing our economic relation to the future, using terms like stock price, interest rate, the advance of technology, and economic growth. But none of these terms explains the source of unearned income, to show how those coming later pay the bill. Nor do they explain how lives in the present are encumbered by previous extractions from the future, or how such privatised relations to the future, in which tomorrow's forms of life become assets bought and sold in the present, contribute to the destruction of a viable collective future. In fact, the language of finance blinds us to this relationship, persuading us that future human livelihoods are not the source of the gains, but our beneficiaries.

In the face of the climate crisis and other anthropogenic disruptions to the balance of Earth systems, including the destruction of river basins, the collapse of habitats, the accelerating extinction of species, and the poisoning of land, sea, air, and human bodies with synthetic plastics and agricultural biocides – and aware of the very unequal vulnerability of different human communities to these disruptions – we need to understand how this blindness is produced. Having already exceeded safe and just limits to the human modification of the Earth, we are seeing everywhere a belated acknowledgement that the future liveability of the planet must today be taken into account.[3] So we coexist with two contradictory ways of grasping the future in the present: one apocalyptic, recognising in current disasters and disruptions the immediate signs of a catastrophe to come; the other a mechanism of blindness that has brought the calamity upon us, by arranging the future as a calculable source of extraordinary wealth, enriching the wealthiest in the present by imposing debts on the vast majority. Governments appear unable to take account of this contradiction, while their actions often seem to be powerless against the agents seeking the further control of future assets, or to be working on their behalf. Even if it were possible to overcome these difficulties, the consequences would still seem unworkable. Capitalism, whatever its costs, claims to have given us growth. How could we survive under a different temporality, in which the future was not defined by a principle of economic expansion?

For as long as we have organised collective life around the principle of economic growth, there have been efforts to point out its limits: that growth is unsustainable, is mismeasured, or comes at too great a social and ecological cost. Those are important criticisms, but there is another way to see our relation to the future. Growth is not the logic of capitalist modernity, but its alibi.

In the past, we often spoke of modernity in terms of physical expansion. Historians described capitalism as a process that began in Europe and gradually expanded to encompass the world. We now see this as a partial account of changes that were never isolated in one place. The reworking of patterns of trade and credit, the intensifying exploitation of labour and the soil, and the destruction of populations and ecosystems were always occurring on a transcontinental scale. To describe this as the spatial 'expansion' of the West reflected some aspects of the reworking. But as so much postcolonial scholarship has shown, it was a product of ways of analysing change that, in measuring movements assumed to originate in Europe and move outwards across a passive geographical space, obscured as much as they reported.

Is there a similar way to revise our understanding of time? Not just to be as critical towards conceptions of history-as-growth as we are towards geography-as-expansion, but, as it were, to develop a similar kind of postcolonial perspective? To include the perspective, not only of those whose lands and livelihoods have been colonised, but of those whose future has been taken from them? To achieve this, we need to understand the diverse mechanisms of extraction from the future that have operated under the alibi of growth.

Value Over All

Capital is not something saved up from the past. As others have shown and this book will further explore, it is a capture from the future.[4] We can start thinking about this relation to the future through the straightforward case of the way a modern shareholder-owned firm acquires its value. When a company is floated on the stock market, the shares offered for sale represent a claim on the ownership of its future profits. Since the revenue is not available immediately, the value of each year's prospective income is adjusted downwards, or 'discounted', to compensate for the delay in time until it arrives. Adding up the 'present discounted value', as it is called, of the years of future profits, produces the firm's valuation.

Let us return to the example of Uber. Although most of this book is concerned with longer histories, wider cases, and other lands than those of a contemporary US tech firm, the case of Uber can serve here to illustrate some of the broader issues about capital and how we understand it. Thanks to its expansion across many of the cities of the Global South, in addition to Europe and North America, Uber will be familiar to most of the readers of this book.

At the time the firm went public, Uber had not yet made a profit. The company had been setting the price of car service rides below their actual cost, to drive competitors out of business. These subsidised operations were losing billions of dollars every year. To value the firm, financial analysts assumed that Uber would continue to expand until it achieved 'market dominance'. By eliminating alternatives, Uber and its one US rival, Lyft, could continue to claim a portion of every fare their drivers earned, taking on average a 20 per cent share, while using their growing dominance to limit the part paid to drivers and increase the cost to passengers. These assumptions suggested that Uber would stop losing money six years after turning public, and, within ten years, would be earning annual profits of almost $5 billion.[5]

An investor-owned company provides not just a claim on future profits. It is a mechanism for acquiring that promised income in the present. In offering shares for sale on the stock market, the investors who own a firm are selling a form of property, the ownership today of assets acquired from the future. This is the process known as 'capitalising' a future revenue. The windfalls the promoters earn from the sale come not from thin air, but from the political robustness of capitalisation – the method of monetising and marketing a private claim to the future.[6] The term 'political robustness' here refers to all the forms of authority, law, policing, economic reasoning, and disregard of claims for social justice or planetary futures on which the extraordinary value of a future claim depends.

The windfall represents the value of an encumbrance imposed on the firm's future customers and workers and on the communities and ecologies to which they belong. The company's profits, and thus its shareholders' dividends, depend on maintaining and indeed increasing this burden. The value of the share, and the dividend on which it depends, takes priority over any demand from workers for fairer wages, from customers for lower prices, or from communities for the protection of common goods – a priority that reflects the greater strength of the company compared to its workers, customers, and communities. This strength is the power indicated by the misleadingly narrow term 'market dominance'. The encumbrance is not a necessary cost of running a business, but a surcharge that the dominant position of the company allows it to impose. The $82-billion valuation of Uber represented the present value of such a power arrangement. The firm's drivers and passengers, and the wider populations affected by its impact on public transport and other collective goods, would repay this value, over time, from their pockets.[7]

The shareholder corporation is thus an apparatus for colonising time. It provides a means of enriching a group of entrepreneurs and financiers in the present by imposing an additional charge on tens of millions of users in the future. The windfall acquired today by those who set up the control mechanisms and arrange the credit lines out of which the apparatus is built will be paid from the incomes of those living months, years, or decades from now – in fact, as far into the future as the apparatus of capture can be extended.[8] Besides enriching its founders, the shares in a business firm can also provide a source of gain to the retail investors and investment funds that purchase and trade them, to those who charge fees for such trades, and to those who speculate in the rise and fall of their price.[9] Indeed, as the extractions imposed on future incomes increase, even the moderately well off turn to these modes of capture, relying on private retirement funds, property investments, and other appreciating assets to protect their standard of living. This compounds the imposition of costs on the future, in particular upon those increasingly unable to purchase housing or other assets.[10]

Nothing about this capture of the future is unique to the contemporary era of tech firms, venture capital, and asset-management companies. For centuries before the rise of modern business firms, there were means of placing populations in debt, typically through merchants providing credit to those experiencing sudden hardship or burdened with tax obligations. But merchants' profits arose more often from taking advantage of differences in price across geographical space than from the temporal postponement on which capitalisation depends. The scale of encumbering the future is more recent. When the modern corporate method of capturing revenues emerged over the last century and a half, the shareholder corporation quickly became what the great Norwegian American economist Thorstein Veblen called in 1923 'the master institution of civilised life'.[11]

There are other methods of extracting payments from the future, and of living under the weight of past encumbrances, several of which this book will explore. When colonising corporations began to provoke colonial warfare, the monarchs who chartered them drew on credit from the same large merchants to fund the costs of war, creating ruinous public debts. The cost of war was repaid not from current revenue, as rulers had done in the past, but by pledging the tax revenue of future years, creating what became known as the national debt – inventing, in the process, the modern 'nation' as the body accountable for this imposition.[12] Government bonds and other kinds of public debt became the largest instruments of credit creation, turning the power of taxation into an expanding apparatus for extraction from the future. Militarism continues to be a principal means in many countries of forcing populations into long-term debt, shrinking the share of public resources available to maintain healthcare, education, and other common benefits. In many parts of the Global South, the forms of national debt, swollen by dependence on international creditors, both recently and in the colonial past, have repeatedly turned countries as a whole into debt machines, enriching those who organise the supply of credit.

While military debt, foreign loans, and corporate stock markets impose costs on a population in general, there are numerous devices for creating encumbrances on specific individuals and households. About a decade after Veblen wrote about the shareholder corporation, a second 'master institution' emerged for realising future revenue in the present: the mortgage bank and the housing market. Little used in the United States, Britain, or elsewhere before the 1930s, home mortgages converted housing into another widely used mode of capitalisation. In many countries of the South, they are still used sparingly today, although strangely, turning land and housing into financial assets has been touted in recent decades as a magical solution to global poverty. The speculative development of housing has a much longer history, especially in countries like England where the land for urban expansion was often monopolised by large private estates. But new housing was typically leased or rented, with the lease of the building calculated separately from the lease of the land, and based simply on the cost of construction.[13] Housing mortgages (loans secured by the property) were not widely used, typically covered less than half the value of the home, and were usually paid off in a lump sum after a handful of years. The invention of long-term mortgages in the West, in the years before and after the Second World War, subsidised with government guarantees and repaid in monthly instalments over decades, transformed housing into an expanded apparatus for the capture of payments from the future. Speculative builders could now sell homes not at the cost of their material construction, but at the capitalised value of occupying a residence over thirty or more years.[14] In the second half of the twentieth century, as housing became by far the largest vehicle of assetisation and debt across the North Atlantic world, between 75 and more than 90 per cent of the increased price of housing was attributable not to the cost of construction but to this mode of capitalisation.[15]

As the price of housing began exceeding the cost of construction by a large factor, the real estate and mortgage industries grew to rival the joint-stock corporation as apparatuses for indebting the future and capturing a prospective revenue in the present. This unearned increment led a New York architect to describe a new building as 'a machine that makes the land pay'.[16] The idea of a payment for land reflected the fact that the increased cost of housing appeared as the higher value of development land, even though the value came not from any change in the nature of land but from the enhanced mechanisms for securing decades of prospective rent and mortgage payments, including the machinery of zoning, racial segregation, and foreclosure. The cost was charged even for a used property whose land purchase and production expense were paid off many years before. The payments came not from the land but from those needing a place to live or work. The profits of banks and property firms thus added an extraordinary financial burden on the incomes of ordinary people. In later decades, the automobile loan, the credit card, the college education, the medical bill, and many other devices emerged for converting the course of human lives into repayment schedules.[17]

Today, almost any arrangement of prospective payment can be capitalised. Any encumbrance reliably placed on the future becomes an asset that is open to being bundled, marketed, and acquired in the present at a discount, from corporate shares and bonds to credit card debt, from housing rents to infrastructure fees, from water and electricity supplies to data streams, from trading platforms and cloud services to musical and writing royalties.[18] Sold at a discounted price to investors and often traded in secondary markets, the payment stream then carries the burden of repaying this advance at full price, creating an unearned increment that the investor enjoys as 'interest'. The burden of capitalisation is not simply a charge added to the cost of capital or to the value of an asset. Capital itself comes into being through this process – not as an asset saved up from the past, as we usually imagine, but as these mechanisms of extraction from the future.

Driving Profits

The surprising thing about our relationship to the future is that we have become so blind to the way in which we impoverish it, living in the present under the burden of previous impositions. A century ago, it was quite clear to an economist like Thorstein Veblen how this method of sabotage, as he called it, operated. Business made its profits, he explained, not through innovation or efficiency, but through powers of disruption and obstruction. His younger and better-known contemporary, Joseph Schumpeter, whom we will first meet in chapter 1 as a youthful economist in Cairo, where he helped rescue a sugar company and develop a small sugar plantation, had a different but related view of capital as disruptive and an understanding of its dependence on a command of the future. Throughout this book, we will pay attention to histories of such sabotage, encumbrance, command, and disruption. Today, economists have available a different language. They transpose the method of living at the expense of those who will repay the debt into what is called 'growth'.

Two moves are made to portray the imposition of encumbrances as growth. In the case of a business firm, the first is to attribute the increasing value of the business not to its power of extraction from the future and the 'dominance' on which this depends but to an improvement in technology. The second is to measure both the windfall gained in the present and the charges through which it will be repaid in the future as equivalent contributions to a larger good – the growth of what we call the economy. While these arguments are routinely offered to explain the source of an increase in the value of a business, they reflect a wider way of thinking about the development of capitalism and the sources of unearned gain.

The first move ascribes the gain to technology. Is not the high valuation of successful companies due to innovation, which produces increased efficiencies? Are not the windfalls enjoyed by investors simply their reward for fostering innovation? Will not a firm's future employees and customers be the beneficiaries of these efficiencies and cost savings? Is not the progress of technology a wider principle governing the growth of economies and the difference on a global level between rich nations and poor?

Let us return once more to the simple but useful example of Uber. Its success, as Hubert Horan explains, cannot be attributed to any technological breakthrough.[19] Its business model, replacing company-owned taxi or limousine fleets with private cars whose drivers now carried the costs of fuel, ownership, insurance, and repair, was borrowed from rival companies. Its smartphone app was a rudimentary program 'hacked together' by outside contractors.[20] Uber did not invent the smartphone, the lithium-ion battery, the internet, GPS, cellular networks, electronic payments, or any other technology used by car firms. In fact, vital components of these systems were provided through public funding, including global positioning satellites (maintained by the United States Air Force), internet protocols (developed as a common resource by peer groups of scientist and amateur programmers), and automated clearing houses for payments, provided by the Federal Reserve Banks and other national banking services. The miniature element running most of these systems, the silicon-based semiconductor – invented in the 1950s by an Egyptian American mechanical engineer, Mohamed Atalla, and his colleagues, and said to be the most widely manufactured device in history – emerged from the need of the US military for components light enough to use in rocketry.[21] By the time Uber was founded, the coordinated use of smartphones, cellular networks, GPS, electronic payments, and other systems was spreading in almost every area of urban life, from ordering food to shopping to catching a city bus. Their use in private transportation, borrowed by Uber from earlier start-ups, was soon adopted by most car service companies and taxi firms.[22] Uber's innovation, if any, lay in transferring the cost of using these devices onto individual drivers, and in crowding the streets with so many empty vehicles cruising for rides that competing firms would struggle to survive. Uber itself acknowledged that a main threat to its business was not the arrival of new technologies but the presence of rival companies in every country offering the same service at better prices.[23]

As so often in the long history of large business firms, Uber's expansion was based not on technological innovation but largely on its predatory pricing and the manipulation of political support, intended to break down municipal regulation and to force rival firms and local companies out of business.[24] Its venture capitalist investors equipped it with a fund of $13 billion, including $3.5 billion from the Saudi Arabian sovereign wealth fund, which it used to set the price of fares well below the cost of each ride. Initially, the firm was paying $1.50 in costs for every dollar it earned carrying passengers. Later it reduced the loss, but only by increasing the share it claimed from each fare and forcing down the income of its drivers, increasing its average 'take rate' from 20 to almost 30 per cent. In some cases, the portion of the fare that Uber took could be 50 per cent or more.[25]

Rather than a new technology, this was the 'shock of the old'.[26] As often happens, the core of the new business was something surprisingly unoriginal, the century-old machinery of the private car. No machine was more important than the private automobile for building the unsustainable worlds of the twentieth century. The motorcar made the production of petroleum the world's largest industry, contributing more than any other device to the growth of carbon emissions and the threat of climate collapse. Private cars had a parallel effect on how people lived, accounting for up to 50 per cent of land use in cities for roads and parking and enabling the creation of suburbia, with its carbon-intensive modes of housing, land use, and privatised transportation. And the individual ownership of cars, by far the most expensive item that most households might purchase, generated the first and largest forms of corporate consumer finance. The car industry pioneered the creation of widespread personal debt, through which everyday lives and labour became an expanding system for funding the payment of fees and interest to banks.

Instead of developing a novel technology, the new transportation firms had found a different way to earn payments from the use of private vehicles, slotting in alongside oil companies, property developers, motorcar manufacturers, petroleum-based sovereign wealth funds, and the banking industry. A handful of global car service corporations could now promise a future in which they would extract monopoly rents from every vehicle journey.

For many decades, economists and economic historians have been appealing to the idea of 'technology' to explain the creation of great wealth.[27] There are many examples of this way of accounting not just for the profits of particular firms or entrepreneurs but for large transformations in the organisation of collective life. From Europe's colonisation of the non-European world, to the British industrial revolution; to the re-engineering of ecologies through river control, dam building, and land clearing; to the construction of railways, electrical grids, oil pipelines, data centres, and other large infrastructures of transportation, storage, communication, and power: again and again, the idea of technological advance offers the explanation for the power of capital to expand. For almost as long, historians of technology have been questioning the power attributed to the abstract force of technology. The question is not whether many devices have improved the human condition – although not always for everyone, and not without costs, not least to the very prospect of the continued viability of a planet hospitable to human flourishing. Rather, as the case of Uber shows, technical developments, even where they make life more liveable, are often jury-rigged from older components and are usually quickly shared.

What Economists Would Like the Economy to Look Like

Technical innovation does not easily account for unearned increases in wealth. Attributing these windfalls – and the encumbrances on which they rely – to technology operates as an aspect of what I call the alibi of capital, in which the relations of obstruction and control that allow the future to be encumbered are misidentified or overlooked. In this overlooking, economics plays a special role. It operates as a field of knowledge with no tradition of studying how technical objects and political relations are coproduced; or, more significantly, how the boundary between what is technical and what is political or economic is continually negotiated; or how this negotiation helps to establish both relations of power and the meanings of concepts. Instead, economic thought depends on a simple idea of technology, not only to understand the current era – captured by the very image of 'tech' – but to explain capitalism itself. Rather than seeing capital as a relation to the future, the alibi substitutes an old view of capital as funds that have been saved up from the past, to be loaned to entrepreneurs, typically to invest in the machines or technology required to make more money. In the popular imagination, and many economists' models, this technical equipment comes to stand for capital itself. The windfalls and interest payments accrued by investors and entrepreneurs can then be described as the reward given to those whose apparent 'savings' have been used to grow more wealth – rather than the charge imposed by investment funds and financial institutions that, thanks to their state-guaranteed privileges and protections, organise the command of payments from the future.

In other chapters of this book, we will introduce a number of terms for understanding the place of economics in both facilitating and obfuscating the capture of the future. These include the idea of 'technopolitics', meaning an approach to studying politics and political economy that treats technical and material processes not merely as the object of political and economic analysis but as the arrangements through which our politics takes shape; 'capitalisation' and 'economisation', referring to the ways in which goods and activities are captured and made calculable; the 'EconoCon', referring to the production of 'the economy', by analogy with Michel Foucault's reading of Jeremy Bentham's Panopticon, as a device for ordering desire and managing the future; and 'oikodicy', by analogy with the theological concept of 'theodicy', for understanding the role of economics in the justification of human suffering. Adopting a term from the study of religion is intended not to suggest that economics has come to replace the role of religion, but to draw attention, as we will see, to the parallel emergence of 'religion' and 'economy' as modern concepts, working as connected practices of justification.[28]

For now, however, let us stay with the simpler idea of Uber. Despite the evidence to the contrary, economists continue to attribute the extracting of future rents to supposed improvements in technology. In the case of Uber, the firm employed its own economists to describe the increasing command of surplus as a customer benefit. The company's monopoly provided the private data through which such claims could be made. In setting passenger fares and drivers' wages, Uber benefited from exclusive control of the information gathered from every ride taken. This enabled the firm to adjust charges according to an algorithm that calculated how low drivers' wages could be pushed, or how high passenger fares increased, to maximise its own share at every moment. Known as 'surge pricing', this evasion of fare regulation and minimum wage payments was promoted as the technological source of new value. The proprietary data from millions of private fare payers was used to construct the argument.

Economists at Uber published an academic paper in 2016 estimating the value of the benefit. Every dollar paid for Uber rides, they claimed, produced $1.60 in value, generating what they labelled imaginatively as a 'consumer surplus' of $6.8 billion a year. This figure was the difference between the fare Uber charged and the highest fare passengers might have been willing to pay, estimated from their responses to surge pricing.[29] In other words, Uber's failure to fully deploy its surge-pricing algorithm and extract the highest possible price at every instant became a benefit to those dependent on its car services. The business press and the blogs of economists promoted these findings as evidence of the novel forms of value that Uber's technology was creating. Uber paid university economists fees as high as $100,000 to author academic articles favourable to the company's claim that its method of calculating fares for riders and payments for drivers produced benefits for both.[30]

Many economists work hard to make the economy embody the truth of their ideas, including the idea that a more efficient or coercive pricing method can create value. Their writing often exemplifies the 'technologies of hubris' that deploy mechanically proficient but conceptually narrow forms of analysis to confirm and celebrate some technical improvement.[31] A prominent University of Chicago economist, who co-authored the Uber paper on surge pricing, described the firm as 'the embodiment of what the economists would like the economy to look like'.[32]

What appears as a technological breakthrough can instead be the source of new costs and inefficiencies.[33] Uber and Lyft differed from local car service firms in one important way: the new companies did not own the vehicles. Requiring drivers to use their own cars made the vehicles more expensive to own and manage, as owners could not benefit from fleet discounts for buying and insuring them or from supervised maintenance programmes. The companies set drivers' wages and terms of work but refused to class them as employees, with rights to a minimum wage or employment benefits. Owning no cars allowed the new firms to evade the laws with which cities regulated the car service industry. Municipal regulation, however imperfect, had allowed for the screening and licensing of drivers and for rules ensuring public goods, such as the requirement to accept riders travelling to poorer neighbourhoods or to accommodate those with disabilities, and for fares that assured drivers a minimum wage.[34]

There were wider costs to this undermining of municipal government, which offset the benefit of any possible technical improvements. Uber's long-term goal was to destroy not only urban car service companies but also public transportation.[35] A study for the National Academy of Sciences found that the companies' subsidised rides drew passengers away from mass transit, depriving public services of income.[36] One of the authors of that report found in another study that 60 per cent of car service users in large, dense cities 'would have taken public transportation, walked, biked or not made the trip' if the new car service companies had not been available, causing a 160 per cent increase in driving on city streets. To build their monopoly, the new firms promoted the instant availability of vehicles, which relied on surplus drivers cruising the streets waiting for rides, clogging roadways with cars at the expense of pedestrians and cyclists and increasing air pollution and the burning of fossil fuels. The report summed up the impact as 'more traffic, less transit, and less equity and environmental sustainability'.[37] A further study showed that the introduction of ride-sharing services in American cities also increased the rate of vehicle accidents and fatalities, including pedestrian deaths, helping to reverse the long-term decline in American traffic fatalities. Using what the US Department of Transportation calls the Value of a Statistical Life (VSL), the annual cost of the increase in fatalities associated with ride sharing was estimated at $9.48 billion per year – a cost included nowhere in the calculation of company profits or technological improvement and a dramatic curtailing of futures for the tens of thousands who were injured or lost their lives.[38]

The future profits of the new companies would arise not from technical efficiencies or improvements in collective welfare but from the opportunities for building monopoly power and from political campaigns to protect it. Uber attempted to expand its monopoly into general transportation services, launching a food delivery service in 2015 called Uber Eats. But delivering food was harder to monopolise than delivering people, and, within five years, the company was abandoning the service in many markets, including Egypt, the Gulf, and India. Expansion offered no technical innovation and no means of turning a loss-making company into a machine of future profits, except where the firm's political power enabled it to extract a rent payment from future drivers and customers. What Uber eats is the future.

Growth

What about the second move for misapprehending our mode of living at the expense of the future: describing our relationship to the future as growth? The capture of prospective revenue in the present reflects the expectation of later profits. As business grows, it will create 'economic growth'. The income may come from the future, but it is a future that will surely be larger and more prosperous. The windfall in the present is a reward, this view explains, for the entrepreneurs who engineer growth and create greater prosperity for all.

Like technical improvement, economic growth is an alibi of capital, a mode in which our relation to the future is misrecognised. We can distinguish two aspects of the alibi: the growth of the individual business firm, and what appears as the growth of human society as a whole. We have come to reckon the second in terms of the first, measuring the human collective as if it were a collection of business firms. The unusual name we give this collective firm, as we will explore later in this book, is 'the economy'.

The first aspect of the alibi is the individual firm and its investors. The original entrepreneurs sell shares to other investors, who acquire ownership of the company's future revenue. They are unlikely to enjoy the windfall gained by the founding investors, and risk losing money if the stock market decides the founders' estimates of future earnings have been exaggerated. But, to attract them as buyers, they are offered the future rents at a discount.

The discount is calculated by considering what the buyers of stock might have earned by purchasing shares in another business. By convention, the value of that foregone earning is assumed to be the amount that banks would charge for extending credit to a firm – the rate of interest. The concept of 'interest' is a modern way of describing the so-called time value of money, a value that we now take to be a natural property of money. As this book will show, it is better understood as a product of modern arrangements, such as the joint-stock company, that reliably postpone income payments and a banking industry that bundles the future payments into credit money available in the present.[39] Without such mechanisms of reliable postponement and aggregation there would be no time value of money. In fact, given the way most money is created today – through the supply of credit by banks – there would be virtually no money.

The practice of discounting the future is a political technology. It has a long history, to which we will return in chapter 3.[40] The method of devaluing and purchasing a future revenue is usually described in reverse. The ordinary investor understands it not as the purchase of money at a discounted price, but as an 'investment' in the present that somehow 'grows' in value over time. The term growth suggests some kind of material expansion. But nothing is required to increase in physical size or complexity for such growth to occur. If anything, it requires something to shrink: the future revenue is acquired at a fraction of its value. This shrinkage is produced by organising the power of postponement. Money does not possess this ability to purchase future income at a discount by nature. The ability is derived from the fact that time can be controlled – by the construction of a political apparatus that will reliably capture and colonise the future. We have come to inhabit a world governed more and more by such arrangements. We diminish the value of the future by developing mechanisms to acquire it cheaply in the present, then experience the path to that purloined future as 'growth'.

The other problem with the idea of growth is that it is seen not just as a feature of individual business firms, but as the collective trajectory of society. To manage the control of time, a larger political frame has been constructed that helps stabilise the sometimes unsteady temporality of the business firm. Invented only in the mid-twentieth century, this supporting armature is called 'the economy'. We usually think of the economy in spatial terms, as the sum of all monetary transactions within a given geographical territory. But the economy is also a kind of time machine, a way of arranging our relation to the future – indeed of producing the future, as a frame inside which our lives are to be managed. Like the value of a business firm, the nature of the economy is to appear to 'grow' – to expand year-on-year. As with the firm, such growth hides the fact that more and more future income has been acquired at a discount. Its subsequent repayment, at full price, turns a mode of consuming the future into what appears as an increase in size.

When households purchase and consume material goods, this consumption is measured as part of the economy. But when they pay the encumbrances imposed by monopolistic firms, the interest payments charged by banks and mortgage companies, the debts incurred for education and healthcare, and every other expense and fee imposed for increasingly privatised and monopolised services, those escalating payments all count towards the measurement of 'growth'. In fact, in the United States and many other countries, a significant part of so-called growth is accounted for by rents, fees, and surcharges of this kind, which are now an increasing part of the cost of creating goods or providing services.[41]

We live in a world organised to place the future in debt, with the income discounted to the creditor and later repaid in full by those encumbered or indebted. The difference between the initial discount and the subsequent full repayment is measured as a growth in the economy and mistaken for an improvement in collective well-being.

These contemporary forms of finance and debt are sometimes described as 'financialisation', a term that contrasts such costs with the needs and productivity of the 'real' economy – the production of actual goods and services. The idea of financialisation is not a term we will use here, for two reasons. First, credit money and debt are never conjured 'out of thin air'; they require building a real apparatus for the capture of future payments, the control of which is the source of all credit money. Uber's attempt to generate financial profits through a material control over the way millions of people move physically around cities is nothing unusual in this respect. Second, this opposition between the financial and the real is precisely the misleading distinction whose production we have to understand. An important study of financialisation defines it as 'the tendency for profit-making in the economy to occur increasingly through financial channels rather than through productive activities', adding that the term productive refers to activities that involve 'the production or trade of commodities'. As an example, the fact that Ford and other car makers were now making more profit from their financial arms, providing loans for the purchase of vehicles, than from manufacturing them, suggests that we can distinguish profits realised on the loan from profits made on the sale of the vehicle.[42] However, although the distinction between finance and production is a real distinction, it is nothing new. As this book will show, capitalists have always derived their profits from financing and producing the future forms of life on which their products depend. Creating the distinction between a future form of life and the 'product' that requires it, and hence between 'product' and 'finance', between real things and mere money, was as important to, say, the colonising of the non-European world, carbon-powered industrialisation, or the rise of the modern joint-stock company as it has been to more recent developments. In fact, as chapter 3 proposes, what we call the industrial revolution can be seen as a detour. It happened as a disruption to the mechanisms of credit creation, albeit a disruption and detour that tipped so-called financialisation onto a new trajectory. Credit apparatuses are always constructed out of real forms of life – including, as we will shortly see, such projects as the reorganisation of a river system or other imperial projects of terraforming and ecological destruction.

Growth appears to us as the unfolding of a human trajectory through time, driven by the forces of modernisation. We use terms like capitalism or globalisation to name the power of what propels us forwards. These terms can make the idea of growth seem like something natural and inevitable. They also make it difficult to see beyond. Operating with a naïve sense of historical time, in which the future is no more than the unfolding forwards of the past and the present, we fail to understand the future as a political and material framework – a framework constructed not from natural increments of time but from the techniques of capture and technologies of calculation through which we live at its expense. Escaping the problem of growth seems to require reversing the very movement of history.

Many accounts of the climate question, whose origins we will consider in chapter 1 and to which we will return towards the end of the book, blame our predicament on the phenomenon of growth. They point out, correctly, the inadequacy of most efforts to reduce the burning of fossil fuels and other actions destructive of the biosphere. But we often attribute these failures to a general relationship to the future that we identify, misleadingly, through the alibi of capital, as the very logic of our history.

There is no denying that some processes have unfolded at accelerating rates, such as the extraction of coal and oil. But we also know that other things decrease, such as the extent of rainforests or the amount of most people's leisure time. Perhaps it is time to use a word like growth in more limited ways, to capture some changes and not others, and therefore not as a general term for our relationship to the future.

In doing so, we would discover that the most widespread use of the idea of growth in contemporary politics, derived from finance and economics, denotes not the collective movement of society but the obscure conventions of business accounting. Those conventions describe a mode of living at the expense of the future. Through the distraction and misdirection of an alibi, they blind us to the way that future is diminished.

Capital is not money or wealth, despite the way that popular accounts often portray it, nor is it the assets or machinery that can be deployed to produce wealth, as economists usually describe it. Capital is the power to encumber the future. The term 'alibi of capital' refers to ideas and devices that distract, divert attention from, or disguise the encumbering. These include the concepts of growth and technical improvement, explored in this introduction, but also money, interest, economy, and several more that this book will examine. An alibi is a (mis)direction. Literally an 'elsewhere', an 'in another place', it can have both a temporal and a spatial aspect. These include, as we will see, a misunderstanding of how we produce the future in relation to the present, but also the geographical distraction of 'the West'. However much we globalise the history of capital or try to provincialise the place of Europe in that history, this history recentres itself in the Euro-American world.[43] Throughout this book, at different instances, we will consider this 'elsewhere', including at key moments histories of the Indo-Islamic world. In fact, more than half the chapters will draw on the experience of Egypt, a place where histories of the capitalising process and its alibis can be vividly discerned. The 'alibi of capital', finally, is also just that: capital itself as an alibi. Instead of grasping an uncertain process of domination, the capitalising of the future, too often we substitute a structure, 'capitalism'. We treat capital not as the modes of commanding the future, but as a thing, an asset, a value. In distracting us from understanding the multiple ways it comes into being, capital has become its own alibi.

1

Groundbreaking

Somewhere in the recent past, the people of the North Atlantic world lost their sense of time. They began to imagine the future as an infinite resource, and to organise collective life at the expense of those who would come later. Strangely, they always seemed aware that their relationship to the future was flawed. Every decade of the last century brought new warnings about the exhaustion of resources, unsustainable rates of economic growth, and irreparable damage to the Earth. But they framed the problem mainly as one of miscalculation. The economic system, they said, failed to take account of nature. In calculating costs and benefits, it largely excluded the damage done to an object outside themselves, the natural world. If they were living at the expense of the future, the appropriation seemed to happen as an inadvertent extraction from the Earth.

There is a different way of understanding how future lives and ways of living are impoverished. The loss of the future is not the result of a failure to calculate. We live within a set of arrangements organised to calculate the future, establish its value, and allow some to extract and enjoy that value – not in a time to come, but in the present. Rather than neglecting the future, our mode of life is oriented towards it. But we are oriented in a particular manner, through specific terms of engagement. In the same way, we do not ignore the physical and biological world. But we relate to it through concrete practices and understandings. We express our relationship using terms like nature, technology, economy, and growth. Those words, which acquired their current use and meaning largely in the twentieth century, blind us to the way we live at the expense of others.

How did we become uncomprehending of the ways in which we are rendering the Earth inhospitable to human flourishing? How did we develop arrangements that cause us to misrecognise our relationship to the future? To find answers, we can start by looking back to the beginning of the twentieth century. While ways of thinking about the natural world, technical processes, and economic forces have much longer histories, at the turn of the last century new forms of political and technical life gave such ideas a different shape. Those forms of life were produced at the high-water mark of European imperial expansion. It is in the events of empire that one can most clearly see them appear.

Civilisation Must Have Its Victims

In 1902, British hydraulic engineers in Egypt, a country Britain had invaded and occupied only two decades earlier, completed the first stage of the destruction of the River Nile. They supervised the building of a dam at Aswan, 500 miles south of Cairo, intended to hold back most of the river's summer floodwaters. The aim, the engineers said, was to replace an ancient mode of irrigation based on capturing the flood in field basins, which allowed only a single crop each year, with year-round water storage that would allow perennial irrigation and the growing of up to three crops a year. Storing the water behind the dam would remove the threat of exceptional floods, which caused crops to fail when floods were low and devastated the fields when extremely high. Regulating the uneven flow of the river would also reduce the financial uncertainty caused by the unpredictable extent of cultivation. Combined with new canals bringing marginal lands into productive use, the expanded cultivation would enable the feeding of a growing population, while allowing the country to meet the demand for industrial crops like sugar cane and cotton. Even today, histories of the Nile and accounts of the country's modern development repeat this narrative of necessary improvement introduced by the hydraulic engineers.

Like so many accounts of the benefits brought by colonial modernity, almost every element of that story is misleading or wrong. Year-round irrigation did not replace the growing of a single annual crop with multiple cropping, but accomplished almost the reverse, replacing three-season agriculture with the widespread cultivation of a single crop. The risk of catastrophic droughts and floods was not an ancient problem solved by new engineering but mainly a modern threat exacerbated by irrigation works. Financial uncertainty was not an unevenness to be removed but an effect to be engineered, to create the possibility of extracting in the present payments from the future. Rather than bringing unproductive lands into cultivation, the new works altered the way marginal lands were used and rendered every field in the country less and less productive, as the symbiosis of flood, soil formation, crop rotation, macrofauna, and microbial life was disrupted and destroyed. Rather than increasing in yield, most crops declined, as the flood-borne silt no longer fertilised the land, with the exception of one plant, water hyacinth, a mat-forming aquatic weed from the Amazon introduced as a garden ornament by colonial Europeans, which took advantage of the unusually still waters created by the dam to gradually clog with dense matting every small canal and waterway north of Cairo. Rather than providing food for a growing human population, whose diet had worsened as food crops gave way to cotton, the elimination of the flood turned humans themselves into hosts, feeding an expanding population of parasites including those carrying hookworm disease, malaria, and bilharzia. But the most misleading claim is the main one, that damming up the river eliminated ecological risk by introducing the year-round storing of its waters. The dam and reservoir that the British built at Aswan did not initiate the storage of the waters of the Nile. They stored them differently, reducing and eventually destroying a much larger, more diverse, but less visible means of storing water and managing the uncertainties of climate, rainfall, and flood.[1]

Some of these negative consequences have long been recognised, although usually as unavoidable side effects. 'Civilisation', remarked Lord Cromer, the man who managed Britain's occupation and celebrated the collaboration of the financier and his 'powerful ally', the hydraulic engineer, 'must, unfortunately, have its victims.'[2] However, no modern account has explained the working of the previous system of water storage in field basins – and, as we will discover, beneath them – or how it was destroyed. For that one must turn back to earlier sources, including the writings of medieval Arab historians on the agriculture of the Nile. Part of the reason why modern accounts fail to understand the ecocide of the river is that they participate in another view of the world, a view shaped by the same events. That view is based on an alternative practice of storage, and an alternative practice of uncertainty, forming a different relation to past and future, a difference fundamental to building a colonial-modern state and modern capitalist enrichment. The engineering of the Nile helped give rise to a novel way of relating to the future, one that came to be misrecognised as 'economic development'. In fact, in an unusual intersection of the history of hydraulics and the history of finance, the destruction of the Nile played a minor but telling role in the very birth of modern economics and of its powerful method of misidentifying our mode of living at the expense of the future.

To uncover these connections between ecocide and economics, between the flows of irrigation water and modern financial flows, let us start with what was different about the engineering of the dam. How was a world to be ordered around extraction from the future through its capitalisation, or the ability to realise that unearned prospective income in the present? How was this extraction, and the breaking of the earth and destruction of life on which it rested, misrecognised and presented, through the birth of economics, as the 'development' of 'capitalism'?

Durability

All forms of life modify the elements around them as they interact with one another and appropriate the materials they need to survive. Plants take nutrients from the soil, which in turn they help create, both through their own decay and through their interaction with the microbial life on which they feed; soil fauna such as nematodes (roundworms) can alter root structures, transport water, and manufacture tunnels and air passages through the soil; larger animals consume plants or other animals, and in many cases make shelters from the same materials.[3] The nests of harvester ants become large granaries, in which they store the seeds that they gather, providing seed dispersal for the plants from which they harvest, and thereby altering local ecologies. Some animals create new ecosystems, as with beavers constructing dams or corals manufacturing underwater reefs. Humans do the same, on what seems at first to be an altogether different scale.[4]

The mass of human construction and modification of the world is astonishing. Large dams alone have redistributed the volume of so much water that building them has changed the speed and axis of the Earth's rotation.[5] The weight of sediment that dammed rivers once transported is now dwarfed by the human removal of earth and rock to construct roads and buildings and extract minerals. The 'overburden' removed to break apart the earth and access coal, iron ore, and other resources greatly exceeds the volume of the mineral itself, in the case of coal by as much as twentyfold. This anthropogenic transport of sediment is estimated to be twenty-four times the weight of sediment the world's major rivers now carry to the sea.[6] Geologists suggest we recognise the geomorphological driving force of human construction and modification of the lived world as the making of a fifth sphere of the Earth – alongside the four classical spheres of atmosphere, hydrosphere, lithosphere, and biosphere – for which they propose the name 'technosphere'.[7]

The technosphere refers to the large-scale arrangements of technical equipment, human sociality, and material life, including the apparatuses that allow energy, information, and materials to flow, encompassing power stations and transmission grids, roads and houses, offices and warehouses, industrial plant and machine tools, coal fields and oil wells, farms and commercial fishing grounds, plantations and managed woodlands, and the dammed and rerouted channels of great rivers. While the word technosphere suggests industry and infrastructure, human terraforming occurs mainly in farming and forestry, which together have modified 38 per cent of the Earth's land area, compared to 1 per cent occupied by buildings and infrastructure.[8] A preliminary estimate suggested that the mass of the technosphere might be as much as 30 trillion tonnes.[9] Humans themselves are said to have a biomass of about 300 million tonnes – no small figure, representing about twelve times the biomass of all wild terrestrial mammals, although humans are outweighed in turn by their domestic animals – but representing a miniscule part of the technosphere.[10] The mass of the technosphere, in these estimates, is five orders of magnitude larger than the human biomass. Humans have built or modified a system of cities, roads, houses, factories, canals, commercial forests, dams, farmland, and re-engineered floodplains that might be 100,000 times their own size.

Still more astonishing, however, is the difference between the extent and mass of the technosphere and the miniscule term with which – for just the last seventy-five years or so – we most often try to grasp its workings: 'the economy'. Conventionally we think of the economy as a macro-object, analogous to the other seemingly large objects that the social sciences describe, such as state, society, culture, and even nature. The economy purports to be a way of measuring and understanding the world's industrial processes, infrastructures, food systems, housing, and other determining aspects of human material life. But the impression of something large-scale is misleading. The economy came into being thanks to the spread of modes of payment, pricing, and estimating that made it possible to exclude most of technical and natural life from consideration in the definition and measurement of our material existence. The economy, as we will later explore, was measured by estimating monetary payments, without even asking how the mechanisms of payment were built or locating where in the future the funds for those payments were manufactured. In comparison to the technosphere, we should think of the economy not as a macro-object, but as something remarkably small.

Other terms that try to capture the force of human impact on the Earth, such as the environment, are not much better. They imply a natural world that human action endangers or seeks to manage. But humans do not exist apart from the technosphere, as the image of a human organism and its environment suggests. The technosphere forms a mixed assembly of elements that are human and nonhuman, biotic and abiotic, material and political.

While the idea of the technosphere depicts processes that are much larger than the system of prices and exchanges that we measure as 'the economy', even here to focus on its scale can mislead us. In colonial Egypt, for example, the new dam and irrigation works operated mainly on the surface, ignoring the interaction between surface water, waters stored beneath the soil, and the complex life forms of the soil and subsoil itself. The calculation of the mass of the technosphere assumes that in agriculture on average only the top four to six inches of soil should be included in its measurement.[11] The technosphere refers only to a small part of the 'critical zone', the sphere of animate and inanimate exchanges that encompass and support forms of life, and itself a thin layer forming just a skin around the surface of planet Earth.[12]

Rather than its scale, what matters most in the human technosphere for our purpose of understanding modes of extraction from the future is something different: its peculiar form of durability. The modifications made by other organisms to the elements with which they interact are seldom stable: things are consumed, undergo decay, revert, transform, or develop further. The resulting recycling of materials, in which one organism's waste or decay provides nourishment for others, creates self-sustaining feedback loops that have supported forms of life over millions or billions of years.[13] Much of the technosphere, on the other hand, is assembled out of more stable combinations, designed to reduce exchanges with other elements in order to resist ageing, decay, and recycling. It consists of arrangements and composites – concrete, steel, and plastics, for example – that often have the capacity to minimise degeneration for years or decades. This gives the technosphere a distinctive temporality.[14]

A large part of the technosphere has been produced or shaped in just the last 150 years, helping to define the experience of modernity. The change is particularly apparent in the period of roughly two generations before the First World War.[15] Those years saw a gradual shift in the production of composite materials such as steel and concrete, and the introduction of other materials and machinery associated with their use, such as steam shovels for excavating canals, high explosives like dynamite for rock blasting and dam construction, synthetic chemicals for use in agriculture and munitions, and the steam turbine, whose steel blades allowed the production of electric power at scale, transforming cities through the provision of lighting and expanding them upwards and outwards with electric elevators and electric-powered trams and light-rail systems.[16]

The new composites played a noticeable role in the peculiar durability of the technosphere. Take the example of steel. This alloy of iron was not a new material, having been known to the ancients and used for centuries in making weapons and tools that required a durable blade or resistance to corrosion. But from the 1860s, minor changes in the steel-making process made it more affordable. As late as the end of the American Civil War, in 1865, only about 1 per cent of pig iron (the ingots of cast iron) in the United States was converted to steel; by 1906, the conversion rate was 100 per cent.[17] Steel then played a key part in building fossil fuel systems (mining and drilling equipment, oil pipelines, refineries), electrical power generation, chemical processing plants, steel-frame buildings of enormous height, bridges of unprecedented span, weaponry such as the Maxim self-loading machine gun, ocean-going shipping, and railways.

The building of rail lines, which accounted for the largest use of steel in this period, illustrates the effects of this durability. During the early years of steam railway travel, before the 1860s, rails were made of wrought iron, which carried heavier traffic than the perishable wooden tracks used by horse-drawn railways but was liable to fracture under repeated loads and needed replacing every six to twelve months, and later every thirty-six months. Fractured rails were a frequent cause of train accidents, and their inspection and replacement were a major cost. With the switch to steel after the 1860s, the replacement period increased threefold, the rails lasting as long as a decade before replacement was required. The problem of rail breaks also spurred the development of machines and standards for testing steels, which further improved durability.[18]

The persistence of technical forms and synthetic materials made possible a different relationship between periods of time, one in which the promoters of new schemes in the present could develop arrangements to profit in large and novel ways from lives in the future. These arrangements, while not entirely new, represented a significant expansion of modes of human enrichment and impoverishment. It is a change that most accounts of capitalist modernity fail to understand.

A Device for Taxing the Future

The conventional way to explain how transformations of the technosphere altered the organisation of economic life starts from an increase in 'demand'. Large structures and technical systems provided the opportunities, it is said, to meet expanding human *needs*: for faster travel, communication over greater distances, the increased transportation of materials, the building of taller and larger cities, and the colonising and farming of undeveloped or less intensively exploited lands. These demands, in turn, provide the explanation for a profound change in political-economic organisation in the period after the 1860s. Grand engineering and transportation projects required an expanded managerial control to handle their complexity, we are told, and great quantities of capital. Those requirements brought forth a novel apparatus, one that was to become a powerful political force over the following century and a half: the large, investor-owned business corporation.[19]

Expansion of the technosphere – and the rise of large business – was typically explained by the promoters of such schemes in terms of demands to be met. They promised that railways, ship canals, energy infrastructure, land development, and other projects would allow the movement of great new quantities of people, food supplies, and manufactured goods. In fact, however, many of them – the transcontinental railways of North America, even the Suez Canal when first opened – often struggled to find users and pay their way. They were typically over budget and often detrimental to human communities and to wider systems of life. The projects were in fact often built for a different purpose: not primarily to move people and goods, but, as a study of the American transcontinental railways shows us, to move paper.[20] Let us consider more closely how infrastructure moves paper – by becoming the instrument for the manufacture of debt, creating vehicles not for the movement of people and goods, but for the making of unearned income.

The long-distance, steel-formed railway, the densely built and electrified city, the re-engineered river system, and other large, durable arrangements created not so much a need to raise capital to pay for them, although such payments had to be found, typically through further systems of credit money: the building of US railways, for example, was financed not from savings but through debt created by banks, including via mortgages on land grants, to be repaid by future users of the land; in Europe, those building railways, mines, and other carbon-intensive infrastructure often paid their contractors and suppliers in 'vendor shares', giving them a claim on prospective profits; and in Egypt, as we will see, the government paid for the Aswan Dam by allowing its chief promoter to issue a new kind of money, whose value was based on future tax receipts.[21] Rather than requiring significant funds accumulated from the past, the durability and scale of such projects created a reliable and plentiful stream of revenue from the future, in the form of passenger fares, freight charges, shipping tolls, urban rental incomes, mortgage payments, and increased rents and taxes on agricultural land. The promoter of the project, armed with the proper political and legal guarantees, could offer to sell shares in that prospective income in the present. The opportunity to sell future claims to income in the present, expanding rapidly with the development of steam-powered transportation and coal-fired electric power, represented not a *need* for capital, but a means for its carbon-heavy *creation*. All our conventional ideas about capital, that alibi imagined as a fund accumulated from the past, collapse once we start to study how capital is created.[22]

The sudden explosion in the number of joint-stock firms, commercial banks, and land development companies in the late nineteenth century, in Europe and North America but also in places like Cairo and Istanbul and other areas outside the North Atlantic world, represented an expanding method of capitalisation. The modification of the technosphere had become a system not just for the human colonisation of the common world, but for the command of revenue from the common future. The creation of capital, and the relation between capital and labour, which had always operated as claims upon the future, could take on an extended temporal structure, thanks to the wider possibilities now available to build more durable apparatuses for generating and controlling in the present a stream of revenue obtained from the future.

Adopting a term from Deleuze and Guattari, we can think of the joint-stock company as an assemblage that operates as an 'apparatus of capture'.[23] The phrase draws attention to the multiple forms of material, power, justification, measurement, calculation, and coercion on which a capture of the future depends. The apparatus is neither fully public nor private but combines aspects of both; is made up of both materials and ideas; deploys both law and violence; and depends upon both careful calculation and the imaginative construction of prospective worlds. Financiers and railroad economists, especially in the United States, developed the tools of calculation, and the statistical data on miles of track, consumption of coal, and volumes of passengers and freight carried, that were used to estimate the prospective income, and calculate its present value. Publicising and puffing a venture through newspapers and advertising campaigns, promoters discovered they could sell shares at many times the cost of building the apparatus. They were no longer selling just a present arrangement of assets and equipment. Thanks to its claim on later revenues, the apparatus had become a machine – technical, legal, and political – for the capture and transfer of future income.

The payment of a dividend to share owners became a charge carried by the enterprise into the future. The charge was now a cost of doing business, to be considered in calculating – in the case of railways, for example – future rates for transporting passengers and freight and future wages and benefits of employees. The carrying of those costs became what the great German social economist Max Weber diagnosed as a tax on future users. In a detailed investigation of the Hamburg and Berlin stock exchanges and other European financial markets, published in 1894, he described how both workers and passengers, as well as the purchasers of goods shipped via rail, were now 'taxed' by the extraction of these contemporary forms of 'tribute-duties'.[24] As an apparatus that organised claims to this revenue, the joint-stock company operated as a machine for obtaining tribute from the future, moving prospective income from the general population into the hands of entrepreneurs and investors in the present.

Time Machines

Investor-owned companies had existed in earlier forms, but their proliferation in association with new coal-driven infrastructure projects and the expanding colonisation that accompanied them was something rather different. From the sixteenth to the eighteenth centuries, numerous large joint-stock companies were established in Europe, usually by legislative or royal charter, as an instrument of state-sponsored trade and colonisation.[25] Compared with earlier companies of merchant adventurers or leagues of European cities, which established for their merchant membership trading monopolies in adjacent regions like the Baltic and the Levant, or the even older forms of merchant partnership that had managed trade across the Indo-Islamic world for centuries, the chartered European joint-stock companies built an unusually mobile and lethal apparatus of capture, controllable over long distances.[26] Adopting the versatile lateen sail from Arab seafarers, along with Arab ship design and navigation instruments, and combining them with weaponry, discipline, and a revived Crusader ideology, they built transoceanic armed trading vessels, along with the fortified overseas harbours where ships could take refuge, trade, and refit.[27] Before the age of coal and steel, ocean-going ships were the largest and most costly technical devices in regular production, while violent, monopolistic colonial trade was a principal field of profiting from the future, through offering shares in armed voyages, discounting bills of exchange, and selling insurance contracts. The shipping of enslaved Africans across the Atlantic and of sugar from the Caribbean plantations in which they were put to work created additional and longer-term opportunities for capitalising future revenues.

These earlier forms of joint-stock company did not endure.[28] Their coercive and monopolistic ventures tended to be costly and unstable. Their expansion led to frequent warfare among the European states that chartered them and with the Ottoman and Mughal Empires and other powers outside Europe. The opportunities to profit from trade – from the difference in price of the same good in one location and another – were limited by competition from overland routes and the lack of European goods that were valuable enough to be shipped in return. The shares in the companies, through which the claims on future profits from trade were converted into tradeable assets, were held initially among a limited group of large merchants, who traded the stock mostly among themselves at company meetings. Only at the turn of the eighteenth century did a wider market in shares develop, and crises soon followed.[29] The ability of colonising companies to sell shares on the promise of future revenue led to intensive rivalries and spectacular financial crashes, most famously the collapse in 1720 of both the South Sea Company in England and the Mississippi Company in France. By the end of the eighteenth century, even as their numbers proliferated, many of the colonising joint-stock ventures had failed, or were being re-engineered as something stronger: the governments of imperial dependencies, as in India, the Dutch East Indies, southern Africa, and the islands of the West Indies; or independent settler-colonial federations such as the United States. Meanwhile, most domestic European businesses continued to operate as family-owned firms or partnerships. Joint-stock companies reappeared in brief frenzies of speculation around a new infrastructure, such as the building of toll roads, inland shipping canals, copper and coal mines, and the early railways. These joint-stock ventures tended to be closely held and short-lived.

The modern business corporation did not evolve naturally from such earlier forms. It emerged unexpectedly, in the later nineteenth century, initially for one main purpose: to profit from building large railways. The great majority of shares traded on the stock exchanges of London, Paris, Hamburg, and New York were in railway companies, along with businesses and investment banks associated with railway construction.[30] Telegraph firms, steamship lines, electric power and light companies, and iron and steel companies also traded shares in future profits. The large business firm, the politico-legal apparatus that would dominate the politics of the twentieth century, had its origins in the novel mode of extraction from the future inaugurated by carbon-fuelled, steel-built, terrain-commanding infrastructure.

What distinguished these businesses was not so much their size or complexity but their mode of persistence in time, allowing the capture of prospective revenue over extended periods. The longevity arose in part, as mentioned, from the creation of infrastructure from repurposed materials, such as steel and concrete, which promised years or decades of service. However, the durability also arose from new forms of political strength. The scale and extension of transportation lines and energy networks, along with urban development, irrigation projects, and other immense schemes, depended not just on material durability but upon durable arrangements of law, authority, compulsion, and justification. The resilience of new assemblages was a question not merely of technics but of technopolitics.

The political transformations accompanying the rise of joint-stock companies had several components. There was the development of novel legal forms. Two changes in law are often noted: the gradual removal of the requirement for an act of the legislature to establish a joint-stock company and, around the 1860s, the 'extraordinary expansion of the principle of Limited Liability'.[31] Two other legal innovations, however, were more significant but are often overlooked. First, the share itself became a form of property. Previously, to purchase stock in a company had been to acquire a share that represented the part ownership of, say, a vessel and its cargo or a railway company's yards, sheds, rolling stock, and lines. The share was not itself a piece of property in law, but, rather, a way of recording a claim on this material asset. By the later nineteenth century, the share had changed into a form of property in its own right and of a new kind. It had become a means of owning part of the company's future earnings, its value bearing no direct relation to the material assets of the business.[32] The stronger the forms of policing, political power, imperial threat, financial calculation, insurance, and economic justification that protected it, the greater the promise of future revenue and thus the more valuable the share.

Second, remarkably, this fundamentally new form of property was made perpetual. Other 'paper' property rights established through analogous legal means, such as the patent on an invention or the copyright on a literary or artistic work, were limited in duration, lasting a fixed number of years. Before the mid-nineteenth century, the same was true of a joint-stock company and its shares. The right to incorporate as a public company was, in most cases, an arrangement that expired. Corporate charters were awarded for a limited period, typically twenty or thirty years, with a shorter limit in the case of banks. The charter of the Bank of England, a joint-stock company, expired every eleven years.[33] Other charters expired once the shareholders had recouped their initial investment. In Massachusetts, for example, the Turnpike Corporation Act of 1805 allowed the legislature to dissolve the corporation when earnings from the tolls on road users reached the cost of construction plus 12 per cent, after which the turnpike became public.[34] By the late nineteenth century, the share had acquired the possibility of perpetual life.[35] While it carried the risk of losing value or even, with the collapse of a company, expiring, as a right to future earnings or yields it became analogous not to a lease, copyright, patent, or other time-delimited claim, but to a perpetual right of ownership, similar in this way to the freehold of land or other real property.

The expanding apparatuses for the capture of the future helped produce another novel form of property: the ownership of risk. The spread of capitalisation processes was creating new ways of measuring risk, managing its production, and profiting from (to use a term we will explore later) its 'economisation'.[36] The destruction of the River Nile, as we will see in the next chapter, was to be justified by the production in new ways of the 'risk' of its flooding and the capitalisation of that threat. The danger of losing one's land or one's housing in foreclosure for debt, emerging widely in the same period in Egypt and across the world, provided a novel means of organising material compulsion and moral obligation. But, as we will explore throughout this book, risk was equally the source of new methods of profiting from the future. The building of railways and other large ventures in this period helped trigger the development of accident insurance corporations, which could assign a value to the risk of personal accident and transform it into a tradeable asset. Previously working only in the specialised sphere of insuring long-distance sailing vessels and their cargoes, insurance companies now grew into large-scale traders of this new financial asset.[37] The development of commodity futures markets in cities such as Chicago (for grain) and Cairo (for cotton), dependent on the development of the telegraph, the grain elevator, and other technical mechanisms, generated another tradeable asset derived from separating the risk, in this case of price fluctuations, from the object to which it was attached.[38]

Accompanying these legal and technical changes was the development of coercive power. Police forces, mobilised militiamen, and troops extended their powers as both users and protectors of rail lines and other infrastructure, whose security was an increasing justification for their deployment.[39] Imperial warfare and colonial occupation could be rationalised by the need to secure new communication routes, such as the Suez Canal, or the coaling stations and later the oil installations on which those routes depended, or to colonise the upper reaches of a river re-engineered as an irrigation system, as with the British argument in the 1890s for its invasion of Sudan.

The age of large technical assemblages was also the 'age of empire', the period in which a handful of European powers, along with the United States and Japan, began transforming older modes of military and commercial domination into the formal annexation and political control of most areas outside Europe and the Americas.[40] What was the connection between an expanding technosphere and expanding imperial power? Again, we must be careful about the direction of the explanation. It is easy to assume that imperial states always favoured larger infrastructures of transportation, communication, and control. But that explanation for the rapid expansion of the technosphere seldom survives close inspection, as shown by such projects as the Suez and Panama Canals, the trans-Ottoman railway connecting Vienna to Istanbul, the

Indian railways under British rule, or the transcontinental North American railways.[41] Although their promoters usually claimed that these projects were serving the needs of Western commerce and empire, it was seldom the case that imperial trade or communication created the demand for them. The reverse may be closer to what happened: the opportunity to build exceptionally large structures for the capture of payments from the future tended to be colonising in form. The command over places and people, and the political capture of the future, were more easily organised and justified as an imperial project.

Vulnerability

If these forms of technical, legal, and political arrangement all contributed to the durability and scale of the new modes of capitalisation, their scope and duration was also the source of vulnerability. They introduced extraordinary instabilities – instabilities that define our current collective predicament.

First, the capture of the future through the development of railways, river networks, canals, urban expansion, and other schemes caused the modification of Earth systems on a scale that caused repeated ecological crises, undermining entire forms of human and nonhuman life. We have already noted how the re-engineering of the River Nile led to the elimination of the flood pulse on which the ecological balance of an entire country had depended, a destruction to be explored in detail in the next chapter. Similar damage was caused to the Bengal and Orissa deltas in eastern India, to the Indus Basin in the northwest of the subcontinent, and to other great rivers wherever Europeans extended their colonial rule.[42] Railways could bring equally significant disruption, both directly, through the modification of the landscape, and indirectly, through the destruction of grasslands and old-growth forests, the spread of pathogens, and the forms of genocide and mass killing that accompanied the elimination of subsistence and the more rapid extermination of native peoples, especially in the Americas.[43]

Second, in the longer term, there was another destabilising consequence to the new scale of breaking of the earth, as serious as the destruction of habitats, species, and human populations. Since durable structures and synthetics required making materials resistant to interaction with other elements and thus to ordinary decay, their construction or synthesis typically required the use of great quantities of process heat.

By forcing chemical bonds to break down and re-form, high temperatures could harden, split, purify, or combine elements, as with the heating of limestone in a kiln to make cement, iron ore in a smelter to produce iron and steel, atmospheric nitrogen in a reactor to create fertiliser and munitions, or crude oil in a refinery to manufacture fuels and petrochemicals. The carriers and haulers that used the new durable infrastructures – rail traffic and ocean shipping, followed by mechanised road transport – also depended on combustion heat. The heating created waste products. Unlike other parts of the Earth system, most of this waste did not become the feedstock for other processes and thus was not recycled.

The principal waste product of combustion was a gas, carbon dioxide, which simply accumulated in the air, in quantities that gradually exceeded the ability of other living and non-living elements to recycle it. Within less than a century, by the 1960s, measurements using an infrared gas analyser were able to show that this accumulation was changing the balance of gases in the atmosphere. Two decades later, it was increasingly clear that the resulting greenhouse effect was warming the surface of the Earth at a rate not seen in millennia. By the end of the twentieth century, it was widely understood that this production and accumulation of unrecycled waste would bring to a close the relatively stable, self-regulating balance of life processes that, since the ending of the last glacial period 12,000 years ago, had supported the conditions for the flourishing of human and nonhuman life.[44]

Third, the great expansion of the technosphere after the 1860s gave rise to a further destabilising force, for which writers of the period introduced a new term. They called it 'capitalism'.

In later decades, the word capitalism was used in a more general sense, to trace a longer historical development, with its origins in the reorganisation of labour in manufacturing, if not earlier commercial changes.[45] Those previous events were significant; we will explore them further in chapter 3. But the expression 'capitalism' was first used in the late nineteenth century to characterise a more specific process, one whose sudden expansion in that period appeared to leading economic thinkers like Alfred Marshall and J. A. Hobson in England and David Ames Wells, the most cited American economist of the period, to be a more significant change than earlier mutations, including the first wave of industrialisation.[46] The new term referred to the expanding activities of the capitalist, meaning, at that time, the financier or speculator, the person who organised the creation of credit and profited from the flow of credit money.[47]

While the decisive shift to modern capitalism is often located earlier, in the factory system and industrialisation, the joint-stock corporation was a new and more effective device for the appropriation of surplus. If the factory was one kind of technopolitical time machine, operating partly by extending and then speeding up the working day but also – as we will see – through the credit device of 'wage labour', then large technical structures, along with the joint-stock companies and joint-stock banks to which they gave rise, represented a different machine, one that extracted surplus from labour not by the compression of production time or the working of individual credit and debt through the wage, but by wider methods for the transfer of the future into the present. Let us look more closely at this new figure, the 'capitalist', and his role in this process of destabilisation. We can begin with the case of a particular individual, one who was to have a pivotal part in, among other schemes, the destruction of the River Nile.

Self-Loading

Sir Ernest Cassel, a German-born British banker, was to profit more than any other capitalist from the colonisation of Egypt.[48] The relationship between durability and instability in these years, and thus between the colonial and the catastrophic, can be seen in his career. Alongside his contemporaries the Rothschilds and the Barings, Cassel became one of the three largest London-based financiers of the late nineteenth century. Unlike his peers he did not arise from a merchant banking dynasty, nor did he go on to form one, so his name is now largely forgotten. For many years, however, he was considered the strongest financier in Europe.[49] The shifting focus of investment projects that he and his partners undertook charts the changing forms of large technical systems through which finance houses created and profited from credit – and their dependence on processes of colonisation.

More than many of his peers, Cassel combined two roles: capitalist and entrepreneur. That is, he was both an assembler of credit capital, arranging the debts or promises to pay that were enabled by large infrastructure schemes, and an assembler of those schemes, bringing together engineers, business firms, bankers, colonial governors, and others wielding imperial power.

Cassel began his career in the 1870s with an investment firm in Paris, Bischoffsheim and Goldschmidt. The firm supplied credit to Khedive Isma'il, the viceroy of Egypt, including a large loan in 1870 to develop sugar cane plantations on the Nile. Bischoffsheim created the loan via a mortgage on the royal sugar estates, which Cassel was later to control. With Bischoffsheim's son-in-law, Maurice de Hirsch, Cassel helped to rescue the financing and oversee the construction of the Oriental Railway, an Ottoman project to connect Istanbul by rail to the capital cities of Europe. Baron Hirsch sold his shares in the completed rail line to Deutsche Bank in Berlin and used the proceeds to establish what became for a while the world's largest charitable organisation, the Jewish Colonisation Association. The association planned for the mass emigration of the Jewish population of Ukraine, Lithuania, and other parts of the Russian Empire to settlements in Argentina. Reflecting the importance of railway building for both imagining and undertaking colonisation, Hirsch called the Argentina scheme a 'business like that of constructing and operating a railroad line'.[50]

Mentored by Hirsch, Cassel undertook the financial rescue of failing railroad firms in the United States, working with a leading German American financier, Jacob Schiff, who became a lifelong friend.[51] In the 1880s, Cassel turned his attention to Europe, where he made a fortune through the control of iron ore for steelmaking. He took over the struggling railway companies of Sweden, the country with some of Europe's largest iron deposits, connecting rival lines to ship the ore, after a minor modification to the new Bessemer process for the mass production of steel made it feasible to use ores containing phosphorous, a deleterious element typical of most European supplies, including those of Sweden.[52]

In the 1890s, he turned back to opportunities in Egypt, which became his main field of activity. Egypt's default on the 1870 Bischoffsheim loan and other debts had allowed the European creditors to take control of the government's finances, reinforced in 1882 by the British invasion to put down an army-led popular revolt and establish Britain's military occupation. Cassel arranged the financing to construct the Aswan Dam, and at the same time acquired the sugar estates that had been foreclosed by the Khedive's European creditors. The dam multiplied the prospective value of the estates, producing a significant profit when the lands were divided and sold to investors. He also developed a sugar plantation, near Aswan, at the village of Kom Ombo, which was settled with farmers and became a company town.[53] He combined the irrigation works and sugar plantations with a third element, helping to form the

National Bank of Egypt, a private financial consortium for investment in the new land development made possible by the dam's restriction of the Nile flood. The National Bank also acquired the exclusive right to issue Egyptian bank notes. With French and Egyptian partners Cassel then invested in a sugar-refining monopoly, built from the repossessed assets of the Khedive. The sugar monopoly, as we will see, was to have a curious role in the history of economic thought.

These projects relied upon the power of Britain and other imperial states, helping in turn to extend that power. Many of the schemes depended on the opportunities that railways and canals created for settling and resettling populations. They also took advantage of the synergies that arose from crossing between different fields and technologies, such as railway building and iron-ore mining, or river control and land speculation, or arms manufacture and the financing of imperial warfare. In fact, the capitalising of these projects depended on such interconnection, difference, and imperial scale; these features allowed the bundling of risk on which credit creation depended. Cassel took an interest in the Maxim machine gun, the world's first self-loading weapon, and financed its development, then helped form an arms company, Vickers, Sons and Maxim, to profit from it.[54] At the same time, he extended credit directly to governments expanding their colonisation projects, including the French in Morocco and the British moving south from Aswan to occupy Sudan. It was there that the Maxim gun demonstrated its new powers, in the overthrow of the government of Muhammad Ahmad bin Abd Allah, the Mahdi or redeemer, who had defeated the earlier Ottoman-Egyptian colonisation of Sudan.[55]

At the end of his career in finance, towards the close of the first decade of the twentieth century, Cassel returned to the scheme that began his career, the Vienna-to-Istanbul Oriental Railway, and worked on a more ambitious expansion. He tried to obtain from the Ottoman government a concession to extend the line to Baghdad, at the centre of a region now suspected to contain significant reserves of petroleum.[56] With his friend Jacob Schiff, with whom he had often wintered in Egypt, he supported a plan to use the railway project to settle colonies of Jews from the Russian Empire in Iraq, in association with irrigation and land development schemes. The British engineer responsible for the Aswan Dam project and for Cassel's sugar plantations in southern Egypt was brought to Baghdad to help with plans for river development.[57] The project for Jewish rural settlements in Iraq was a successor to Baron Hirsch's Argentinian scheme and a similar colonisation plan that Schiff had supported in the southwestern United States. Combining railway building, river development, and land settlement, the Iraq scheme was designed to integrate Jewish colonisation into an Ottoman political order, as a means to thwart the rival plans of the recently established Zionist Organisation for a racially segregated Jewish state in Palestine, another part of the Ottoman Empire (successful imperial financiers like Cassel had no interest in the vision of the antisemites, a rising political movement in Europe that saw Jewish communities as a distinct race or nation).[58] In combination with the short-lived Iraq scheme, Cassel also formed an investor group that attempted unsuccessfully to acquire monopoly rights to all of the Ottoman world's oil reserves, including those of Egypt and the Arabian Peninsula. One of Cassel's final acts, shortly before his death in 1921, was to endow a fund that supported the teaching of economics at the new London School of Economics and to provide a loan of £5,000 to John Maynard Keynes, rescuing the young economist (and future rescuer of economics) from a failed speculation in the Indian rupee and other foreign currencies.[59]

Let Us Go into the Stock Exchange

When we return to the history of Egypt and the destruction of the Nile, we will look again at Cassel and his role in financing the Aswan Dam and associated schemes to produce sugar, to better understand these methods of capturing the future. But, first, let us consider another aspect of these transformations in the technosphere and the durability of means of capitalisation. Trading shares in that future helped create another element that was both stabilising and disruptive, the enterprise in which Cassel himself invested at the end of his life: the discipline of economics. As a professional academic field, economics emerged around the 1870s, in the same period as modern stock markets, modern imperialism, and the modern 'capitalist'. This was no coincidence.

There are two points of connection between the colonising transformation of the technosphere and the birth of modern economics. First, the technical arrangements that produced the stock market also produced the possibility of the science of economics. The joint-stock company, as we have seen, created a new kind of asset, the share, which now appeared to be separate from the material structures and processes from which it acquired its value. The share was an abstraction, and a real one, manufactured through legal practice, pricing mechanisms, the policing of labour protest, and the imposition of colonial power. This abstracting made the share into something liquid, a so-called 'financial' asset—one that, seemingly unconstrained by the future forms of life that produced its value, could be traded instantly for other financial instruments. It had a price. Economics was a product of the same abstraction. The new discipline substituted for the study of colonisation, capital, and the technosphere the analysis of the constantly moving prices on which it built its particular mode of abstraction.

Second, economics attempted to explain, or provide an alibi for, an extraordinary phenomenon that seemed characteristic of the new 'capitalism': unearned income. The expanding power to colonise the future was somehow generating unprecedented windfalls in the present. This novelty required some innovative system of knowledge to explain and appear to resolve it. What kind of science might serve to justify exceptional forms of wealth and inequality, by finding a way to distract attention from the very mechanisms that created them? While other aspects of the history of economics will be considered in a later chapter, let us examine first how the emerging discipline dealt with the problem of unearned income, or the phenomenon of finance, and found its alibi.

The modern theory of finance emerged at the same time, just over a century ago, in a period of credit expansion, windfalls, and periodic crises – a period that echoes the events of the present, including not just the Ubers and other tech firms of today but the global economic crisis of 2007–8. The late nineteenth century experienced a wave of stock market booms, which crested at the turn of the twentieth century and crashed in the great Wall Street Panic of 1907. Like their counterparts a century later, the economists of the day struggled to make sense of these events. The problem was to explain the astonishing power of money to make money. Stock exchanges, commercial banks, bond dealers, and mortgage companies had multiplied, not only in the financial capitals of the North Atlantic world but in places like Buenos Aires, Cairo, Bombay, and St Petersburg.[60] Large financiers made (and sometimes lost) fortunes, but even small investors could purchase bonds or shares and, with no effort on their own part, benefit from the strange ability of money to grow.

What explained this fecundity of money? What was the secret of unearned income? For economists at the turn of the twentieth century, the puzzle was framed as the problem of interest – or the relationship between money and time. Today, we are so accustomed to the assumption that the passage of time allows money to grow that it is easy to forget the strangeness of this idea. A century ago, it was perhaps the central puzzle of economics. Older ways of explaining money's powers of reproduction no longer seemed appropriate, or even necessary. Previously, money had appeared to be one of those unusual living forces that can expend itself and yet not lose its original power, like human labour or a renewable natural resource. Two centuries ago, the political economists of the early nineteenth century had explained that capital, like land and labour, possessed its own 'productive agency'. Jean-Baptiste Say, the great French political economist whose *Treatise on Political Economy* of 1803 had built upon the work of Adam Smith and surpassed it in influence, offered the most persuasive account.[61] Smith himself had said nothing systematic about interest, understanding it simply as a charge imposed on the product of labour by those who supplied the producers – the actual source of wealth – with materials and equipment.[62] Say developed a different view, in which capital itself was productive. When capital is loaned for a fixed period, he wrote, 'its agency is transferred for that space of time' into a 'material substance' such as the 'works, machinery, utensils, provisions, and stock in hand' of a manufacturer. At the end of the loan period, the original value of the capital 'is returned to the full amount, and emerges in a perfect state from its productive employment'. The payment of interest was similar to the rent paid for the use of land or the wage for the employment of labour. It was the rent charged by the owner of capital for the temporary use of money's peculiar productive power.[63]

By the late nineteenth century, the large expansion of so-called finance capital had made this explanation seem inappropriate. On the one hand, there was no clear relationship between the earning of interest on a loan and whether it was put to a more productive or less productive use. The rate of interest had become a standard charge set by the money markets – stock exchanges, bond traders, commodity futures markets, and other places where financial instruments were now rapidly bought and sold. On the other hand, those markets had given birth to a new kind of thought, the science of economics, from which there emerged a different explanation of the power of money to earn money.

'Let us go into the stock exchange of a large investment centre like Paris or London,' wrote the French economist Léon Walras, in his 1874 text that was to establish general equilibrium theory as the basis of an economic science. 'We shall see, now, how competition works in a well-organized market.'[64] The introduction of the telegraph with its rapid transmission of price information, the development of the stock ticker tape with its visual display of the continuous movement of prices, the transformation of the share from a representation of material equipment and physical structures into an abstract monetary instrument, and the practice of calculating and publishing a daily closing price for each share: all these material developments had made the stock market appear an almost perfect model of buyers and sellers in action.[65] As with all markets, the creation of a 'price' was a product of technical and political arrangements. Those who controlled the market mechanisms – the telegraph, the ticker, and the electrical wiring and battery power on which they depended – managed access to information and even membership in the exchange.[66] But these material devices and legal arrangements generated an abstraction around which economic theory could now unfold.

The money markets facilitated an abrupt change in the understanding of wealth. 'The common expression Money Market denotes no locality', wrote William Jevons, a cofounder with Walras of the new science. 'It is applied to the aggregate of those bankers, capitalists, and other traders who lend or borrow money, and who constantly exchange information concerning the course of business.'[67] This aggregate, this seemingly placeless abstraction, allowed political economy to abandon its attempts to study the wider world – countries, populations, industries, resources, transportation, infrastructure, empire – and become a science of the price mechanism.[68] The novel way of thinking, later referred to as 'neoclassical economics', produced the understanding of capital and interest that went on to govern economic analysis for the next century.[69] It is the story of money that today, as we will see in chapter 8, seems unable to make adequate sense of our predicament, in particular the problem of climate crisis and so-called economic growth. This inability can be traced to the fact that money markets – the very place where it seemed that money could effortlessly make money – provided not just the problem to be explained, but the laboratory within which the new science produced an answer.

Seeking an alternative to attributing a 'productive agency' to capital, economists found the solution to the question of how capital could earn interest by attributing this payment, as with everything else in neoclassical economic thought, largely to the operation of a human preference. If people were willing to pay a premium to have money now rather than at a subsequent date – in other words, to pay interest for the temporary use of capital – that willingness must reflect a natural inclination to consume today rather than later. As the American economist Irving

Fisher explained in *The Rate of Interest* (1908), the work that established the new understanding of capital, the power of money to earn interest had nothing to do with the transformations in the technosphere and the expansion of imperial power that had made possible the capture of revenue from the future on an unprecedented scale. It was simply a consequence of human impatience.[70]

At first, this shift from the agency of money to the preferences of humans had its critics.[71] How could the theory of capital and interest pay no attention to the productive process? Surely the earning power of capital must have something to do with the technical processes into which it was put. An answer to this objection was soon found, by introducing another human agent, and a more reliable source of alibi: the figure of the entrepreneur.

The Entrepreneur

Among the earlier, classical economists, a similar move had been made. In fact, it was Jean-Baptiste Say who introduced the term *entrepreneur* to explain how the production of wealth – and the earning of interest – depends on this pivotal figure. The first edition of Say's *Treatise*, in 1803, said little on the subject. But, over the following decade, forced out of public life by Napoleon Bonaparte's seizure of power, Say tried his own hand at organising a productive process. With funds from a wealthy aunt and other investors, and the assistance of his younger brother Louis, he canalised a small river above the village of Auchy in northeast France and set up a water mill for spinning cotton fibre into yarn. The venture faced a recurrent technical difficulty: the parts of the wooden machine that transferred motion from the *moteur hydraulique* to the spindles on which the yarn was wound repeatedly broke. Replacing the wooden parts with wrought iron failed to end the breakages, and replacing those with parts of cast iron, which is stronger but more brittle, also failed. Like many would-be industrialists, Say found the attempt to introduce mechanisation too much of a challenge. The wastage of material damaged in production, which amounted to 25 per cent of the raw cotton, added to the difficulties. Unable to make a profit from his hydraulic machine, Say dissolved the business in 1812 and became a rentier, investing in a number of other ventures. In particular, he helped his brother take over a sugar-refining business. A refinery was less prone to breakdowns than a textile mill, as boiling and purifying raw sugar was a simpler process than spinning cotton fibres into fine yarns. The Louis Say Refining Company went on to become the most successful sugar business in France and one of the country's largest industrial enterprises – although, as the firm's manager was to discover with the fingers of his own hand, even in the refining of sugar industrial accidents could have their consequences.

Meanwhile, Jean-Baptiste turned back to political economy, and in 1814, with the defeat of Napoleon, he was able to publish a revised edition of the *Treatise*. The new version broke with the view of Adam Smith and David Ricardo that the value of a good is determined principally by the amount of labour required for its production, emphasising the equally important contribution of a new figure, the entrepreneur. This person coordinates the use of labour and machines, improves the organisation of work, supplies or draws in capital, and, above all, uses his understanding of the principles of supply and demand to regulate the pace and direction of production – including, for example, deciding whether it is more profitable to produce cotton or sugar. In other words, the 'productive agency' of capital is effective thanks to the combining of technical know-how and economic science in the person of the entrepreneur.[72]

A century later, in the opening years of the twentieth century, the figure of the entrepreneur reappeared. Once again, this character would provide an alibi, offering the theory of capital and interest a way to explain the ability of money to make money. As economics narrowed its focus to become a science of the price mechanism, it would find a means to readmit into the explanation of the nature of money every other piece of machinery and process of production that gave money its reproductive power, via the tenuous human link of the entrepreneur. The role of this figure, however, was redefined. Rather than a vital coordinator, the entrepreneur was now imagined as an occasional disruptor – in fact, as a groundbreaker.

In 1911, three years after the appearance of Irving Fisher's *Rate of Interest*, a young Austrian scholar, who would later move to Harvard University and come to rival Fisher for the title of the most influential American economist of the first half of the twentieth century, published his most important work. Joseph Schumpeter's *Theory of Economic Development* proposed an account of capital and interest in which the very ability of money to earn money turned decisively on the role of the entrepreneur. For Schumpeter, however, the entrepreneur was not a coordinator but an agent of disorder.[73] The entrepreneur set in motion the processes of innovation, upheaval, and technical reorganisation that he later termed 'creative destruction'.[74] Today, more than a century later, the Schumpeterian account of capitalism as a process driven by a unique force of disruptive innovation remains, among critics of the power of capital as much as its defenders, perhaps the most persuasive explanation for the very experience of temporality that defines capitalist modernity.[75]

While Schumpeter referred in places to the writing of Jean-Baptiste Say, the impetus for placing the role of the entrepreneur at the centre of the new account of money did not come from Say's *Treatise*, nor from other texts he encountered as a student in Vienna. It arose from his first job after completing his education. That employment was in Cairo.

Schumpeter's ideas were formed in Egypt not by Say's writings but in his encounter there with the powerful sugar business created by Jean-Baptiste's brother, the Louis Say Refining Company. It was the collapse of the Company's scheme to make a fortune from sugar in Egypt that shaped the most influential twentieth-century account of how money is made from money.

Nowhere better exemplified the sense of capitalism at the turn of the twentieth century – the power of money to make money – than the place that Europeans called Egypt. Cairo and its countryside had seen a first financial boom in the 1860s. As the government constructed railway lines, expanded cotton plantations, and established sugar cane estates, taking advantage of the interruption to the supply of slave-grown cotton and sugar from the American South, European financiers drove an expansion of credit that created a 'Klondike on the Nile'.[76] However, after the defeat of the American slave states and the resumption of US cotton exports, the price paid for cotton collapsed. The international financial crisis of 1873 created further pressure and led the Egyptian government to default on its debt payments and fall under European financial control. A popular uprising that followed was ended by the British military occupation of 1882. The occupation laid the groundwork for renewed investment in cotton and sugar production, which by the turn of the century created a second credit boom. Land itself became a speculative asset, its value fluctuating, to the astonishment of local observers, with the rise and fall of the price of cotton and sugar.

By 1905, this expanding world of credit organised around speculation in cotton, sugar, stocks, and land became a topic of continuous debate in political circles in Cairo. A leading writer, Ya'qub Sarruf, serialised a novel that year in a literary supplement to the monthly journal he edited, *al-Muqtataf*. The suspense from the appearance of one instalment of the story to the next replicated, Elizabeth Holt suggests, the uncertainty of a world in which the wealthy were simply 'those who win from the loss of another' and where 'no one is sure of his tomorrow'.[77] The materiality of the new modes of capture – its buildings, exchange mechanisms, and information systems – were duplicated and refracted in new forms of print culture and imaginative life. 'In all the capitals of Europe there is a huge building on which is written "Stock Exchange",' a member of Egypt's ruling elite remarked to a French American visitor in 1905, complaining about the 'plague' of speculators sapping the country's life. 'Here the whole country is one vast Stock Exchange.'[78] The collapse of the boom that same year was to trigger the Great Crash of 1907.

The events that took Joseph Schumpeter to Cairo, in the wake of that crash, were set in motion by a suicide in Paris in 1905.

Schumpeter's Sugar Mill

The death of Ernest Cronier in August 1905 seemed to epitomise the financial fragility of those days. A servant found his body on the bedroom floor. The autopsy, carried out in the adjoining bathroom, revealed that Cronier had tried to poison himself with potassium cyanide. Unable to keep the poison down, he then shot himself in the heart.[79] Curiously, the hand with which he pulled the trigger had two prosthetic fingers, the consequence of an accident twelve years earlier inspecting repairs to machinery at his sugar refinery.[80] Those repairs, and the missing fingers, provide a clue to which we will return.[81] Known as the sugar king, *le roi du sucre*, Cronier had built the kingdom he controlled, the Louis Say Refining Company, into a business large enough to corner the French sugar market. Hired first as an engineer and manager, he had taken control of the firm – founded by Jean-Baptiste and his brother Louis – following the death of Louis's grandson, Henri Say, in 1899.[82] But a large speculative investment Cronier had made in sugar futures had failed. Then a letter reached him in Paris from Cairo suggesting that his fraudulent management of the sugar industry he controlled there, which had funded the speculation, had been uncovered. Within days, he took his life.[83] On the news of his death, on 27 August 1905, the value of the Company's shares collapsed. The Paris stock market fell, causing huge losses, and traders felt the shock over the following days in the exchanges of London, New York, and Hamburg.[84]

In Alexandria and Cairo, where the money markets were among the most active in the world at that time, the shock was even greater. The Louis Say Refining Company had only recently taken control of the Egyptian sugar industry, in partnership with local investors and Sir Ernest Cassel, the British financier we met earlier, creating the country's largest industrial enterprise. The firm had faced an uncertain future. The completion of the Aswan Dam at the turn of the century had led farmers unexpectedly to decrease rather than expand the growing of sugar cane. Instead of profiting from processing the cane, the firm had become a business for producing debt. Its directors had prospered by issuing fictitious bills of exchange, with no underlying sale, creating the credit money that funded Cronier's speculations in sugar futures in Paris. The manager's suicide forced the Egyptian company to suspend dealing in its collapsing shares and reorganise, closing half its refineries and selling off its sugar estates and agricultural railways.[85]

Perversely, the sell-off fuelled a new round of speculation in Egypt's colonial future, creating a boom whose collapse eighteen months later, in April 1907, was to reverberate around the world. Over those eighteen months, buyers bid up the price of agrarian land development companies, and property firms promoted a frenzy of construction in Cairo and Alexandria, as did the banks and mortgage companies organising the creation of debt.[86] 'People were apparently mad,' according to a later report to the shareholders of the National Bank of Egypt, the financial firm that Cassel had helped create. 'They seemed to think that every company that came out was worth double its value even before it had started business.'[87]

Suddenly, however, the long-term future of Britain's colonial occupation of Egypt, a political future that had, along with the dam, produced the value of those shares and mortgages, seemed less certain. In the summer of 1906, protests against the British occupation intensified, in public condemnation of the hanging of four men and the imprisonment and flogging of others in the Delta village of Dinshawai, forty-five miles north of Cairo, after they resisted British soldiers shooting their domestic pigeons for sport (housed in dovecotes, pigeons were raised for their droppings, which were used as fertiliser, providing nutrients of increasing importance following the deleterious effect of the new irrigation schemes on soil fertility). Anxious to reduce political opposition, the following April the government in London accepted the resignation of its proconsul in Cairo, Lord Cromer, who had administered the colonial government since 1883.[88]

A financial crisis, the threat of which had been clear since January, then broke. An emergency shipment of £600,000 in gold from London failed to stop the run on the banks. 'Egypt went first', the most persuasive contemporary account of the worldwide financial crisis of 1907 reported.[89] The panic spread to Japan, Germany, and Chile, and then by October to New York, where the collapse on Wall Street triggered a financial contraction, whose severity was surpassed over the next century only by the Great Depression.

Meanwhile, the Egyptian courts had appointed a group of experts to reorganise the sugar company. Their plan was to expand the refining of local sugar cane, rather than trying to profit from processing imports of subsidised Austrian beet sugar. To assist in the reorganisation, those in charge hired a young Austrian lawyer – the future Harvard economist Joseph Schumpeter. Trained in economics (still taught as a branch of law in European universities at that time), Schumpeter spent most of 1908 in Cairo, and in his spare time developed the ideas for the book that he completed writing soon after his return to Europe.[90]

Schumpeter's *Theory of Economic Development* addressed the question of how to explain this expanding world of money that makes money. The question was posed, once again, as the issue of how capital earns interest. He wished to understand what he called 'the suction apparatus' that enabled the capitalist, meaning the supplier of credit, to draw 'a permanent stream of goods from profits'.[91] To explain the apparatus, the book drew on insights he obtained while reorganising the sugar business. In Cairo, his biographer notes, Schumpeter 'acquired practical business experience and firsthand knowledge of how business and commerce worked in the real world', in helping to reorganise the sugar company and implement different methods of production. 'His participation in this innovation made an indelible impression on Schumpeter. He had witnessed firsthand the innovation process and its consequences. It would later become a centrepiece for his theory.'[92] Just as the modern stock exchange had provided a previous generation of economists with a simplified site from which to derive an abstract model of the price mechanism, a brief encounter with the business of sugar production in colonial Egypt provided Schumpeter with the laboratory in which to generate, in the figure of the entrepreneur, an entire theory of capital.

The widely influential argument of the book was that the power of money to expand – the ability, in other words, of the capitalist to earn interest from the creation of credit – arose not from the properties of money, nor from the powers of the capitalist who supplied it, but from the distinctive role of a different figure, the entrepreneur. The role of the entrepreneur was not to coordinate business, as Say had argued, but to disrupt it.

Interest as a Reward for Disruption

The models of economic life that men like Walras and Fisher devised, based on the principle of equilibrium, where the mechanism of a market price is assumed to continually produce a balance between the demand for goods and their supply, had no place for deliberate disruption. These economists were therefore unable to account for the earning power of money, except by positing a universal 'time preference' for present over future goods. This preference, they surmised, allowed those supplying credit money to a borrower seeking a good immediately to exact a premium (an interest charge), forcing those unwilling to wait for a good to pay more than its value. As Schumpeter pointed out, there was no evidence for this universal preference. Most businesses funded their continuing or expanding activities through retained profits, with no need to pay a fee for the use of money (just as today, stock markets played very little role in providing capital for business firms).[93] Only those in distress, such as a debtor facing dispossession or a business facing an unexpected crisis, were forced to borrow money at interest. To explain the fact that the payment of interest on money had recently become a general feature, one had to understand the role of the entrepreneur.

Entrepreneurs are those who introduce 'new combinations', Schumpeter wrote, which reorganise an existing productive process. The rearrangement makes a business briefly more profitable than its rivals. Carried out repeatedly across the whole of industrial life, this continual rearranging causes the circular flow of goods between production and consumption, as envisioned in the equilibrium account of economic life, to be repeatedly disrupted. The disruption gives rise to a dynamic movement. As a contrast with the image of economic processes tending towards an equilibrium, Schumpeter termed this disruptive movement economic 'development'. His account was not just a sociological explanation for the working of a capitalist system. It was a solution to the problem of explaining the power of capital to earn interest, of money to accumulate money. What distinguished the entrepreneur from the ordinary operator of a business was that, as an innovator, the entrepreneur was the only agent unable to fund operations out of retained or expected earnings. Valuing the access to present funds over future money was not a general human trait; it arose from the special needs of the entrepreneur, who was required to operate as a debtor. This need established a market for money, for which the price was the rate of interest. The demand for money among entrepreneurs in different fields was transformed by the market into the general rate of interest for all capital.

For Schumpeter, interest was not an intrinsic property of money, nor a reward for patience. Rather, it was simply the special kind of profit earned by an entrepreneur, as a temporary reward for reorganising a process of production. The rearrangement briefly makes an enterprise more profitable than its rivals. As soon as other businesses adopt the same methods, competition forces down the prices of goods and the entrepreneur's profit comes to an end. To illustrate this process, he gave an example drawn directly from his own experience in Cairo. 'Let us assume that a planter who has previously cultivated sugar cane changes over to cotton-growing, which until recently was more lucrative than it is now. This is a new combination; the man thereby becomes an entrepreneur and makes a profit.' Then, competition forces down receipts, for example through an increase in the rent of the land – 'as has actually happened', he notes – and at that point 'the entrepreneurial function of this man disappears'.[94]

Jean-Baptiste Say had chanced upon the importance of the entrepreneur by trying his hand at cotton manufacturing, failing, and then collaborating with his brother on a simpler scheme for refining sugar. Schumpeter had had a similar revelation, with the same materials, by describing a plantation owner who shifts their cultivation from cotton to sugar cane. And he did it while rescuing the same French sugar company.

While in Cairo, Schumpeter may have experienced exactly this choice between planting sugar cane or switching to the cultivation of cotton. Besides rescuing the sugar company, he also had a hand in the organisation of a new agricultural estate, belonging to a princess of the Egyptian ruling family. The sources do not tell us the name of the person. However, this was likely the estranged wife of the Khedive, an Austrian like Schumpeter, who had recently acquired her own palace and estate on the northern outskirts of Cairo. She had met the Khedive through her brother, who had been his classmate at the Theresianum in Vienna, where the political elite of the Habsburg Empire was educated and where several Ottoman-Egyptian princes were sent to study. Schumpeter was a graduate of the same school. Fond of recalling other achievements besides his business success, and in the habit of 'charm[ing] his way into intimacies with young women', Schumpeter later boasted of having an affair with the princess.[95] Rather than any personal liaison, what is interesting but never mentioned is the nature of the enterprise in which Schumpeter had a hand. Located at Mostorod, on the northern outskirts of Cairo (today the site of Egypt's largest oil refinery), the estate was both typical and unusual for the forms of entrepreneurship emerging there at that moment. It was built on lands made open to the summer cultivation of cotton by the completion five years earlier of the Aswan Dam. It formed part of a group of model farms that the Khedive had recently acquired, including a neighbouring estate of his own, to experiment in the expansion of cotton cultivation. An English writer had just described the model in detail, in a book aimed at encouraging English settlers 'disposed to enter upon a business career in this favoured country' to move to Egypt and buy up land to develop as cotton plantations.[96]

In his short and eventful stay in the booming city of Cairo, Schumpeter believed he had discovered, through his own brief experience of the role of the entrepreneur, the secret of capitalism's distinctive relationship to the future.[97] The entrepreneur makes his appearance here, however, not as the key to the dynamic of capitalism, but as the one who helps to conjure into being its alibi.

The alibi of capital emerges in this combination of what appears as both groundbreaking and the source of growth. If capital has acquired its mysterious ability to expand, enabling money to make money, an explanation is to be offered in the disruptive force misrecognised as innovation. To name his understanding of capitalism, as a system characterised by this new power of money, Schumpeter presented in the title of his book a novel term: 'economic development' (*wirtschaftliche Entwicklung*). Development, or the introduction of new combinations, was the process triggered by the entrepreneur. Distinguishing the static, circular process of exchange studied by so-called neoclassical economic theory from the dynamic process of business, or the capitalist system in which money had an expanding power to make money, depended on this emergent concept of development.

The term 'economic development' in Schumpeter's title did not yet have the wider meaning it would acquire in subsequent decades. By the mid-twentieth century, as we will see in chapter 5, the invention of the economy, an object conceived and measured in terms of its 'growth', would transform the meaning of development. In turn-of-the-century Cairo, the British colonial regime had as yet no conception of economic development. It was not a term they used. They were preoccupied with development, but their use of the word had an earlier and more specific sense – it referred not to the development of an economy, nor even to a vaguer economic development, a term they did not use, but to 'land development'.[98] It described the focus of British policy at that time, the reorganisation of the Nile, the expansion of irrigation works, and the transformation of agriculture into a field of extensive private and corporate debt creation. This was precisely the process from which Schumpeter abstracted his theory of capital: converting land to more profitable purposes, or rather to purposes more easily capitalised, thanks to irrigation schemes and the spread of industrial crops such as cotton and sugar cane. Land was remade as a more powerful apparatus for the 'suction' of income from the future.

There was nothing innovative about the cultivating of cotton or sugar cane, or the decision to switch from one to another. Both had been cultivated in the Nile Valley, as we shall shortly see, for more than a millennium. Nor did they have any simple relationship to growth, since the yields of both crops were set to decline.[99] In introducing here these agricultural schemes, our aim has been to ask about the new relationship to the future that they engineered, while avoiding the implication that this represented 'improvement' or a general process of 'growth' or 'development'. To grasp a different understanding of the engineering of the future under capitalism, the starting point is to replace a history conceived as economic development with the question of how the control of future revenues is engineered, and how the future itself is produced as a frame of collective life, through organising combinations of the durable and the liquid.

Schumpeter's Egyptian sugar mill allows us to connect together several questions that this book has begun to ask. In the age of empire, meaning roughly the five decades of rapid imperial expansion prior to the First World War, there occurred both an accelerated breaking of the Earth, in the rapid expansion of the technosphere, and a simultaneous lightness or evanescence of the arguments and ideas with which economists and others tried to explain the extraordinary capture of wealth that accompanied this ground assault. The earthly, material destructiveness of the groundbreaking appeared as the very opposite of the immaterial, ideational, imaginative work of the entrepreneur – although in practice the two were inseparable. Groundbreaking made it possible to extract from the future on a large scale, and in the process to generate instabilities and uncertainties personified in the new figure of the groundbreaker, the 'capitalist'.

Schumpeter and his wife both contracted a serious fever while in Cairo, causing them to cut short their stay in the country after less than a year. In the next chapter, we will linger in Egypt a little longer. Leaving the gates of the sugar factory and the model farm, it is time to open up wider questions about how the capitalist relationship to the future, and the modern power of money, were organised. As we will see, the choice between growing cotton and sugar cane was not simply the enterprising decision of an entrepreneur. To produce the entrepreneur and his decision, an entire country had to be remade, its ecology re-engineered, its people impoverished, and its river destroyed.

2

On Rivercide

Egypt in the nineteenth century endured a catastrophic reordering of its geography and ecology in the service of the production and repayment of credit. The catastrophe allows us to look more closely at the connection between colonial occupation, ecological destruction, and the capitalising capture of the future. Egypt became a place, perhaps more than any other site in the world, that exemplified the power of money to make money. Joseph Schumpeter used his brief experience in Cairo to develop an explanation of capitalism as 'creative destruction', an understanding of capitalist modernity that still shapes the way it is imagined today. One of the few concrete examples that Schumpeter offered of this destructive creativity was drawn from his brief involvement in the agricultural transformation of Egypt. But what exactly did that destruction involve? What reordering of the technosphere was required? And on what producing of the future, reimagining of the past, reorganisation of know-how, arranging of alibis, and expansion of ignorance would it be based?

'Every hydraulic engineer considered himself a financier', wrote William Willcocks, reflecting on his career in Egypt, 'and every financier considered himself a hydraulic engineer.'[1] Willcocks produced the first designs for the Aswan Dam, built in 1899–1902, and later managed the sugar estates of the man who financed it, the English capitalist and entrepreneur Sir Ernest Cassel. It was the collapse of Cassel's sugar investments, as we saw, that had brought Schumpeter to Cairo. In Egypt, hydraulic engineering was the basis of a new financial order, enabled by a new degree of creative destruction. The apparatus of credit creation, of the suction of payments from the future, was being built not just by entrepreneurs like Ernest Cassel and his partner Ernest Cronier, but out of canals, sluices, dams, barrages, and steam pumps. The acquisition from the future would require, in turn, a rewriting of the past.

The British had occupied Egypt in 1882 in response to a political-financial crisis. European bankers had deployed their expanded tools of credit creation to float successive government loans, each establishing a hold over a different source of future revenue. The taxes of individual provinces, the state railways, the salt tax, customs duties, and eventually the ruler's private plantations were each mortgaged in turn to the bankers. An increasing part of every new loan was devoted to repaying interest and fees on the previous debts.[2] Each loan combined the durability of a future flow of income with the liquidity of the arrangement that converted the control of those prospective revenues into a credit payment in the present. The international financial crisis of 1873, precipitated by the collapse of speculation in railway shares in Europe and the United States and leading to a worldwide depression, along with the lack of any further public revenues to mortgage, forced the Egyptian government into bankruptcy. A popular revolt against the regime and its creditors led to the creation of a nationalist government and caused the country's largest landowners and financiers to seek external support to restore the old order, precipitating Britain's military occupation. In 1884, the success of another popular revolt – the Mahdist overthrow of Egyptian rule in Sudan – and another round of financial crisis in Cairo persuaded Britain to turn its temporary 'intervention' into an open-ended imperial administration.[3]

A Volatile Nature

There followed a five-decade experiment in remaking a country as a repayment machine. The building of this machine required both the work of irrigation engineers to alter the flow of the world's longest river and the work of historians to rewrite one of the world's longest recorded histories. To understand the working of this machine, the standard account of how Britain built a colonial regime through the re-engineering of the Nile needs to be reconsidered.[4]

The standard account began to be written almost immediately. In 1890 the American engineer David Ames Wells, whom we met in passing in the previous chapter, published what was to become the most influential US economic text of the period. He placed the transformation of Egypt at the centre of his account of the revolutionary 'disturbances' brought on by technical innovation since the 1870s, anticipating the ideas of Schumpeter two decades later. He started with the example of the Suez Canal and ended with the transformation of the Nile by the work of British engineers.

> In 1884 Britain virtually took possession of Egypt, and from that moment there was initiated, under the management of a body of skilled engineers and practical men, a renovation of the water-supply and irrigation system of the country upon which the life and prosperity of its people depend, and which has already been attended with results of extraordinary beneficence. In Lower Egypt land reclamation is going on at the rate of fifty thousand acres per annum, and in other sections of the Nile Valley at double that amount; giving to a down-trodden and impoverished race better opportunities than they have had for centuries of supporting themselves by their own labor.[5]

One of Britain's first steps, the conventional story goes, was to abolish the so-called 'traditional' system of forced labour, the *corvée*, on which the old agrarian system was based.[6] Next, the occupying authorities addressed the persistent problem facing the development of agriculture in Egypt – the volatile forces of nature. 'The history of Egypt is the story of the use man has made of the Nile', wrote the author of what became the leading economic history of modern Egypt. 'Left to herself, the river would be a capricious and changeful mistress, bringing prosperity one year, famine the next. Ever since man settled down in the Nile Valley, he has endeavoured, with the varying degrees of success, to tame the river and turn it to his service.'[7] This history of volatility, uncertainty, and famine, as we will discover, was a fabrication.

All cultivation depended on the annual inundation of the river, whose extent, historians claimed, was unpredictable. To remedy this, the British engineered a new system of what they termed 'flood control' to alter the flow of the river.[8] First, they supervised repairs to a barrage north of Cairo, raising the level of the river where it supplied the main canals of the Nile Delta. They then built the much larger dam in the south at Aswan, together with a third barrage midway between the other two, at Asyut. These made it possible to store the water of the Nile in the reservoir behind the dam, regulate the annual inundation in the summer, and convert most of the irrigation system from flood basins to permanent canals. The new storage mechanisms allowed an increase in agricultural productivity, it was said, by switching most of the country's agriculture from a system based on a single annual crop, sown in the catch basins after the flood waters receded, to a system of perennial cultivation, in which two or three crops a year could be grown in the same field. Many parts of the Nile Delta and small areas of Middle Egypt had already introduced perennial cropping, lifting water from the river or deep-lying canals constructed earlier in the century, via animal-powered irrigation machines and, increasingly, steam pumps. The new barrages worked differently: rather than deepening the canals to carry water when the river was low, they raised the water level in the river and the canals throughout the year, allowing the conversion of more than three-quarters of the country's fields from the seasonal use of flood basins to year-round cultivation. This in turn enabled a large expansion in the production of the two main industrial crops, cotton and sugar cane. The area devoted to growing cotton for export, which had already increased tenfold by the 1860s and 1870s from its modest levels in the 1820s and 1830s, more than doubled again in the two decades after 1882.[9] Its cultivation became the mainstay of the country's economic growth, according to standard accounts, and integrated it into so-called world markets.

Thus, the history of Egypt is told as the encounter with nature, in the form of a river whose unpredictable flood made her a 'capricious and changeful mistress' in need of being 'tamed'. The men taking charge of her taming had mastered the technical means to regulate and manage her capriciousness. A parallel argument was made about the administering of untamed social forces. The *corvée*, together with the *courbage*, the whip made from the hide of the Egyptian hippo and used to enforce the compulsory labour regime (even as the hippo was being driven extinct by the very irrigation works its hide enforced), epitomised a corrupt and despotic government that was to be rendered juridical and humane.[10] Just as the forces of nature were to be controlled, the natural despotism of the East was to be replaced with a more rational, more humane, administration. The role of technical reorganisation, in this account, was to improve, to advance, to rationalise, and to make more productive.

This history appears differently when the focus is placed not on the problems of ordering, improving, and extracting a surplus from an untamed nature (whether physical or human), but on how the making of a technosphere built out of irrigation works a new apparatus for the capture of the future.[11] We should not assume that technical changes were improvements, but ask how the problems were defined and for what the solutions were useful. The main threat faced by Egyptian farmers in the nineteenth century was not the problem of high and low Nile floods. That 'risk' was easily exaggerated and the engineering works themselves made it worse. Farmers faced a different, recurrent ecological threat, one that was the cause of recent, devastating crop failures and economic calamity. It may also have increased the likelihood of excessive inundation. The new threat affected an existing arrangement for storing surplus Nile waters for use after the inundation and draining of the floodplain, an arrangement that allowed cultivators to manage the problem of high and low floods, but which was now to be destroyed.

Unlike the irrigation works carried out earlier in the century, the storage system built by the British did not recognise the existing mechanism, did not consider its strengths, and did not deal with these other vulnerabilities. The new arrangement addressed not those issues, but the problem of capitalisation: how to convert the fields from arrangements that produced and stored wealth, allowing only limited forms of credit creation, into an apparatus producing flows of prospective revenue from the future, which could be realised in the present and made into unearned income for those separated from the fields in both space and time. None of this was driven by a particular *need* for raw materials, such as cotton or sugar, nor by a flow of 'surplus' capital from Europe, nor by a demand to integrate Egypt spatially into a world market. Those are the usual stories. Nor did it represent a new, late-nineteenth-century stage of financialisation, for the creating and profiting from credit construction had been the purpose of European involvement in irrigation and agricultural schemes since at least the middle of the century, and in trading schemes long before that.

Can the Waterwheel Squeak?

During the 1990s and over the following two decades, I lived for several periods in a village in the south of Egypt, learning how people farmed the land and made their livelihoods. Next to the house where we lived, against the mudbrick wall of an animal shelter, stood the disassembled parts of a wooden irrigation machine – the large, toothed wheel of a *saqiya*, or animal-driven waterwheel. The machine, I learned, was once used to raise water from a well that had been dug adjacent to where our house now stood, to irrigate the surrounding field. The men of my age remembered helping with the wheel in their youth, before the completion of the High Dam at Aswan in the late 1960s. From older men I learned something remarkable. In the old days before the dam, once the flood waters receded each year, waterwheels like this would be assembled across the fields, each one placed over a large, brick-built underground cistern or well. The wells and waterwheels constituted an extensive system of water storage and irrigated farming about which I had never read.[12] Even today, thirty years later, this method of storage, and the world it organised, has never been described.

The question of storage is important to how we think about capitalism. In many accounts, pre-modern states governed through the control of storage. A large grain store in a city such as Cairo could serve as the recipient of government tax payments, a machinery of credit, and an instrument of price control. It could also be used to prevent disorder, by supplying grain in periods of shortage. Under capitalism, it is said, storage was transformed into a different method of organising the control of flows and populations. Most goods could be acquired on demand through market exchange, so now only certain strategic items were stored, mainly those available in liquid form – water, petroleum, and above all money.[13] The largest of these, money capital, is imagined as a reserve accumulated from the past, but one that loses value if stored. It must be quickly put in motion and dispersed, to replenish the store and indeed expand it. So, most accounts of capitalism in fact retain a simple storage model: what is piled up from the past governs the future. Here we are exploring a different understanding: capital as apparatuses that capture, or assetise, the future, allowing a few to enjoy that future income in the present and imposing on the majority the burden of its subsequent repayment. The acquisition from the future operates not only through financial engineering, but through the engineering of industrial life, human livelihoods, and ecological systems. The history of water storage and hydraulic engineering in Egypt can offer a different account of capitalist modernity. Storage neither disappeared nor increased. It was reorganised into new hands, relocated in new locations, and vastly simplified in operation. This terraforming transformation, in which the very life of the river was destroyed, was to serve not the storing up of a reserve but the constructing of a new command of the future.

For the British, irrigation engineering dealt with what was visible and measurable. It ignored what could not be directly seen and calculated. Colonial engineers addressed the Nile using the science of hydrology: in terms of the flow of surface water. They paid little attention to what would later be known as hydrogeology, dealing with water that was not visible or measurable, stored beneath the surface.[14] This inattention was the product of agnotology, not ignorance. They knew that water was present beneath the ground and, as we will see, widely used, but chose to ignore it.[15] By estimating what could be seen and easily measured and controlled – the area of a cross-section of a river channel or canal and the velocity at which its water moved – they were able to calculate the flow rate, or the volume per unit of time. The flow rate allowed them to estimate the 'duty of water', a term coined in the use of water power to run industrial machinery in Britain but taken up by irrigation engineers in India and then Egypt to calculate the area of each crop that a given supply of surface water could irrigate.[16] The aim was to make 'the Nile supply . . . do the maximum work that could be got out of it'.[17]

To understand the transformation that occurred, we need to consider the Nile differently, in terms of both visible and invisible water, bringing into view a different ecology. Historians refer to the previous arrangement as basin irrigation but mention only one-half of the system. No modern scholarship has described how farmers used the river as a year-round storage system, supporting perennial irrigation and multiple crop seasons.[18] The dams built by the British to impound the Nile did not introduce the storing of its waters and did not initiate perennial irrigation and year-round agriculture. Rather, they stored the waters differently, and less effectively, in a narrow channel on the surface. The engineers instigated, not perennial irrigation, but a trapping of waters on the surface of the land, where they had previously been dispersed and stored beneath it.

The Nile Valley runs downhill, both longitudinally (like all rivers), in this case northwards towards the Mediterranean Sea, but also laterally: the floodplain of the river is highest at the point closest to the river, on the levees adjacent to the main channel. Over millennia, as the river, swollen in late summer by the monsoon rains of Ethiopia, overflowed its banks and deposited a layer of silt, it dropped the heaviest particles of silt first, closest to the channel, creating a berm higher than the rest of the floodplain, as shown in figure one. Beyond the levees the land slopes downwards towards the outer edges of the cultivable valley, which ends at the limestone escarpment of the desert.[19]

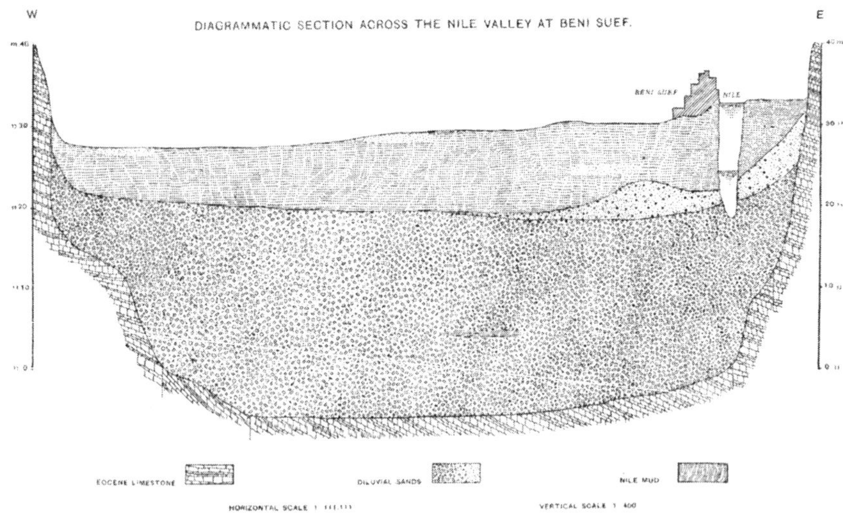

Figure 1. Cross-section across the Nile Valley at Beni Suef.[20]

The method of basin irrigation had been in use for millennia, through periods of development, disuse, and restoration.[21] Under this system, the main irrigation channels were cut through the berm to carry the summer inundation to the fields beyond. Earthen dykes divided the floodplain into large catch basins, each of a thousand to tens of thousands of acres, subdivided by smaller dams into manageable sections. Sluice gates or temporary openings in the dykes and dams made it possible to fill each basin in turn to a depth of one to three metres, store the nutrient-rich inundation for about forty-five days, and then release the surplus into a lower basin downstream and eventually back into the river.[22] Under Muhammad Ali, who ruled from 1805 to 1848, engineers had improved the arrangement of catch basins, enhancing the sequence of upper and lower basins and adding canals drawing flood water from further upstream to supply the elevated fields on the berm close to the river.[23] After draining the flood waters, as the ground began to dry and crack, farmers sowed the two main cereal crops, wheat for the bread of the well-to-do and barley to make ordinary bread and to feed horses and make beer (barley was preferred to wheat when the Nile flood was weak, as it required a shorter growing season and less water).[24] They used the same method for the leguminous crops, principally clover to feed cattle and water buffalo, alongside lentils, chickpeas, fenugreek, lupins, and broad beans. Unlike the cereal crops, the legumes added further nutrients to the soil, working symbiotically with bacteria in the root zone that absorb nitrogen from the air and process it into useable form. The interaction of water, air, silt, soil, bacteria, and human labour produced these crops without need for artificial fertiliser or further irrigation, nor even for the labour of ploughing, stone removal, pest control, or weeding.[25]

Every account of Egypt's agrarian history describes this system. But there was a second method for storing water, which was used to produce a second crop, in early summer, after the main winter harvest – and even a third crop, in the autumn.[26] The improvements introduced under Muhammad Ali were designed in part to enhance this arrangement.[27] The water was stored not in the field basins but in a substrate beneath them. Convinced by the colonial irrigation engineers that the old system allowed only one main crop a year, 'the land lying fallow during the rest of the year for want of irrigation water', modern historians have never described the old system that later-nineteenth-century irrigation schemes destroyed.[28]

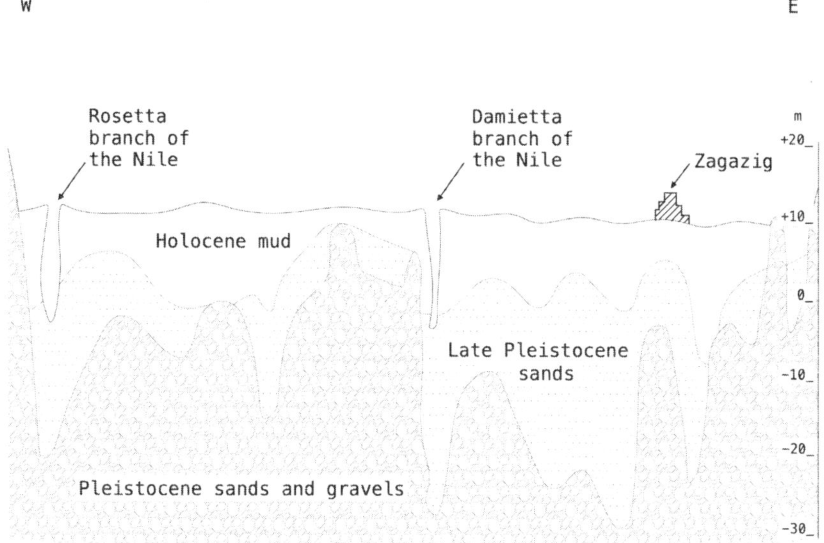

Figure 2. Cross-section across the Nile Delta at Zagazig.[29]

The soils of the Nile floodplain consist of two layers, as shown in cross-section in figure 2: an upper level of alluvial sands and clays, whose relatively impermeable material forms the basins that stored and slowly absorbed some of the flood waters; and a substrate of gravel and sand. This more coarse and granular material could hold large amounts of water trapped beneath the clayey soils above. As the flood waters inundated the surface of the fields via openings in the dykes and canals, they also recharged the gravel layer below, seeping down through the clays, or percolating directly from the sides of the riverbed as the river rose, or via the flooding of irrigation pits dug in the fields.[30] Even today, despite depletion by surface reservoirs upstream and consequent seawater intrusion downstream (exacerbated by rising global sea levels), this underground storage forms one of the largest groundwater reservoirs in the world.[31]

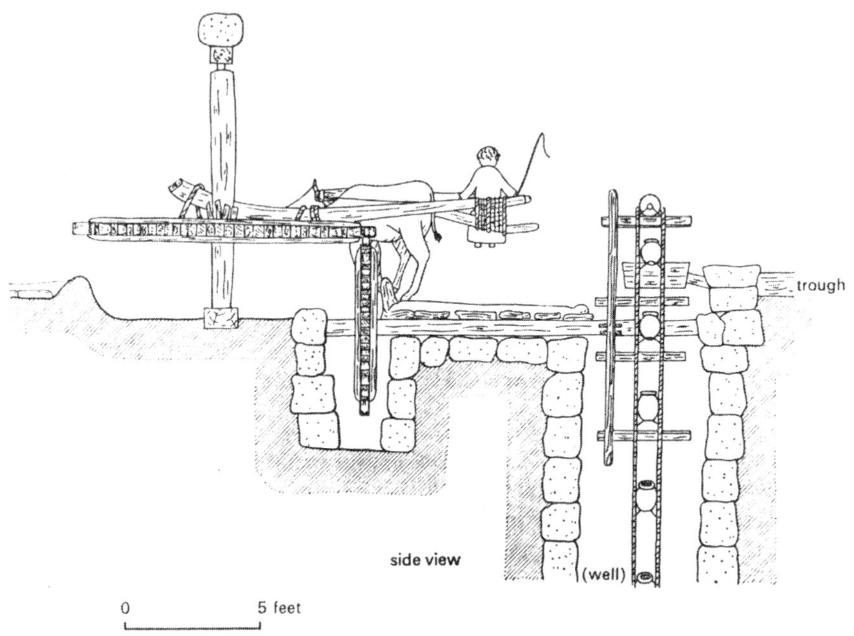

Figure 3. *Saqiya* raising water from a well.[32]

The second crop was irrigated from this subterranean reservoir, using *saqiyas*, or animal-driven waterwheels, placed over large, brick-built wells.[33] The *saqiya* is familiar to every reader of Egyptian history, but is usually assumed to have lifted water from the Nile or from a canal. Before the twentieth century, however, cultivators used it to raise water mainly from wells. It consisted of a horizontal wheel driven by an ox, turning two vertical wheels connected by an axle (see figure 3). The larger of the vertical wheels lifted the water from the well in a chain of pots. The well over which the *saqiya* was placed is less familiar. It was a large chamber, constructed by excavating a pit about ten metres in diameter, to a depth ranging from three metres to ten or fifteen metres, depending on the level of the ground water. The sides were lined with a circular wall of clay bricks, produced on site and fired to impart strength and resistance to water (see figure 4). To sink the well, as a medieval text relates, bricklayers at the surface laid courses of masonry on a ring-shaped wooden base made of acacia, while other men used buckets to excavate silt and water from the bottom of the pit, digging beneath the frame.

'As the soil under it loosens', the text explains, 'and as its own weight increases with the brick courses laid upon it, it goes down into the ground.'[34] When the well reached its required depth, the bricks were arched inwards to form a vault to support the machinery above, leaving a rectangular slot at the centre of the roof through which the vertical wheel of the *saqiya* and its chain of buckets could descend.[35] The brick-built wells were a permanent feature of the landscape, although except for the slot on the surface largely invisible. The waterwheels were typically disassembled each autumn before the flood waters arrived, then reassembled the following season.

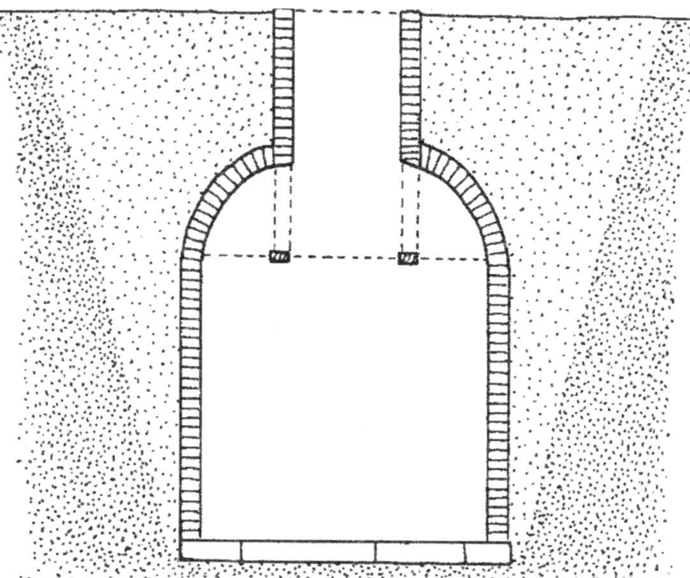

Figure 4. Cross-section of a *saqiya* well, showing annular wooden base, brick walls, vaulted roof, and rectangular opening.[36]

As the *saqiya* rotated, day and night through spring and summer, its wooden gear wheels groaned and squeaked. For any visitor, this was a distinctive sound of the countryside. In March 1845, at the order of

Muhammad Ali Pasha, the Egyptian engineer Joseph Hekekyan spent several nights in a village, investigating hemp cultivation. Among 'the noises and inconveniences of the quietest part of an Egyptian village night', which had greatly annoyed him, he listed 'the creaking of the heavy waterwheels'.[37] Foreign visitors usually travelled through the country by boat on the river, seldom spending time in the villages and fields. They apprehended rural life through its soundscape, often describing the characteristic noise of the Egyptian countryside in summer as the doleful squeaking of the waterwheel. 'Along here, seven or eight miles below Assouan, there is no vegetation in sight from the boat, except strips of thrifty palm-trees,' recorded one of the most popular accounts, 'but there must be soil beyond, for the sakiyas are always creaking.'[38] Much of this soundscape was soon to disappear.

According to the medieval agrarian almanacs, such as the one reproduced in the *Khitat* of Maqrizi, in the month of Tuba (9 January to 7 February), the fifth month of the Coptic calendar, as the main crops were maturing, farmers would restore the wells in the middle of the field basins, then bring out and assemble the waterwheels.[39] While the wells might be several metres deep, in many cases the gravel layer rose higher at the edges of the floodplain, both along the riverbed and at the outer margin. The pressure of the raised water table at the edges created an artificial head, forcing the supply up towards the top of the well.[40] Each well, Maqrizi tells us, could irrigate four to six acres; or up to ten acres, the almanac of Ibn Mammati specifies, if the land was close by.[41] The wells were gradually depleted during the growing season, reaching their lowest level on the fifteenth day of Abib (9 July), according to the almanacs, at which point the rising Nile would begin to replenish them.[42] The wells and waterwheels were so important to medieval agriculture, especially for producing sugar cane and other summer crops on large estates, that the sources often refer to such estates as 'waterwheels' (*sawaqi*).[43]

When European land speculators discovered these wells around the turn of the twentieth century, they assumed the pressure must come from reservoirs deep below and set up several artesian water companies to invest in exploiting this new resource. No artesian springs were found and the companies failed.[44] The recharging of the wells by the flood 'came as a surprise to us', the irrigation engineers later acknowledged.[45] They brought to Egypt an Oxford-trained geologist, Hartley Travers Ferrar, a veteran of Captain Scott's first Antarctic expedition, to investigate. His report recorded that at Qift, a town in Upper Egypt about twenty-five miles north of Luxor, 'a stream of flood water was seen flowing into a sakia-pit without raising the level of the water in it'. Noting that there were still an estimated 40,600 *saqiyas* in field basins, he added that 'if this takes place at every sakia which is submerged during flood time, the quantity of water which passes into the diluvial beds by this means must be very great'.[46]

The gravel beds extended beneath the entire area of the floodplain. However, in years of stronger floods, this underground storage could be supplemented with another reserve by allowing water to collect in low-lying ground, including land at the outer edges of the floodplain, or even adjacent wadis in the surrounding desert. Being the slowest to drain, low-lying, waterlogged lands were planted with flax, a valuable crop that prefers wetter soil and provides, when processed into linen, as we will see, another storage option.[47] Or the lowlands could be deepened by excavation to form pools that stored irrigation water for the following spring and summer. Numerous such *khaznas*, or storage ponds, were still in use in the nineteenth century.[48] In fact, along great stretches of the Nile floodplain, prior to the irrigation works of the nineteenth century, such pools formed a second river channel along the desert edge. Fed from ground water and surplus flood water, and filling as the main catch basins were emptying, this surface network supplemented the wells as a source of stored water.[49] (In fact, the deep 'summer canals' built by Muhammad Ali and his successors were, in some ways, an extension of this system, for they filled not only from the rising Nile but from the seepage of groundwater.)[50] In Middle Egypt, an ancient canal, the Bahr Yusif – formed, like most older canals, from one of these secondary channels of the river – allowed surplus flood water to escape to a much larger depression, the Fayyum basin, lying about fifty miles north of Cairo. In the ancient and medieval periods, Fayyum became one of the country's most productive agricultural regions and a centre of its linen industry. Managing the use of these escapes further reduced the risk of excess floods downstream.[51]

The principal crop produced from the well irrigation was *dhurra*, or sorghum. The cultivation of this summer-season crop was so extensive that sorghum served as the staple food of Egypt prior to the elimination of flood-basin agriculture.[52] Native to the semi-arid regions of Africa, sorghum is a grain adapted to the problems of light or variable water supply, typical of the flood-recession farming practised for millennia in the large river basins of the Sudanic belt – the Nile, Niger, and Senegal rivers.[53] Compared with cereals like wheat, it has much deeper roots, and its leaves are able to suspend photosynthesis during drought, entering a kind of hibernation.[54] Sorghum also tolerates waterlogged soils and flooding, in part because it is moderately salt-tolerant, withstanding the higher salinity of over-saturated soils.[55] It is a multipurpose crop, its leaves and stalk supplying forage for livestock, fuel for bread ovens, and thatch for housing. In years of low floods, if the winter harvest was poor, farmers expanded the summer cultivation of sorghum using additional well irrigation. To further compensate for a poor winter harvest, in the third growing season, the autumn, the area planted with sorghum would be 'very considerably increased', enabling small farmers 'to refill their granaries'.[56] The improvements to the basin system under Muhammad Ali facilitated this cultivation of sorghum in the third season.[57] Thus sorghum was suited to the challenges of using stored subterranean water and to variations in water supply. At the same time, it was a resource that could remedy the shortages to which such variability might lead.

The wells were also used to cultivate smaller areas of orchard fruits and other valuable summer crops, including indigo, sesame, cotton, and sugar cane, and to grow the nutrient-rich Colocasia (*qulqas*), or taro, a root crop that flourishes in moist soil and in fact grows better when submerged in water, allowing it to remain in the ground even into autumn when the next season's flood began to rise. Prized as a delicacy by the country's Ottoman-Turkish elite, taro produced twice the income per acre of cotton or sugar cane in half the growing time. A large household with little land, one source tells us, might live off an acre of taro.[58]

Public Storage

Storing water beneath the surface had several advantages compared to the system of river barrages, permanent surface reservoirs, and so-called perennial irrigation with which the British were to replace it. First, the old system allowed the inundation of the fields and depositing of nutrient-rich sediment. Soils were enriched both by the nitrates, phosphates, and trace metals in the sediment and by the enforced resting of the land during inundation. The new barrages and reservoirs were to destroy this self-moving process of flooding and fertilisation, creating a need for the heavy labour and expense of year-round water lifting from canals and mechanical fertilisation. Every historical account notes in passing this consequence of the new system and the advantage of the old. The further benefits are seldom noted.

The second advantage was that this self-moving enrichment flowed laterally in both directions, first the flood and then the ebb. The annual expansion and contraction of the stream allowed the two-way exchange of nutrients and biota between the river channel and the floodplain, sustaining terrestrial, avian, and aquatic life.[59] Destroying the seasonal 'flood pulse', as hydrologists now call it, eliminated interactions that reproduced life on both land and water, including birds and waterfowl and an abundance of fish and their food sources, whose cycles of nesting, spawning, and migration depended on the flood cycle and the nutrients it supplied.[60] When the flood receded, fishing nets placed at the mouth of the channels that carried the waters back to the river yielded an abundant catch.[61] Fish were also caught by hand in the fields, using reed traps, or from the wells into which the waters drained – allowing the underground cisterns to serve a second role, as fish tanks.[62] Cured fish, which could be stored in earthenware pots for months, was reportedly the animal food that ordinary people consumed most often.[63] Young fish also preyed on mosquito larvae, preventing the spread of malaria, which was endemic in Nile Basin countries farther south.

Third, with subsoil storage, little of the stored water was lost by evaporation. Under perennial irrigation, large volumes of water evaporated from the new reservoirs and from the expanded surface area of the river. At the same time, inundating the fields each year, then storing the flood waters below the surface, from where they were slowly released back into the river channel as it returned to its minimum level, automatically drained the land and flushed harmful salts to below the root zone. The flood and ebb of the river also flushed and dried the branch canals that carried waters to and from the fields, cleaning them of weeds and pests. Perennial irrigation using barrages and dams raised the average river level and water table by several metres, leading to waterlogged and increasingly saline soils, declining crop yields, and the need for extensive drains and pumps. Transformed into still, permanent ditches, the branch canals became a manufactured haven for water hyacinth – the floating plant introduced as an ornamental by Europeans – whose fibrous root mats, developed to resist the fast-flowing rivers of Amazonia, began to clog the waterways and lakes of Lower Egypt (as it did the major rivers of colonial India, the southern United States, and later the entire Nile Basin), doubling the loss of water by evaporation and killing off other forms of life.[64]

Fourth, the permanent canals and their banks also became breeding grounds for the snails and nematodes that carried bilharzia (schistosomiasis) and hookworm disease. As Jennifer Derr explains, these two debilitating parasitic infections became endemic to rural Egypt, slowly colonising the bodies of a significant part of the population.[65] Those bodies were also weakened under the new irrigation regime by increasing malnutrition, in particular the spread of pellagra. The canal system facilitated the widespread cultivation of cotton, which encouraged a switch to the consumption of maize (American corn), whose short summer growing season synchronised with the two-year rotation of cotton, and whose straw supplemented imports of coal from south Wales as fuel for the boilers of the steam-driven irrigation pumps.[66] A great transfer of workers from the sorghum-growing regions of Upper Egypt to the labour-intensive cotton plantations of the lower Delta contributed to this change in diet. Unlike other grains, maize does not supply niacin (vitamin B3), unless boiled in an alkaline solution, the method used in making tortilla flour in its native Mexico. Caused by lack of niacin, pellagra results in painful damage to the skin and digestive system, and in dementia and sometimes death. The British became alarmed at increasing rates of dementia.[67] The elimination of the flood and the exponential increase in cotton and cane cultivation also allowed other infections, parasites, and pests to flourish.[68] Rats, attracted to the juice of sugar cane, whose tall stalks protect them from birds of prey, infested cane fields. By gnawing the base of the canes, they could destroy up to a quarter of the crop.[69] The most destructive pest was the cotton worm, a moth (*Spodoptera littoralis*) whose larvae thrived on the increased cultivation of cotton. They could destroy entire fields of cotton if not checked by picking off the infected leaves one by one.[70]

Water stored beneath the soil had one further advantage over the new storage in reservoirs and canals. Unlike water held on the surface, it was beyond the control of powerful plantation owners, who might monopolise access to the new canals, and of the irrigation engineers who served their needs. 'Here we see basin irrigation at its best', noted a survey of the province of Sohaj, a region in the south that had not yet suffered the damaging engineering work of neighbouring provinces, in particular the extensive sugar plantations created further north. 'The land is rich, the crops luxuriant, the people well off and independent, and the subsoil water well utilised.' Farmers 'whose flood waters no engineer can cut off, and whose plentiful subsoil water at a convenient level is their own property', enjoyed a notable independence and prosperity.[71]

The waterwheels themselves were the focus of a complex interaction among humans, animals, soils, and crops. They were usually turned by oxen, whose fodder – mostly Egyptian clover – could be grown in the same fields, supplying three or more cuttings from a single planting and improving the soil. Cattle produced their own young, thus reproducing the country's main source of motive power. As in most agrarian societies, including those of Europe, rearing them was a principal means of storing and increasing household wealth (in English, the word cattle derives from chattel or capital, meaning stock; its sense narrowed to *live*stock only in the sixteenth century).[72] Using oxen as irrigation labour gave value to male calves; their replacement with mechanical pumps would later render the males largely useless, forcing their slaughter for meat and an increase in meat-based diets. Cattle dung, formed into flat cakes, dried in the sun, and stored on rooftops, provided fuel for baking and cooking.[73] The wells also became important to the country's silk production. Previously reliant on imported thread, Egypt developed a domestic silkworm industry at the turn of the nineteenth century. By the 1820s, the number of mulberry trees, whose leaves provide the food for the larvae of the silk moth, reached an estimated three million. The silk moth, a domesticated insect (the only one, in fact, besides the honeybee), produces eggs that require a cool place to incubate; they were therefore stored in the irrigation wells.[74]

Building a well required brickmakers to make fired bricks and bricklayers to construct it. The well machinery needed carpenters to fabricate the wheel and assemble and disassemble it each season, sustaining the largest non-farming livelihood in the villages.[75] To turn the machine, up to eight oxen worked in shifts, so a number of houses would typically share in each wheel, which distributed the water to individual plots via a network of shallow field ditches.[76] The collaboration kept the wheel turning day and night, and included the labour of those who dug and maintained the ditches and grew the crops that fed humans and animals alike.[77] Collaboration does not imply harmony or the absence of compulsion; as in any community, hierarchies within and among households would shape the allocation of tasks.[78] Twentieth-century accounts describe surviving instances of this cooperative arrangement as a '*saqiya* ring'. Often linked by kinship, the farmers each provided their own draft animal, sometimes participating in more than one ring.[79] In northern Sudan and Nubia, where the rising Nile usually remained within its channel and seldom inundated the wider floodplain, using the *saqiya* was so universal that the word came to refer to the village itself, a community formed through the collaboration around the wheel.[80] In Upper Egypt, the main summer harvest feast was celebrated not when the sorghum was cut but when the *saqiya* completed the tenth and final watering of the crop. After sharing *kunafa* or other sweetmeats brought out to the waterwheel in the field, then returning the empty plates piled with heads of sorghum, the owner would cut four more heads and fasten them to the foreheads of the two oxen that had turned the wheel, before driving them home in gratitude and celebration.[81]

On lands where a share of the produce was claimed by a tax farmer or other holder of rights to revenue, an arrangement widely used in the Mamluk and Ottoman periods, prior to the re-emergence of individual proprietorship in the late nineteenth century, the collaboration around the waterwheel would include the 'assistance' (*'auna*), meaning obligatory labour, required to run the wells of the person claiming those rights.[82] The periodic expansion of the *saqiya* system in pre-modern times may have been associated with the development of sugar plantations, requiring compulsory labour, both human and animal, on a larger scale.[83] From the 1820s, as the regime in Cairo began to recolonise the south of the country, the first orders to officials creating large agricultural estates were to secure cattle and 'erect waterwheels'.[84] Later in the century, especially after the great cattle plague of 1863–4, the ruling household and other owners of large estates began to replace *saqiyas* with British-built steam pumps, which initially were too heavy to disassemble and remove before the annual inundation but were robust enough to survive being 'completely buried under water every flood'.[85] Engineers then adapted the locomobile, a steam-powered plough, to serve as a mobile pump that could be wheeled into the fields and placed over wells when the flood receded.[86] Meanwhile, most farmers continued to rely on *saqiyas*, although, in the twentieth century, they gradually replaced them first with diesel pumps and later with small electric pumps (which today can be solar-powered).[87] However, their use was increasingly restricted to lifting water from the new irrigation canals. The water was now stored remotely, behind barrages and large dams upstream, especially after the height of the Aswan Dam was raised, in 1912 and again in 1933, and a second dam was built in 1937, at Gebel Aulia, on the White Nile just above Khartoum, and more so after the completion of a much larger structure, the Aswan High Dam, in the 1960s, and of a second giant barrage, the Grand Ethiopian Renaissance Dam, on the Blue Nile in Ethiopia, in 2020. With the elimination of the Nile flood and its benefits and with irrigation water now available by pumping it year-round from canals, the subterranean storage system was largely abandoned.

Silencing the *Saqiya*

The waterwheel and the well have disappeared. They have been eliminated both from the way the river is used and the way the colonising of Egypt is described. Their elimination from the history of the river can offer us clues as to why they were removed from the landscape. With them have vanished from sight the very ecology of the river and the economy of storage and modes of life they enabled. This motivates the question we have just posed: Can the waterwheel squeak? If we can never adequately recapture the subaltern voices that colonialism silences, there are other silences, other sounds, and other agencies, human and nonhuman, that are equally difficult to recover, but that an account of apparatuses of capture can nevertheless try to acknowledge.[88]

There are three reasons for the disappearance of the *saqiya*. Each tells us something about the way surplus was captured and how its capturing was to be reorganised. First, the best accounts of the history of the Nile prior to European colonisation are based largely upon sources that were concerned with a specific mode of governing the river. Their focus is the large-scale works through which authority was exercised and revenue extracted.[89] A principal source is a government register known as *al-Jusur al-sultaniyya*. Compiled by the Ottomans after they conquered Egypt in the sixteenth century, and based on older records, the register is a compendium describing the channels on the surface through which the flow of the river was managed.[90] The term *jusur* refers to the embankments or dykes whose upkeep ensured that the flood waters were properly contained, encompassing both the banks of the river and the canals that carried its water to field basins and escapes. The registers were still available in the late nineteenth century, when they appear to have been a source for the most comprehensive account of the country's irrigation network, Ali Pasha Mubarak's *Khitat*. A leading engineer and public official, Mubarak devoted volumes 18 and 19 of that work to describing village by village the infrastructure of irrigation.[91]

Writings that draw on these registers, directly or indirectly, to describe the irrigation system make only occasional mention of wells and the *saqiyas* that worked them. One could suggest a straightforward reason for this. As large-scale infrastructures, the canals and embankments required and enabled the development of a wider structure of authority, reflected in the maintaining of registers. As discrete, local equipment, the waterwheels and wells did not. A temporary apparatus assembled and disassembled each season, owned and maintained by households, the waterwheel provided no infrastructure for building the powers or claims of government. They were not, therefore, 'registered'.

The same relation between a technical apparatus and the exercise of authority is seen in a second major source for histories of the Nile, the records of the legal system. Court cases dealt disproportionately with disputes among strangers; they were perhaps more concerned with the problems produced by the same large-scale infrastructure. Canals created conflict in ways that waterwheels did not, for they traversed multiple properties and communities and were easily subject to interruption and dispute. The oxen that drove the wheels might be traded among strangers, or break loose and damage another property, so, if the *saqiya* appears in the sources, it is often in relation to disputes over animals.[92] The legal system expanded around such issues, and these were the main topics around which the history of the river could be written.

By contrast, sources that reflect the resolution of disputes within the community, such as the legal manuals used by local jurists (*muftis*) to settle conflicts without resort to the court system, can reveal more about the social world of the *saqiya*. 'If a person borrows another's ox to use with his *saqiya* and it falls into the well', one of the most widely used manuals explains, 'then he is liable for it'.[93] The replacement of the *saqiya* by a system of canals and reservoirs went hand in hand with the displacement of local juridical opinion by the system of courts and government registers that an expanded infrastructure enabled and depended upon. In other words, the presence or absence of the waterwheel in historical sources may reflect not just alternative means of irrigation but expanding or receding forms of justice and power.

The disappearance of the *saqiya* was further caused by the reliance of many modern accounts on European sources, especially the French engineers who helped develop Nile irrigation works under Muhammad Ali and his successors. The focus of the engineers, reproduced by more recent historians, was those schemes. Many contemporary scholars depend on the classic account by Roger Owen, who relies on Jean Mazuel's work of 1937, who drew on Linant de Bellefonds's 1874 study, to suggest that only about an eighth of the agricultural land in Lower Egypt was used to grow summer crops, and even less in Upper Egypt. This figure was a principal source for the erroneous idea that most of the country's land produced only a single crop each year. But Linant was referring only to the area fed by the new summer canals. He noted that powerful proprietors monopolised the canals, while ordinary farmers, 'who have dug the canals, are obliged to supply their summer plantations with water from wells, raised by means of sakiehs'. The *saqiyas*, he indicated, irrigated a far larger area – but were not recorded in the estimates of land devoted to summer cultivation.[94]

The main reason for the disappearance of the system of well irrigation, however, is that the British worked to eliminate it, both from how the river was managed and how its history was written. Egyptian engineers, forced aside after the British occupation, and aware of the deleterious effects of the large summer canals, especially the Ibrahimiya Canal built in the 1870s to serve the sugar plantations of Khedive Isma'il, the country's ruler, had argued in favour of retaining the system of catch basins and well irrigation. The engineer Ali Mubarak served as Minister of Public Works in Khedive Isma'il's final government in 1878–9, just before Britain and France pressured the Ottoman sultan to remove the Khedive from power. In 1879, while out of office, Mubarak published a book on the management of the Nile, *Nukhbat al-fikr fi tadbir nil Misr*, which made a series of proposals for improving the existing mixed-crop system and valley-based (rather than exclusively river-based) water storage: increasing the length of time water was stored in field basins, and releasing it more slowly; building additional storage in side valleys and depressions; and adding small barrages to raise the level of the river in parts of the Delta. He opposed the increased cultivation of cotton and sugar cane, in part because of the quantity of forced labour they required for ploughing, planting, tending, and harvesting and for digging and clearing the summer irrigation canals. He produced tables of figures to demonstrate that the mix of food and fibre crops Egypt cultivated under flood-basin farming in 1800 was more valuable than the extensively grown cotton and cane of the 1860s. He also argued that the greater diversity of crops supported animal husbandry and a range of local industries based on the processing of wool, flax, and other fibres.[95]

British officials had to work hard to discredit these views, especially when they were supported at first by the most knowledgeable of the colonial irrigation experts, William Willcocks himself. The man who would become director of reservoirs in the Egyptian government and the architect of the first Aswan Dam, Willcocks was the author of *Egyptian Irrigation*, a compendious study on whose third edition, published in 1913 and 900 pages long, every later historian in this area relies. In the first edition of that report, however, only 350 pages in length and published in 1889, just seven years after the British occupation, Willcocks argued for retaining and even re-expanding the system of catch basins and irrigation from wells. He compiled figures showing that the flood-basin system was more productive and profitable than converting land to canal irrigation, in part because 'the cost of raising crops on an acre of land in the basins is insignificant'.[96] While completing the Delta Barrage had allowed the irrigation of lower-lying lands by permanent canals, it was still the case even in Lower Egypt that 'above an 8-meter contour [above sea level], wells are freely used for irrigation', while in the basins of Upper Egypt 'wells are dug everywhere for summer irrigation', allowing a second and third growing season.[97] Other accounts published after the British occupation made similar arguments.[98] Like Ali Mubarak, Willcocks and other engineers preferred schemes for storing surplus flood waters in escapes at the side of the valley, rather than the damming of the river itself, and hoped that one day, when people were 'in despair' at the consequences of the new canal system and barrages, the Delta would be restored to the process of inundation still used throughout the south, reintroducing 'the old healthy system, by which the land is washed, matured, and limed annually by the Nile itself without any other agency'.[99]

Egyptian Irrigation was published in New York as well as London, intended in part for 'persons desirous of investing money in land reclamation schemes'.[100] However, Willcocks's superior, Lieutenant Colonel Justin Ross, added his own introduction to the first edition, saying that the author's argument in favour of flood basins and well irrigation was wrong. One of eight Indian-trained irrigation engineers, including Willcocks, whom the British had brought to Egypt, Ross had recently been promoted over him as inspector general of irrigation. In one of his first acts on becoming his supervisor, he reworked Willcocks's calculations to claim that well irrigation allowed the summer sorghum crop to be grown on only 10 per cent of the land. The other 90 per cent would be unused in summer, 'owing to the scarcity of labour'.[101] This calculation made the old system appear less productive. Leaving aside the fact that steam pumps could replace human and animal labour to lift well water in flood basins, a development already under way, labour was scarce for two reasons, one of which we will come to below.[102] But the immediate cause, as Willcocks himself explained, was the emptying of men from villages in the south to provide labour in the north, to excavate and maintain the system of canals on which cotton cultivation now depended, driven there by the imposition of forced labour or hired on contracts paid from public funds.[103] In other words, the extraordinary need for labour that made the new canal system in the north so expensive was used perversely to make the *saqiya* system appear impractical.

Four years later, in a lecture to the Scottish Geographical Society, Ross conceded that irrigation using wells was suitable, but only for small cultivators: 'To introduce summer crops, such as sugar, cotton, or rice,' he acknowledged, 'we have only, in the first instance, to dig wells and lift up the water by machines.' He insisted, however, that this method 'is suitable to cultivators who have small holdings, and who live in a self-supporting fashion. It is not economical, nor is there a maximum of produce for a minimum of labour expended.' Alluding, perhaps, to Sudan, where waterwheels (as noted) were still widely used and the Mahdist state had defeated the Egyptian occupation, he added, 'For semi-barbaric countries, liable to be upset by war and incursions of semi-savage enemies, it is excellent.' He even admitted that 'There are still many tracts of Upper Egypt cultivated thus, and the well-water is practically inexhaustible, as the Nile-water is never more than 35 feet below the surface even in summer.' However, for summer cultivation on a large scale, he argued, the use of groundwater must be replaced by a system of surface canals and steam pumps.

> We must bring the water of the Nile to the surface, either by a canal or a low-grade slope taking off far up the Nile, or by pumping water by steam power up to the field level or into a large canal, or we must make a dam in the Nile to raise its summer level, so that the cost of excavating the canal may not be excessive.[104]

The later editions of *Egyptian Irrigation* reduced the emphasis on the *saqiya* system, confining remarks on the use of wells mainly to preliminary observations on 'The Soil of Egypt'. When the report moved on to the detailed discussion of irrigation, the focus was on flood control and the construction of canals and barrages.[105] By 1908, Foaden and Fletcher's *Text-Book of Egyptian Agriculture*, published in Cairo by the Ministry of Education in the year of the founding of a national university, stated flatly that 'under the basin system, only winter crops can be grown . . . There is no agricultural work to be done between harvest and flood.'[106] The system of well irrigation using subterranean water was rendered invisible. Egypt's agrarian and economic history could now be written, with half the picture missing.

Animal Power

A third reason why historical accounts have ignored this system, besides its absence from government registers and the British desire to eliminate it, may be that most European writers saw only its vestiges.[107] Like the Spanish conquistadores landing in sixteenth-century Peru and finding an Inca world already collapsing from the smallpox that had reached there ahead of them, the French engineers under Muhammad Ali and the British who later replaced them encountered a country attacked by a series of epidemic and pandemic diseases that had travelled to Egypt in advance of the European forces. Although we have no reliable figures, in 1800 the population of Egypt may have been at its lowest point in centuries, afflicted by periodic outbreaks of plague and by the new threat of cholera. Both diseases spread with the increased warfare that came with Russian expansion into Ottoman lands to the north and British and French military expansion in the Eastern Mediterranean and South Asia, as well as the increasing movement of trade and pilgrimage traffic.[108] Building on earlier Ottoman public health policy, Muhammad Ali and his successors introduced quarantines and other measures that allowed population levels to recover, but epidemics of cholera and plague struck again in the 1830s, and cholera reappeared frequently, including in 1881, on the eve of the British occupation.[109] Europeans tended to see plague and other diseases as a natural aspect of a sickly Orient, part of what defined its difference from the West, rather than as the product of forms of violence, upheaval, and intensification of exchange in which they themselves played a leading role.[110] So the 'scarcity of labour' that, in their view, rendered existing methods of cultivation unprofitable was caused not only by an increase in the more labour-intensive method of canal-irrigated cotton production, but by longer-term effects of imperial expansion and warfare.

Even as measures to combat plague and cholera reduced their incidence, another disease affected more directly the employment of well irrigation using *saqiyas*. It attacked the main motor of the system, the country's cattle population. This was rinderpest, the cattle plague, a contagious disease that targets the digestive tract causing fever and extensive gastroenteritis, producing a 'loathsome stench' and, in most cases, death. In its severest form, it can kill more than 95 per cent of the animals it afflicts.[111]

The first major epizootic of the modern period struck in 1784–5, decimating the livestock of southern Egypt. It returned in Cairo and the north three times in the next four years. The government urged farmers to respond by using water buffalo instead of oxen to turn the waterwheels.[112] In many cases humans had to replace the dead animals, turning the waterwheels with their own bodies.[113] In the 1820s, the government organised the importing of oxen from Kordofan in central Sudan to deal with the continuing shortage.[114] Without animals, the wells fell into disuse. In 1826, Muhammad Ali Pasha complained to his official in charge of Minuf and Ashmoun, two western districts of the Nile Delta, that the district's 'wells and waterwheel structures are in disrepair'.[115] In 1833 the Pasha set up a government office for importing replacement cattle from Sudan, building large barns on the Nile at Qina, where the animals arrived from the south.[116] The virus struck again in 1841–2, arriving from the Russian steppes via the Danube and Trieste, and destroying as much as 90 per cent of the country's herds.[117] In 1861 and 1863–4, there were outbreaks in the south. Local reports speak of some localities losing three or four hundred head of cattle a day, farmers 'unable to water the land for want of oxen', communities too weak to carry off and bury all the carcasses, and dead cattle flowing down the river in their thousands. Once again, 'men [we]re turning the sakiahs' with their own bodies.[118] Further outbreaks were recorded in 1872–6 and in 1881 on the eve of the British occupation. Finally, there arrived the great rinderpest panzootic of 1889, which started in Ethiopia as Italian forces expanded their colonial occupation from the coast of Eritrea.[119] The plague spread north and south along the length of the continent for almost a decade, with devastating consequences in Egypt and the countries to its south.

In Ethiopia, the resulting famine, exacerbated by three years of El Niño–related droughts, was the worst in memory.[120] The loss of cattle removed the main source of heat energy, dried cow dung, putting pressure on scarce supplies of timber. By the last decade of the century, tree cover across the Ethiopian highlands, already reduced by agricultural expansion and further damaged by the invasion of Ottoman-Egyptian, British, and Italian armies, was at its lowest.[121] Denuding the land of its trees contributed to changes in the level and intensity of rainfall and runoff into rivers. The runoff becomes the flood waters of the Nile, which, in turn, saw more extreme high and low floods. So, the cattle plague may have had a second impact in Egypt: as well as repeatedly destroying the system of animal power, it simultaneously made the flood waters that those animals lifted using waterwheels more variable and unpredictable.

When British engineers surveyed and wrote about the irrigation system as they found it in the 1880s and 1890s, and French engineers in the decades before them, they were witnessing a mechanism whose vital parts and main source of energy had been repeatedly attacked and destroyed.[122]

The So-Called History of Famines

Anxious to justify the colonial re-engineering and ecological destruction of the river, the British authorities overlooked the recent history of epidemics and epizootics to which their own imperial actions and those of other European powers had contributed and ignored the system of subterranean water storage that had regulated the supply of water and created a fertile and adaptable system of farming. They ignored the use of marginal lands as an escape, in which excess water could be stored and used in years of high floods. They ruled out the proposals of the leading Egyptian engineer, Ali Mubarak, to increase the use of such reservoirs as an alternative to the damming of the river.[123] They also rejected the Wadi Rayan scheme, the project of an American archaeologist to restore an ancient lake adjacent to Fayyum to store excess irrigation waters, as an alternative to the Delta Barrage.[124]

In addition, as the colonial regime consolidated a more centralised control of the river, based on permanent canals and barrages, a new space of ignorance opened up. The use of the telegraph allowed officials to monitor the threat of flooding from Cairo, becoming fearful of the 'risk'. Calculations and decisions that were previously handled locally were now to be made at a distance.[125] In the past, during high floods, when the rising Nile approached its peak, watchmen would be stationed along vulnerable stretches of the embankment, assisted by gangs of men ready to deal with any breach, while the river was covered with small boats laden with stones and stakes for making repairs, all carrying lanterns. With the lamps reflected in the water, 'the scene is full of gaiety and animation', reported a British district engineer at one such site. 'A stranger might imagine a great fair was on.' Admiring the 'steady business-like manner' in which any threat was addressed, the engineer 'felt that it was almost impossible for a breach to occur, while the officials in Cairo, seated in their offices reading the duplicates of telegrams, were always nervous and timid'.[126]

The alleged vulnerability of the old system was partly a product of the very ignorance produced by the measures to eliminate it. From their offices in Cairo, senior British officials pictured Nile irrigation over the preceding centuries as a system of great human precarity, in which the very survival of the country was permanently at risk from the river's unpredictability and the ever-present threat of ecological devastation and famine. The officials also blamed this unpredictability on resistance to colonial rule, including the establishing of an independent state in Sudan in 1881 under the Mahdi, Muhammad Ahmad bin Abd Allah. 'The Dervishes', they complained, 'prevent any scientific examination' of the flow of the Nile.[127] River control would become an argument for the elimination of such independence. The ecological and political 'risk' of an unruly river, magnified by an unruly population, was a further justification for the river's destruction.

This view of precarity is reproduced in most modern historical accounts, providing an alibi for the act of rivercide. It has been reinforced by a misreading of earlier Arab writers.[128] Histories of medieval Egypt, based on contemporaneous sources, offer a different picture. The flood regime of the Nile was more predictable and more dependable than that of any other world river, thanks to the multiplicity of its sources in equatorial East Africa and the regularity of the Ethiopian monsoon.[129] While the height and timing of the crest of the flood varied from year to year, the variation could be predicted once the water levels began to rise – for example, by recording the interval between surges and the density of the alluvium deposit at measuring stations along the Nile.[130] The area to be cultivated could be extended or restricted based on these predictions, with the further options, as mentioned, of storing excess water in escapes or dealing with low floods by planting additional crops using well-water irrigation in a second and even a third season. If by the third season the wells were running low, the rising Nile of the following summer replenished them.[131] There were years of abundance and years of scarcity, but the extremes were moderated by the ability to store grains and pulses from one season to the next, whether within the domestic dwelling or in the granaries of large merchants and the ruling households, who might hold thousands or in some cases tens of thousands of tons.[132] Charitable foundations (in Arabic, *awqaf*, or *waqf*, singular), whose extensive landholdings developed as a means of shielding rural and urban property from the government's revenue claims, used their food stores and kitchens to support the disadvantaged in periods of shortage.[133] The country's textile production, based mainly on growing flax to weave into linen, an industry that flourished in the medieval period and was revived again in the seventeenth and eighteenth centuries, provided a further storage system.[134] Thriving in the moist soil of low-lying tracts, flax allowed the utilisation of surplus land from a heavy flood, and the storing of the surplus.[135] The flax industry may have expanded partly for this purpose, alongside the profit that merchants made from advancing credit for its production. Transforming a field crop into bolts of cloth provided a means of storing wealth that was more secure, valuable, durable, and fungible even than grain. In fact, rolls of cloth could be used more easily than granaries to secrete wealth to avoid the payment of taxes, and may even have been used to pay those taxes.[136] Thanks to the ability to store grain, pulses, fabrics, and flood water, in medieval Egypt the fluctuation in the flood pulse was never a cause of famine.[137]

The main threat to livelihoods in the pre-modern period came not from the variability of the river but from occasional episodes of political strife, as with the 'great calamity' (*al-shidda al-'uzma*) of 1062–73, or the political crises at the end of the twelfth century and again at the end of the thirteenth.[138] More frequently, the threat came from large merchants, who were tempted to hoard supplies of grain in anticipation of shortages in order to drive up the price. They would sometimes exaggerate the threat of a low crest to the flood, to panic the population into purchases that would justify their raising of prices.[139] The principal historical source on these matters from the Mamluk period, Maqrizi's *Ighathat al-umma bi-kashf al-ghumma*, written around 1405, chronicles every known episode over the preceding centuries of food shortage and distress. It acknowledges the role of climatic factors in the variation of the flood and grain production. It mentions in passing a fantastical account from two centuries earlier of a drought so extreme that the well-to-do began to consume the flesh of children, a taste 'to which they became accustomed'.[140] But the text addresses mainly the responsibility of government authorities and the market supervisor, or *muhtasib*, an office in which Maqrizi himself served for the city of Cairo on three occasions, to control prices and punish fraud. He attributes recent episodes of dearth to the decision of the authorities to cease coining money in silver and rely instead on book credit (money of account) supplemented with token copper coins, a practice Maqrizi blames for debasing the value of money and impoverishing the peasantry.[141]

The rhymed title of Maqrizi's book is difficult to render in English but can be translated as 'Aiding the Community by Examining Its Distress'. The French Orientalists who first noted the work's existence gave it a different title: *Le Traité des Famines*.[142] Calling the book a 'treatise on famines' captured not the meaning of Maqrizi's text and its concern with money and credit (a topic to which we turn in the next chapter) but the colonial preoccupation with agrarian crisis. The supposed 'risk' of famine established the inadequacy and precarity of the old order and provided a further alibi for the planned destruction of the river.

Capitalisation

Focusing on surface water and ignoring the methods of storing and lifting irrigation water had an additional reason. The old methods of animal-powered, multiple cropping, based on both surface and subterranean storage, were too adaptable and too complex for the novel purpose that agriculture was now to serve. The colonising and capitalising of the Nile could not easily coexist with the regime of flood-recession farming, not because the land needed to produce more – as we will see, the productivity of the land was set to rapidly decline – but because it needed to produce differently, in ways that turned inundation from a resource into a threat. Colonialism would now engineer a 'flood-vulnerable landscape', one whose risks were better suited, not to the exploitation of nature, but to extraction from the future.[143]

The Nile under British colonialism was to operate as a means of profiting from – by taxing – the future. Mechanisms for extracting payments from the future were not in themselves something new. For centuries, the large merchants of Cairo had prospered from the ability to extend credit to cultivators, through contracts for the advance purchase of crops that they could sell for a higher price when delivered. The requirement to pay taxes at regular intervals helped compel those who farmed the land to sell their produce in advance and become indebted. The authorities imposing taxes and the merchants advancing credit for their payment depended on one another for their own enrichment. In recent centuries, under Mamluk and Ottoman rule, the right to a share in the tax revenues of specific villages and their lands had been assigned as salaries to military officers and later auctioned to tax farmers. These rights to revenue evolved into rural estates, inserting the holders as middlemen between merchants and cultivators in the mechanism of extraction. In the early nineteenth century, Muhammad Ali replaced tax farming with direct taxation, tried to cut out the large merchants by commanding the delivery of crops directly to government warehouses, and then began allocating future harvests, and gradually entire estates, as credit payments to foreign merchants and as recompense for the military officers and officials of an expanding army and rural administration. Those controlling the estates, many of which were confiscated from the former Mamluk ruling households, were made 'responsible' (*muta'ahhid*) for the taxes owed to the government, which gradually raised and simplified the tax rate.[144]

In the second half of the century, as Amr Khairy Ahmed explains, two crises helped generate a new scale of capitalisation. First, waves of cattle plague, wiping out the animal power that drove the country's irrigation machinery, created the opportunity to introduce large numbers of steam-driven pumps, supplied on credit. While cattle were raised, nourished, and reproduced through collaborative arrangements or partnerships that seldom provided openings for extracting credit payments, the much greater cost and longevity of a steam pump allowed a proliferation of loans, including those provided to purchase imports of coal. Second, the perpetuation of slave-based cotton production in the United States triggered another crisis, the American Civil War and the worldwide cotton famine of 1861–5. Coinciding with the 1863–4 cattle plague in Egypt, the jump in cotton prices offered the country's European and Levantine financiers the means to create credit on a new scale.[145] Steam-powered cotton gins joined the irrigation pumps as machineries of capitalisation. When the cotton famine abruptly ended, Khedive Isma'il, who ruled from 1863 until his deposition at the behest of his European creditors and their governments in 1879, switched to sugar production. He commissioned the digging of the country's longest canal, the Ibrahimiya Canal in Middle Egypt, equipped it with large steam-powered pumping stations to feed extensive cane plantations, and built 'immense sugar manufactories worked by steam, in which the cane is ground and the boiling and refining done'.[146] The size of the machinery, infrastructure, and plantations facilitated an even larger creation of credit. Between half and three-quarters of the loans that Isma'il took on were used to purchase steam pumps and build factories, light railways, canals, and other infrastructure for the new plantations.[147] The increasing independence of Isma'il from Istanbul allowed the financiers to secure the credit through mortgages on government revenues and eventually on entire government and royal estates. By the end of the 1870s these state and royal domains, now seized and administered by the European creditors who had declared Isma'il bankrupt, occupied 10 per cent of the country's cultivated area.[148] The plantations and machinery, the levels of debt they engineered, and the ecological, financial, and political crises to which they contributed triggered the British occupation. What Isma'il and his creditors had pioneered on the country's sugar and cotton plantations was now to be attempted across the country as a whole.

Expanding further the system of extraction from the future required water to be stored differently: rather than using field basins and the substrate beneath them, and the handful of deeply excavated summer canals, which depended on seepage from the same groundwater, storage was to be located in a large surface reservoir, in the raised volume of the river itself, and within the cellulose walls of industrial crops.

To capitalise Egyptian agriculture on a new scale, turning the Nile itself into a machine of credit and time transfer, financiers sought to expand the growing of the two industrial crops, cotton and sugar cane, neither of which was popular among small farmers. Industrial crops are not consumed locally and cannot be set aside for future use. While farming households and merchants can store cereal grains, pulses, rice, and locally processed textile fibres for years, they cannot easily store cotton or sugar cane. At harvest, the entire crop moves quickly out of local hands and is carried to sugar mills or cotton gins.[149] The interruption between the growing of the crop and the processing and manufacturing of its product made it easier to reduce the share of the crop's value retained by the grower. Farmers who conventionally kept a large part of a staple food crop might receive less than half the value of crops grown for manufacturing. The non-farming share could now be made available to investors, via the sale of title to the land, becoming a long-term future income to be capitalised.

For reasons related to their industrial purpose, industrial crops grown on an extensive scale were incompatible with the existing system of water storage, and thus incompatible with the life cycle of the Nile. They required a different period of cultivation, based on a different timetable of access to irrigation water. Most basic food crops grow quickly – the food is typically stored in the fruit or seed, which can be ready to harvest within a few weeks of sowing.[150] In an environment with year round sunshine and irrigation water, from flood water and well water, this allowed, as we have seen, the growing of multiple crops in a year. The value of the two industrial crops, in contrast, lies not in a fast-growing fruit or seed, but in the cellulosic structure of its support – the stalk that stores the juices of the sugar cane or the boll that contains the lint of the cotton plant.[151] This cellulose storage vessel takes much longer to mature. Long-staple cotton occupies the ground for nine months and sugar cane for up to twelve, in addition to the weeks needed for ploughing and preparing the soil, allowing for only one crop per year. The old double-layered system for storing irrigation water required the flooding of most fields for up to six weeks in late summer. The industrial crops, needing up to a year to build their own storage system, could not be synchronised with the flood cycle.

It was always feasible to reserve small sections of land for cane and cotton, along with the third industrial summer crop, indigo, to be irrigated from wells or from the rising river using waterwheels.[152] In the medieval and early modern period, for large merchants, the extended growing season of these crops was not a handicap but a source of their profitability. Cotton was generally restricted to small plots of an acre or an acre and a half, in a countryside where field basins extended for hundreds or thousands of acres.[153] Sugar cane was more extensively cultivated. From around the eleventh century, Egypt developed an important sugar industry using irrigation of small tracts by waterwheel. The sugar was valued as a medicine, and the investors in early cane-pressing factories were often physicians.[154] In the seventeenth century, according to Nelly Hanna, there was a renewed investment in sugar cultivation and, by the end of that century, the refining and trading of sugar was one of Cairo's most profitable businesses.[155] For large merchants, the advantage of cane lay precisely in the length of its growing season. The year-long delay between planting and harvesting forced farmers to enter into debt by selling the crop in advance, to provide funds for tax payments and other needs. For merchants, a significant part of whose profit came from the supply of credit (as we will explore further in the next chapter), the longer the growing season, the greater the potential profit.[156] For the same reason, most farmers avoided planting sugar, due to 'the advances that its cultivation requires'.[157]

Expanding the cultivation of cotton and sugar was inconsonant with the flood-recession ecology of the Nile. To extend their production on the scale first attempted in parts of the Delta in the 1820s, expanded in the Delta and Middle Egypt in the 1860s and 1870s, and envisioned for the entire country by the British colonial administrators, and by financiers like Ernest Cassel and his partners, the life of the river now had to be destroyed.

In fact, the expansion of industrial crops, under way from earlier in the nineteenth century, had already begun to disrupt the river's flood pulse. The crops were incompatible not just with the flooding of individual fields, but with the entire ebb and flow of the Nile. Since the fields that were being converted to cotton and cane were no longer able to store significant flood waters in the summer, the expanding cultivation of these crops left fewer basins available to cope with excess water. More and more of the flood had to be contained dangerously within the banks of the river and its canals. This change increased the risk of devastation in years of a high Nile crest. Moreover, with flood basins the land was usually lying fallow as the waters rose, so an unexpected inundation did little damage to crops, and any flooding of the berms was usually compensated by an increased yield in the following season thanks to the additional layer of silt. But when strong floods or weakened dykes led to the accidental flooding of a field of cotton or sugar cane, the cultivator lost an entire year's production.[158] Conversely, unlike sorghum, the industrial crops were not adapted to deal with water scarcity or drought, so the risk of losses from a low Nile was much greater. The periodic extremes in the level of the Nile flood, and the occasional risk of accidental flooding, became an unmanageable threat. The apparent unpredictability of nature and risk of famine that the British sought to overcome was a product of abandoning a more complex system of water storage and of introducing the vulnerable form of control and durability required for capitalisation.

The spread of industrial crops was also the cause of the other problem the British sought to solve: the widespread and increasingly exacting use of the *corvée* or forced labour. The *corvée* of the later nineteenth century was not an ancient method of deploying labour by a despotic government, as the Europeans often portrayed it.[159] There was a longstanding practice of requiring collective labour to maintain embankments, canals, and other irrigation works, and much larger workforces were mobilised in periods of imperial expansion to construct new basins and canals (usually dug from abandoned channels of the river).[160] Tax farmers who controlled large estates, and the 'tax guarantors' who replaced them under Muhammad Ali, could compel labour from those they taxed.[161] But maintaining the banks as the flood water rose in midsummer and allowing it to flow into the fields in sequence, then cleaning the canals when dry in the winter, was a manageable task compared with the work of keeping the embankments intact around fields of cotton and sugar, using the canals as long-term reservoirs that had to be prevented at all costs from collapsing and flooding the adjacent fields.

A larger problem that the British encountered with the conversion of large tracts to cotton and sugar cane was that with the development of barrages and permanent canals the fertilising silt that used to be carried by the flood waters and deposited across the fields – estimated at 125 million tons per year – was now deposited in the irrigation canals.[162] The ancient source of the fecundity of the fields had been transformed into a waste product. When the British occupied Egypt, the bulk of the *corvée* labour by that point was required for the annual excavation of this unwanted silt (and in addition as unpaid work on large estates in the labour-intensive cotton harvest). Since it was impossible to remove so much silt, the floor level of the main canals rose, raising the level at which they took water from the river and making it more difficult to use the canals in years of low floods and thus rendering the risks to cultivation in such years much greater.

Dams Across the Nile

The solution to the problem of 'waste' was to complete the destruction of the river by building the dam at Aswan and additional barrages downstream and later upstream of Aswan to block its flow. The flood pulse was replaced by engineering a year-round channel of water in the old riverbed, which became a regulated, canalised stream, most of it never even reaching the sea. Although the first Aswan Dam was designed with under-sluices to allow some of the silt-laden water to flow downstream in the summer, much of the 'waste' would now be trapped in the reservoirs behind the dams. The barrages would raise the level of the river, eliminating most of the annual need to dig out the canals. The river and its canals replaced the fields and underground reservoirs as local sites of water storage.

This had unforeseen consequences, several of which have already been mentioned in describing the advantages of the old system of underground water storage: raising the water level of the river and canals year-round caused problems of salination and water logging; perennial canals provided a home for parasites that caused endemic and debilitating diseases; and, above all, reducing and eventually eliminating the flood destroyed the flood pulse on which both the fertility of the fields and the richness of the river's biota depended. In addition, while crops sown in flood basins required no ploughing or harrowing, the land now received extensive mechanical treatment – a minimum of three ploughings in the case of cotton, to bring unexhausted soil to the surface and eliminate weeds.[163] Ploughing, as we now know, destroys the structure and microbial life of the soil, gradually reducing its fertility (and also releases the atmospheric carbon sequestered by plants).[164] All of these factors contributed to the 20 per cent decline in agricultural yields between 1900 and 1920. Farmers were forced to replace the lost nutrients from other sources, mainly imports of nitrates and later of synthetic fertiliser, at an increasing cost, but failed to stem the decline in yields.

A further cost of the river's destruction, and benefit for capitalisation, was that its floodplain was rendered not only less fertile, but available for conversion to roadways and housing. Human settlement had previously been confined mainly to the raised berms along the banks of the current river channel and of ancient, abandoned channels; to the uncultivated escarpment at the valley's edge; and to the numerous 'tells' or human-formed settlement mounds rising above the floodplain, the result of building and rebuilding for centuries or millennia on the same site. With the elimination of the flood, the well-to-do could now expand their housing into the fields below the villages and towns, or build lavish *'izbas*, or estates, in the countryside, including the estate that Joseph Schumpeter took a hand in planning, north of Cairo.[165] In addition, around Cairo and other cities, land development companies began converting both existing and potential farmland into speculative housing schemes, and into factories for brick-making and other industries. So agricultural land became not only less fertile but less available, despite the promise that flood control would extend the area of cultivation. Instead, the cultivated area actually decreased in the years after the building of the dam.[166] Even the much larger High Dam, in the 1960s, produced no increase in cultivated area in the decade after it was built, due largely to urban encroachment.[167] None of that loss of land, however, compared with the loss suffered by the people of Nubia, the country lying between the Nile cataracts south of Aswan. The building of the first Aswan Dam flooded the northernmost hundred miles of their land, while the raising of the dam in 1933 submerged all of Upper Nubia. The High Dam was later to form a reservoir that obliterated the entire country – fields and villages alike.[168]

Finally, ending the flood brought to an end not just a system of farming but the geological process through which Egypt itself had been formed. The annual deposit of alluvium had formed the Delta where the majority of Egyptians now lived and farmed, building out its shoreline as a wide fan into the Mediterranean. Over millennia, the annual increment of a few millimetres of Nile sediment had kept the sea at bay and counteracted the gradual sinking of the Delta, a phenomenon caused by the compaction of the strata beneath. Deprived of the flood, with the land continuing to sink, and with global warming from the late twentieth century causing the sea level steadily to rise, this terraforming process was pushed into reverse. By the early twenty-first century, dozens of metres of coastline were being lost to the sea each year. Concrete sea defences could limit some of the surface loss but could not prevent the sea's infiltration of the groundwaters below. The groundwater of the northern third of the Delta turned saline, increasingly unable to support agriculture, and that of the middle third became brackish, rendering it harmful to most crops. With the Nile turned into a largely enclosed reservoir, less than 10 per cent of its water reaching the Mediterranean, the sea was now salting the earth. All this was occurring even before the much larger rise in sea levels that continued global warming would bring.[169]

The gradual elimination of the flood pulse was accompanied by a transformation in property rights and land taxes. Prior to the re-engineering of the Nile in the nineteenth century, governments in Cairo imposed taxes on agriculture that were adjusted each year according to variations in which lands were flooded and what crops were planted. In the medieval period, there were different tax categories for land planted with wheat, flax, and so on, but also for land planted the previous year with those or other crops that exhausted the soil. These would be assessed differently to allow periods of fallowing, or planting with crops that restored the land's fertility. While the tax was not charged as a percentage of the yield, it was calibrated nevertheless to the variation in fertility, flood pulse, and the nature of the crop. In the nineteenth century, when the government restored direct taxation after the period of subcontracted tax farming, it eliminated this relation between tax payments, crops and soil fertility.

Like the Suez Canal a generation earlier, the dam itself was a project with a durability – an extended temporal and technical scale – that made it available to capitalisation. Scale matters, not because it produces a more efficient or effective irrigation mechanism, but because it creates a durable credit machine. It builds the future that the creation of credit requires.

Since the debt crisis in 1876, Egypt's public finances were under the supervision of a British and French controller. An agency representing private European creditors, the Caisse de la Dette publique (Fund for the Public Debt), received government revenues directly, including customs duties and railway and telegraph revenues, and other sources allocated to debt service. The Caisse operated in effect as a privately run treasury. The arrangement restricted the government from borrowing funds without the approval of the Caisse. The French bondholders, in particular, refused to approve large new loans, including a loan to finance the building of major irrigation works.

The Aswan Dam was financed, not by direct government borrowing, but by creating money on the London money markets. Ernest Cassel, the British financier, created an unusual investment scheme to pay for the dam. This is not just a question of how the dam was financed, how a large engineering work found a supply of credit, but more, how finance *was engineered*, how the creation of large-scale credit found a material structure through which it could come into being.

The contract for the construction of the dam was awarded to a British engineering firm. The work was paid not in cash but in 'irrigation certificates'. This was a form of money created by Cassel and his partners, who set up the Irrigation Investment Corporation in London and issued the certificates through the Bank of England. Investors purchased £100 certificates in instalments, by writing cheques on their bank accounts, starting with an initial cheque for £5, with a £3 fee bringing the total cost of the certificate to £103. None of this involved the payment of funds saved from the past: it was funded by the future tax payments of Egyptian farmers. The banks of the investors buying shares created credit money, by marking each cheque payment as a debit against the name of the investor and crediting the same amount to the Bank of England. The latter, in turn, credited the funds to the contractor's account, over three years, as the stages of construction were completed. Meanwhile, the Egyptian government was supposed to increase the tax on lands that received irrigation waters via the new storage system, although, as noted, the areas of land irrigated, and thus the increase in taxes, were much less than the dam's proponents had anticipated, and yields from the land began a sharp decline. After three years, the Egyptian government began making payments every six months from these tax revenues to bank accounts of the holders of the certificates, through Cassel, with an additional amount equivalent to an interest rate of about 6 per cent. Two-thirds of the interest went to the certificate holders, and one third to Cassel and his partners. For organising a loan of £2.7 million, in which none of his own capital was involved, Cassel, the entrepreneur of the project, earned a fee estimated at £500,000 to £1 million. The dam was a machine for capturing much more than irrigation waters.

This astounding deal was just the start. Cassel and his partners invested in the creation of a sugar plantation on the Plain of Kom Ombo, just downstream from the dam at Aswan. The scheme drew in hundreds of small investors and tenant farmers, who struggled to make a return or a living. Cassel also set up a banking consortium, which he named the National Bank of Egypt, to extend further credit to the Egyptian government. Before the building of the dam was announced, Cassel engineered an equally intricate arrangement to acquire the former Khedive's sugar plantations, the vast landholdings known as the Daira Saniya, or royal domain, with twenty-two cane-pressing mills, and cotton estates, which had been mortgaged to his European creditors. The construction of the dam sharply increased the value of the estates. Some of these were turned over at large profit, again with the credit for their initial purchase created by the financiers involved. The industrial plant and some of the sugar estates were retained by Cassel, in partnership with Ernest Cronier and local investors, who created the Egyptian Sugar Company – the place, as we saw in the last chapter, where Joseph Schumpeter's theory of the entrepreneur was born.

Following this, dozens and eventually hundreds of other entrepreneurs began to realise the opportunities for the capitalisation of sugar and cotton estates. Europeans, sometimes with local Cairo partners, set up land investment companies. These would acquire title to marginal lands that could now be levelled, irrigated, and leased to cultivators. Capitalisation was calculated over fourteen years: land could now be sold, not at the old rates at which parcels changed hand, related to levels of land tax, but (as with the new method of valuing company shares) according to a novel mechanism of calculation: its value was based on estimated future earnings, at fourteen times the annual net profit.[170] Land prices and rents doubled almost overnight. But, to obtain this price, half the land had to be planted in cotton. Cotton requires a three-year rotation, to prevent rapid depletion of the soil. With capitalisation based on keeping half rather than one-third of the land in cotton, soil depletion accelerated and yields began their sharp decline.

Land Development

Meanwhile, much larger futures could be built on the future promised by the dam. The structure engineered not just new possibilities for capitalised agriculture; it implied, for the first time, a long-term future for the British occupation that sanctioned and facilitated the dam's construction. The projected political future provided the frame for other long-term structures. There was an urban land boom, as property development companies were formed to build new neighbourhoods in Cairo, and entire suburbs like Maadi and Heliopolis, typically built in association with monopoly concessions to developers to run tramways and utilities. 'According to the figures given for 1914, out of £E210,000,000 (the total sum of foreign capital investments in Egypt), 166,300,000 or 79 per cent was accounted for by non-productive investments (public debt, mortgage and banks), 26,500,000, or 12.6 per cent, by transport and trade and only 10,500,000 or a mere 5 per cent by industry and construction.'[171]

Note that the powers of capitalisation did not have to be placed in private hands. It would have been perfectly possible to convert the extraordinary windfalls from future payments into public revenue and use them for social ends. Initially, in some cases, this happened. For example, the mainline Egyptian railways, starting in the 1850s, were built by the government. The ruling regime refused to grant the rights to private companies. By 1904, the railways were responsible for 20 per cent of government revenue – although it had been forced by the 1870s debt crisis to commit those revenues to the repayment of debt.[172]

How does capital consume the future? And how are we made blind to this? The consumption of the future appears speculative, flimsy, and artificial. But, in fact, it works through a technical apparatus of capture, which must be engineered. This engineering allows the capitalist, and the coloniser, as we have explored in the case of Egypt, to portray the revenue captured from the future as the product, not of those whose lives later on repay it, but of technical improvements. We have studied the engineering closely, at the level of canals, storage systems, animals, and crops, for the claim of technical improvement and economic progress has such a hold on our imagination – a hold so powerful that we no longer even hear how the waterwheel used to squeak. The engineering does not necessarily produce any improvement in wealth or well-being – in fact, it can make those worse. It was designed mainly to improve the capture of the future.

3

Capitalism as a Detour

Among the many victims of the Covid-19 pandemic, there was a minor, unexpected casualty. The Great Lockdown of 2020 pushed the world into the worst economic recession since the Great Depression. The loss of millions of lives to the coronavirus was accompanied by an even wider loss of livelihoods and economic futures. Yet, amid this hardship, stock markets around the world, after a sharp setback, experienced a year of extraordinary growth. Their expansion followed the decision of leading central banks to create trillions of dollars of new assets, almost doubling the size of their balance sheets, using the method of manufacturing money known as quantitative easing, first widely deployed in response to the global financial crisis of 2008.[1] The multiplying of banking assets saved the financial markets. The casualty was any lingering claim that our common sense, or the textbooks of economics, understood how money works.

Why is it so hard to make sense of money? And what does this difficulty tell us about capitalism? The apparatus of commercial banking, interest rates, equity and bond trading, derivative contracts, quantitative easing, and other parts of the various mechanisms that create credit and debt seem to lie outside both our everyday comprehension and standard models of economic life.[2] Why is the industry of finance, which somehow manufactures all this money, rendered so much more opaque than other branches of manufacturing? Is the financial industry more technical, or more abstract? Surely the design and production of monetary instruments cannot be more complex than the making of a microprocessor, say, or a Covid-19 vaccine? The answer lies not in the complexity of money as a process, but in the relation between the process and the ideas with which we make sense of it.[3]

The difficulty in understanding money arises from the way its manufacturing has shaped the science that explains it. Like money itself, the discipline of economics works by placing out of sight much of the apparatus of its production. Economics is a form of knowledge that operates, again like money, by assuming a remarkably simple distinction: the difference between an object and its price, between the material and the monetary, between finance and society, and between the real world and its representation.[4] These issues concern not just making sense of money. The alibi of capital is produced through these apparently self-evident distinctions. Behind them lies our understanding of how modern forms of life came into being, and how they are governed today. Addressing them requires, among other things, considering the place of colonisation and enslavement in creating modern systems of money, the *detour* into manufacturing and wage labour in the so-called industrial revolution that made money appear to be merely a means of payment or representation of value, and then the rise of the modern corporation and the modern bank in the later nineteenth century as new machines of money creation. Those late-nineteenth-century arrangements emerged in part in the kinds of imperial schemes explored in Egypt in the previous chapter. Such schemes facilitated and disguised both the making of money and the creating of the wider effect of a world that seemed divided into the financial and the real, the fictitious and the material – and the colonising and the colonised.

Money Is No Object

What is money? The answer until recently seemed obvious. We thought of it as an object, in the shape of banknotes and coins, the everyday appearance it took in our hands. We associated its production largely with the powers of the modern state, since governments authorise the coining and printing of currency and establish the legal tender in which taxes and other debts must be paid. Economists conventionally defined its use in terms of three functions. It is a medium of exchange, enabling people to trade goods and services without resorting to barter; it is a unit of account, allowing business firms to record the prices of goods and calculate their profits; and it is a means of storing wealth, in a form that is easy to stockpile, retrieve, and exchange. These descriptions of the creation and use of money, still found in almost every textbook of economics, corresponded to our ordinary experience.

Today, however, the orthodox and common-sense view of money is recognised as inadequate. The financial turbulence of the opening decades of the twenty-first century revealed the need for a different understanding. Financial booms and real estate bubbles were driven by the creation of collateralised debt obligations and other complex monetary products.[5] The banking crisis that followed the collapse of those products in 2007–8 required an unprecedented creation of bank credit, through the practice of quantitative easing. The liquidity crisis triggered by the Covid-19 pandemic in March 2020 led to an even larger expansion of credit money. In these events, money was not just created to facilitate trade or accounting. It was engineered in novel ways on a new scale. Money was a device in whose manufacture extraordinary forms of wealth, instability, and indebtedness were being created.

In light of these events, a different understanding of money has emerged – or, rather, has been rediscovered. In terms of its production, very little money is created by the government. Despite its association with national currencies and the authority of modern states, the production of almost all money is the work not of government treasuries or central banks but of corporate interests. In modern times, commercial banks have acquired an immense power, largely ignored by the textbooks of economics, to mass produce money. In terms of its use, money is not just an object that facilitates exchange or provides a means of storing wealth, but a relation to the future. It comes into being as credit, or the present right to a deferred payment or good. Whatever its further uses, the origin and main importance of money lies in claiming a future income and, crucial to what we call capitalism, capitalising that future claim and making it available at a discount in the present.[6]

As we will see, however, banks do not create money by acting alone. A claim on a future payment would be ineffective if that future was not organised as a reliable source of payment. Banks acquire the power to manufacture money only as part of wider mechanisms for the capture of the future. Those mechanisms include forms of trade, infrastructure, and manufacturing, systems of law and its enforcement, and devices that measure, forecast, and simplify the relation between the present and the future. Finance is often associated with a different aspect of the future, its uncertainty.[7] However, uncertainty is significant only as an element that arises from within, and modulates, an effective power over the future.

Although newly visible, this understanding of money as an effective claim on future goods or payments, and thus a relation of power over the future, is not new. We can turn back more than a hundred years to the early twentieth century, when the modern powers of money were first clearly articulated, and overcome the 'lost century' in economics that followed the forgetting of this work.[8] One of the key forgotten texts is a short essay of 1913 entitled 'What Is Money?'[9] Its author, Alfred Mitchell-Innes, had first learned about modern finance not in New York or London, but in Cairo. He spent the formative years of his career there during the British occupation, serving as under-secretary at the Egyptian Ministry of Finance.[10] As we saw in the previous chapters, Egypt in those years was an important site for the development of new forms of credit money.

After ending his assignment in Cairo, in 1908 Mitchell-Innes was posted to Washington.[11] He published 'What Is Money?' in *The Banking Law Journal* five years later. The following year, after 'many economists and college professors' objected to his views, the journal's editor invited him to publish a second article, spelling out in more detail what he termed 'The Credit Theory of Money'.[12] John Maynard Keynes, who had just completed his own first study of money, *Indian Currency and Finance*, praised the work of Mitchell-Innes for showing that the conventional history of money 'is quite mythical'.[13] Joseph Schumpeter, who as we know had also worked briefly in Cairo in the same period as Mitchell-Innes, had published his own version of the credit theory of money in German in 1911, although its English translation, *The Theory of Economic Development*, did not appear until 1934.[14]

'What Is Money?' exposed two fallacies. The first held that money derived its value from gold or silver. Even though paper notes were used far more widely than coins and bullion, the gold or silver 'standard' on which money was said to be based was thought to determine its worth, and to allow a quantity of one currency to be measured against another. This idea had governed economic thinking in the nineteenth century but was to be largely abandoned after the mid-twentieth.[15] The second fallacy, often attributed falsely to Adam Smith and prevailing for another hundred years after Mitchell-Innes debunked it, explained the origin and purpose of money as providing a means to facilitate exchange. Early societies traded goods by barter. As commerce expanded, people overcame the inconvenience of bartering one good for another by developing the use of money as a medium of exchange.[16]

The available historical evidence, Mitchell-Innes showed, supported neither of these views. First, until modern times, no attempt was made to define the value of a 'pound', a 'dinar', a 'dollar', or any other money in terms of a metallic standard. These names denoted units of account. They were not fixed objects of a determined size and weight. The metal coins that rulers might periodically issue, typically to pay for soldiers or other services, and to reclaim as tax payments, were tokens whose worth varied. Metal-based money was a minor part of any monetary arrangement. It did not determine the value of the sums recorded in the accounts of merchants, which usually represented the largest and most stable system of money.[17]

Second, money emerges not as a convenient substitute for exchange by barter but as a mechanism of credit. The use of credit is found in the records of the earliest civilisations. In agrarian societies it was associated with the crop cycle. Merchants and other powerful households extended credit to farmers against the crop in the field, to be repaid with the harvest.[18] Credit assumed a greater role whenever commercial relations and the power of merchants intensified. In the medieval Islamic world, merchants used the device of *bayʿ bi-salam* (forward buying) to purchase a future crop at a discount, or to cover a tax payment owed to the government, as a means of creating and profiting from credit money.[19] Likewise, both retail and wholesale business were routinely conducted on credit.[20] 'In most cases', wrote a leading Muslim jurist of the eleventh century, 'profit can only be achieved by selling for credit.'[21] The practice of deferred payment for long-distance trade, developed by merchants of the Indo-Islamic world and widely used in the commerce of the Indian Ocean and the Arab lands, spread to Italy and then the rest of Europe.[22] The method allowed the creation of money through a device that in Egypt and the wider Arab world was known by the Persian term *suftaja*. It came to be known in English as a bill of exchange.

In its simplest form, a bill of exchange works as follows. A supplier of goods requires to be paid in advance, to produce or acquire the merchandise and arrange its shipment. The merchant purchasing the goods creates a form of credit, by writing a bill of exchange as an IOU – the process known as 'drawing a bill'. The bill is a promise to make the payment on a future date, nominally the date by which the goods are to be delivered and, by convention, a standard period of days, weeks, or months depending on the type of good and the distance it will travel.[23] The reliability of the promise – the soundness of the bill – rests on the reputation and strength of the merchant.[24] This strength was extended through access to courts and legal remedies. In the Islamic world, judges and their courts played an important role in enabling merchants to extend and reclaim debts.[25] In Europe, where access to judges and the mediation of the law may have been more restricted, the formation of early modern states can be traced to the demand for stronger mechanisms for collecting debts.[26]

The promise of future payment becomes a financial instrument that can be immediately traded. A recipient of a bill who requires the funds right away, to settle an earlier debt or purchase further goods, can sign it over to another merchant as a means of payment. Once the same bill can be signed over, or endorsed, from merchant to merchant multiple times, it functions as a payment system.[27] The more the practice expands, thanks in part to the increasing legal enforceability of claims, the more impersonal it becomes, increasingly detached from specific merchants. The greater the detachment, the greater the 'moneyness' of the payment device. Although bills were associated with long-distance trade, a variant known as a 'bill of accommodation', where the underlying sale of goods was fictitious, provided a further means of creating credit.[28]

Not all credit required bills. Financial credit may often have functioned as an extension of non-monetised relations of credit and debt, which, in ordinary times, created the fabric of reciprocal obligation through which social community was forged.[29] Local suppliers of goods would typically extend credit to retail customers or other businesses by simply recording a debt in the merchant's books, or by oral agreement.[30] Widely documented in European sources, this everyday use of credit is equally evident in studies of the medieval and early modern Islamic world.[31] Unlike bills of exchange, merchants' book credit did not circulate as a payment device and was not always associated with a fee, discount, or interest payment. However, credit balances could be settled by writing promissory notes, which circulated from hand to hand as credit money.[32] In Europe, small businesses might also convert the debts on their books into bills of exchange, especially those obligations that went unpaid for long periods, or they might accept another bill in payment.[33]

Merchants would settle their bills among themselves, with many of the debts and credits cancelling one another out and any outstanding balance converted to a promissory note. Coinage was used only rarely.[34] For local trade, this 'reckoning' would happen on an occasional basis, or simply whenever an account book was full.[35] Historically, periodic fairs provided the occasion for the settling of accounts between merchants from different regions. In the Eastern Mediterranean, this practice continued into modern times. 'Not many years ago at the fairs of Egypt,' Mitchell-Innes remarked, one could still see how large fairs, often associated with religious festivals, served as clearing houses for merchants. Alongside a brisk retail trade, the main business for most merchants at the fair – such as that associated with the feast of Sayyid Badawi held at Tanta in the Egyptian Delta – was 'the settlement of debts and credits'.[36] In early modern Europe, the trade fairs of medieval long-distance trade were transformed into 'exchange fairs', dedicated to the clearing of bills.[37] According to Fernand Braudel, the quarterly exchange of paper and settling of debts at the most important of the fairs, at Piacenza, near Genoa, 'dictated the rhythm of the material life of the West'.[38]

'The West' refers here to a collection of city-states and kingdoms located at the periphery of the great trade routes and commercial hubs of the Indo-Islamic world. The peripheral location gave European merchants, and especially the exchange bankers of Genoa, an unusual relation to money and the opportunities to profit from its creation. As elsewhere, merchant bankers coexisted with princes attempting to consolidate control over their cities and territories by monopolising the right to mint gold or silver coins. Although coinage typically played a small part in monetary systems, in Latin Europe from the sixteenth century, precious metals took on a different role. Merchant adventurers began to use gold and silver not as money but as a principal item of trade, almost the only commodity available to them to exchange for the fabrics, spices, and other luxury goods of Asia. At the same time, princes competing to control this trade and arm its protagonists required increasing access to gold and silver coinage, to pay their administrators and mercenaries. The practice of exchange by bills became entwined with trade in precious metals, allowing those who controlled the exchange fairs to profit – often corruptly – from the proliferating complexity of exchange relations. By the early seventeenth century, the complexity and the corruption helped bring about the collapse of the Genoese system.[39] Europe's armed usurpation of Asian trade routes had set credit creation on a different path. But it would soon be transformed again.

Money Becomes the Discounted Future

As the main routes of Europe's long-distance trade and the source of its increasing wealth shifted from the Mediterranean world, the Levant, and the Red Sea to the Atlantic and Indian oceans, the hub of European money creation shifted from Italy to the north, first to Germany, Antwerp, and Amsterdam in the seventeenth and eighteenth centuries and then in the nineteenth century to London.[40] The largest English merchant houses – firms such as the Barings, who were cotton traders of German-Dutch origin from the English West Country, and the Rothschilds, who grew their business as cloth merchants first in Hamburg and then in Manchester – evolved into merchant bankers, specialising in trading and clearing bills of exchange and underwriting and selling government debt.[41] In the eighteenth and nineteenth centuries, governments began to regulate the issuing of notes, and to designate a national bank or monetary authority as the sole issuer of banknotes, transforming these promises to pay into national currencies. England, emerging as the most successful war-making and colonising power of that period, had pioneered a novel public–private arrangement, in which the government's future tax revenues secured the value of paper money.[42] However, while centrally printed bank notes came to designate a country's legal tender, they were to play only a minor role in the daily creation and supply of money. As private banks took over from merchants the main role in supplying credit, they created money through bills, loans, and ledger balances – forms of debt produced to fund the purchase of future goods and other assets. This became the source of almost all money.

With the development of wage labour in factories, to remain with the case of Britain, most money continued to take the form of privately created credit. Large employers seldom paid wages in cash, which in any case was difficult to find. As we will see, they would pay workers in scrip, shop credit, or other forms of truck, but also with credit notes issued by the growing number of private 'country' or provincial banks.[43] When the government restricted the creation of notes by provincial banks, payment by personal cheque emerged as the main form of paper money. As late as 1913, John Maynard Keynes remarked that in Britain cheques were the principal form of money, except for 'certain kinds of out-of-pocket expenditure, such as that on railway travelling, for which custom requires cash payment'.[44] But, before the consolidation of modern banking in the second half of the nineteenth century, most ready money in Britain, as elsewhere in Europe, took the form of shop credit, promissory notes, local private bank notes, and bills of exchange.[45]

Understanding that money comes into being as credit matters not just for the history of finance. Credit money is a social and technical relation, built upon a power established in the present and extending forwards, to control a debt to be paid in the future. While a monetary system may also provide a means of keeping accounts, its main purpose derives from creating a debt that is paid forwards through a means of reliable enforcement. That enforced future is a world that is constructed and controlled through a technopolitical process, in ways that shape patterns of living and forms of prosperity. Money is not simply an object representing value or lubricating some more real process. It is a technical apparatus of capture. However, the strength of the apparatus rests in part on appearing as only an abstraction, the mere token or representation of a more real object or process.

When organised at scale, this relation across time becomes the source of significant income to the creditor. The gain arises from the practice known as 'discounting'.[46] Money is a present claim on a future asset or debt. Since access to that asset or debt payment lies in the future, the merchant or finance house willing to take a bill of exchange as settlement for a current debt can acquire it by offering less than its face value. The purchase of a bill, or promise of future payment, at a lower price is known as 'discounting' the bill.[47]

This works as follows. A merchant accepting a bill that promises, say, a payment of £100 at a date six months in the future might discount its value by 5 per cent. The discounted value in the present reflects the delay in time until the deferred payment occurs, and the small risk that the payment might never materialise. (The large merchant banking houses, which came to be known as 'discount houses', could minimise the risk in accepting bills by maintaining a large and diverse portfolio.) The owner of the £100 bill who needs to spend the funds right away receives only £95 from the merchant willing to take it. The seller satisfies a need to pay off an earlier debt or make a further purchase immediately, at the cost of losing 5 per cent of the income due. That person pays a penalty in the form of a loss of future earnings. The merchant who acquires the bill at a discount, on the other hand, gains that lost amount, purchasing £100 of income for £95. For the expenditure of little or no effort, a payment of £95 is transformed by the passage of a few months into £100. Repeated on a wide scale, the discounting of bills became the source of significant and almost costless profits.

As we saw in chapter 1, economists in the twentieth century, especially in the United States, came to see this gain as a reward for 'the time value of money', as if time were a natural property of money. But the postponement or delay that creates the opportunity for gain is not a simple phenomenon of nature. The mechanism of delay had to be engineered.

The conduct of long-distance trade provided an expanding opportunity to create credit money. By opening a significant gap in time between a purchase and its delivery, non-local exchange generated the deferral or interruption from which a discounter of bills could make a gain. As the opportunities for delay and discounting proliferated, the creation of credit money became the purpose of trade.[48] The goal of the large merchant or merchant bank was increasingly not to acquire goods and profit from their resale in a different location. It was to organise the movement of goods in order to create opportunities for the creation of credit, and to profit from the delay in time between the drawing (creating) of a bill and its payment. The distance and delay forced the supplier to accept a discounted price for the bill, and the merchant or banking house profited from acquiring future payments at a discount. Rather than organising credit to facilitate the flow of trade, large merchants organised trade to facilitate the creation of credit.

Money is not the mere representation of something more real. It forms a mechanism for acquiring income from the future, in a world increasingly organised, as we will see, to create forms of delay and postponement that set up the future as a place for profitable extraction.

Fabrication

We should think of capital, as previous chapters have shown, not as something accumulated from the past but as what is claimed from the future. It represents not the building up of a store of wealth, but the controlling of prospective payments.[49] This change in perspective suggests, at first, that we are abandoning the study of concrete, material relations. It implies refocusing attention from the material to the speculative, from the real to the fictitious. If capital is a relationship to the future, to something that has not yet happened, it seems to be built on conjectures and fabrications, rather than on the actual production of material life among living humans. Such thinking about our relationship to time, where the future is fictitious and only the past is real, has difficulty making sense of contemporary ways of organising the collective world. It reflects a naïve historical mode of thought, in which the past is a factual sequence of known events and arrangements, while the future exists mainly as prognosis – the range of possible 'outcomes' that might be conjectured from a knowledge of the past.[50] This simple historicism will not get us far. We already know that the past is a fabrication, in the positive sense of that word. It is a present set of references, memories, reports, archives, legacies, traces, and effects whose force we attempt to organise, mitigate, and apprehend. The future is no less present. If it is a projection or fabrication, it is fabricated and projected in concrete ways that are organised, calculated, acted upon, and made available in the current moment. The aim, then, is not to abandon a material account – or better, a technopolitical one. It is to be as attentive to the fabricated effects of the future as we have been to the effects of the past.

A technopolitical account can be distinguished from an account centred in the mere production of material life. The productivist approach to understanding capital actually starts in the future, as we already noted, but only a future of human needs. Those assumed needs create a demand, the demand to provide what is required to sustain human life. The material demand is met by organising processes of production. Capitalism arises from the development of those productive processes. At the centre of this productivist history, one finds not the methods of indebting the future by capturing prospective income in the present, but the transformation of present methods of producing material goods to meet the needs of the future. Its characteristic institution is the large-scale site of the production of goods, the modern factory. Its archetypal form of domination is wage labour, or the abstractions to which this gives rise. And its defining event is the industrial revolution.

That event, however, takes on a different import if we approach it as a shift in modes of capturing the future. The shift decentres the role of Britain and Europe in those transformations, contributing to a better understanding of the place of colonisation, war-making, and slavery in the shaping of the modern era. And it allows us to resituate wage labour, as we will see, not as the source of the abstraction through which domination occurs, but as another apparatus for the production of debt relations through which the future is encumbered and captured in the present.

The story of the industrial revolution in Britain is widely known – and its significance still often exaggerated. Within a span of fifty years, from the 1780s to 1830, textile merchants in northwest England reorganised first the spinning of cotton yarn and eventually the weaving of cloth. Moving away from the putting-out system used in the linen and woollen trades and in finishing imported cottons from the Ottoman Empire and India, they took direct control of large groups of workers, who were assembled in water-powered 'mills' (often repurposed from other uses) to operate spinners and looms that took advantage of fast-flowing streams in the hills around Manchester. The rural location guaranteed a supply of low-cost, forced labour, mostly children, provided by expanding urban poorhouses anxious to move paupers off their books by shipping them to rural parishes. The mills produced rapidly increasing quantities of cotton yarn and later cotton textiles. None of the raw cotton was grown locally, and little of the finished fabric was consumed at home, where the preference continued for better-quality woollens and linens, or the finer Indian and Persian fabrics. The less durable, low-quality machine-woven cotton fabric was mostly shipped abroad, initially to Europe, West Africa, and the slave plantations of the Caribbean and the American South, and later to the Middle East, South Asia, and the rest of the world.

Traditional accounts of this unusual arrangement turn on the role of technology, which functions so often as an alibi of capital. They see industrialisation (and, in some explanations, an antecedent agricultural revolution) as a transformation driven by rapid and continuous technical innovation. The accounts differ mainly in the reasons for this drive. Rival explanations attribute it variously to an accumulation of surplus capital, the competition among producers caused by access to free or propertyless labour, an abundant pauperised workforce of mainly children and women, a new culture of consumption creating a spirit of industriousness, the rise of an effective legal regime enforcing property rights, ecological pressures caused by a rising population and limits to the availability of land, and privileged access to colonies and slave plantations that both produced the raw cotton and consumed the manufactured textiles.[51]

The focus of accounts on the cotton mill and its machinery is no surprise. As the scale of production expanded, the purpose-built mills became large, imposing structures, forming new features in the landscape, some of which survive even today. One of the most well known of the Manchester-area factories, Quarry Bank Mill near Wilmslow in

Cheshire, is now an industrial heritage site owned by Britain's biggest conservation organisation, the National Trust.[52] Built in 1784, it grew into one of the largest cotton-manufacturing plants in the world. Today it is a museum and archive of the industrial revolution, and even the setting of a grim British television drama series whose writers developed the story from documents in the site's archive.[53] Sources from that archive figure in almost every scholarly history of the British industrial revolution. Other sources widely used to construct that history, including the reports of factory inspectors and patents for mechanical inventions, focus the story on technical innovation and the organisation of factory production.[54] Large mills like Quarry Bank were not actually representative of manufacturing, which, in most cases, developed in smaller establishments, but they made an impression on visitors, and benefited from 'the Victorian disposition to reify and magnify the power of "capital"', and from 'the obsessive association of the *Grossbetreib*', or large enterprise, with the age of high capitalism by German writers like Engels, Marx, and Sombart.[55] The raw cotton that the mill processed and the yarn and cloth it produced provide further tangible matter for composing a history focused on the transformation in the production of material goods. The story of cotton brings the wider world into this material history. The Caribbean, the southern United States, India, and Egypt were organised to produce the cotton, becoming incorporated, as the standard accounts put it, into a world-economy centred on industrial production.

This machine-focused, production-centred, British-based account of the genesis of modern capitalism has been questioned for some time. It has several weaknesses. The story of cotton refers to a small enclave, not representative, even in Britain, of how most industry was organised or workers were employed. Even in textiles, the putting-out system long remained more profitable for most fabrics, and non-mechanised production increased vastly; many areas of enterprise continued to use hand labour without large-scale mechanisation, including ship building, brewing, glass-blowing, paper making, leather work, the building trades, and most coal mining and iron smelting.[56] Cotton production was an enclave in another sense. Unlike most British and European industry in this period, the raw material and most of the final product were non-local. As we just noted, no cotton was grown in Britain, while up to three-quarters of the spun yarn and manufactured cloth, a fabric 'too flimsy to be worn by itself' in its country of fabrication, was exported.[57] Operating abnormally in this way as a 'freak' of world commerce, as a Manchester firm described it, the industry had weak linkages with other areas of local life, including the more popular and better-quality textiles made from local wool and linen.[58] Indeed, as Beckert notes, the very abnormality of cotton, as a commodity that was both produced (as raw material) and consumed outside Europe, meant it was suited uniquely to the global process of capitalisation; and specifically, as we will see, to the dislocation required for the large-scale manufacture of credit.[59]

Many other issues have been noted regarding the tradition of explaining capitalism in terms of technical progress. As we saw in chapters 1 and 2, such accounts often overlook the preference of capitalisation for the production of waste. Cotton exemplifies this. The local fabrics it replaced were more durable and mendable; wool in particular required less frequent laundering, since it is able to absorb rather than trap moisture from the body, and unlike cotton it is naturally anti-bacterial.[60] In contrast, the flimsiness of cotton and the disposability of its products favoured frequent replacement and thus the durability of revenue on which capitalisation depends. More generally, the rapid increase in production and wealth associated with the so-called industrial revolution was for the most part not the result of technical transformation. Most of it, in fact, can be attributed to increases in population. Mechanisation did not produce material growth.[61] A key transformation of the period, from water-powered mills to coal-powered steam engines, was driven not by a search for technical improvement, as Andreas Malm has shown, but by the fact that the more cost-effective water power required the shared development of a public resource among a coordinated group of users, whereas the supply of coal could be secured (and capitalised) privately. Steam engines also allowed production to move from rural to urban locations, providing a larger, more vulnerable pool of labour, while the machine process expanded the opportunities for control of a workforce whose concentration in factories had given it an increasing power to disrupt production and thus demand improved conditions.[62] Finally, other regions, such as China, faced similar demographic pressures and opportunities for improvement but took a different path from the mechanisation of production. If anything, developments in Europe were shaped not by material or technical advantages but by a lack of endowments – except for the possession of sub-tropical slave colonies, which provided both a source and a market for the material around which mechanisation was organised.[63]

These issues have already allowed for some time a less Eurocentric account of industrialisation, one that does not seek explanations for what happened in exceptional technical improvement, the peculiar industriousness of the British, or a simple demand for increasing numbers of goods, and one that is more aware of the element of contingency.[64] There is now a strong counter-narrative explaining the industrial revolution in more global terms, in relation to the earlier development of long-distance trade, colonisation, and plantation slavery. Within those explanations, there is an important debate on the sources of credit or capital for financing the industrial revolution, including the relative importance of Atlantic slavery versus the capture of resources of the Mughal Empire.[65] However, to properly decentre Europe and its industrialisation, the terms of those debates need to be altered. Even in the best accounts, commercial or mercantile capital merely prepares the way for industrialisation. The finance it supplies simply 'lubricates' the transition to industrial production.[66]

The question that should concern us is not how industrialisation was financed. If we were to keep with these terms, the question would be how and why *finance was industrialised*. These terms are inadequate, however, for they appear to assume the difference between 'finance' and 'industry', between money that is merely a lubricant and the real 'machine' that it lubricates, and thus between the monetary economy and the 'real economy'. In the following two chapters we will approach these questions from the side of the 'real' and consider the genealogy of the economy. We will examine the emergence of first 'the economic' and then 'the economy' through processes we will identify as 'economisation'. For now, however, working from the side of understanding money, we will continue with the more capacious term 'capitalisation', precisely because it does not start from an opposition between the financial and the real, but explores how that distinction has been invented, deployed, and disguised.

By focusing here, then, not on commerce and 'commercial capitalism' but on capitalisation, where commerce is one mode of capitalising and capturing a future revenue, we can go on to propose two additional arguments. These will place the development of the factory system not as the culmination of earlier commercial developments but as the consequence of a *detour*, a disturbance that unsettled the wider, less-European-centred modes of acquiring future payments in the present. The idea of a detour suggests not just an element of contingency, in which multiple developments combine to create an outcome, but a less linear process characterised by delay, interruption, and discontinuity. And we will see how one notable feature of factory production allowed it to give rise, unexpectedly, to a new mode of extraction from the future – the more localised method of credit creation that operated and was misrecognised as wage labour. In the device of the wage, we will discover, there emerged a modern and very simple conception of money.

Industrial Revolution as a Detour

First, the detour. Why did British merchants, long accustomed to the large profits from financing colonial trade, warfare, and chattel slavery, take the unusual and laborious path of organising workers close to home on a large scale into factories? How was the raising of surplus through a new control over manufacturing connected with the capture of revenue from the future? Instead of resorting to arguments about 'technical improvement' or increasing 'demand', an alternative answer can be found in the worldwide crisis in the processes of capitalisation, dependent on militarised trading monopolies, colonisation, and enslavement, that occurred in the period from the 1770s to the 1820s. Cotton manufacture was the diversion taken to salvage a system of credit manufacture disrupted by global warfare, colonial expansion, resistance, and revolution. One part of this crisis is well-known, the developments occurring within Europe, although it is less often associated with the wider events to which it was connected. But let us start there.

When revolution engulfed Europe in the years after the summoning of the Estates-General at Versailles in 1789, merchant-financiers in London and their business partners in Europe were forced to divert credit into the expansion of textile production. This diversion is explained neither by a sudden consumer demand for more cloth, nor by a 'surplus' of capital looking for new investments. It was the result of a dislocation of merchant finance, requiring a new process through which to complete the circulation that allowed the creation of and profit from credit.

As the French revolutionary wars spread across Europe in the 1790s, London merchant creditors financed war on the continent. The British government paid for mercenaries supplied by the rulers of German states, extending a system of hired soldiery developed during the American Revolutionary War. The scale of these payments for warfare was unprecedented. By the end of the Napoleonic wars, military spending was accounting for 58 per cent of the British government's revenue.[67] Borrowing against future tax revenues, the government transferred the payments not in cash but via letters of credit drawn on London merchants, whose agents or partners in Germany made the funds available to recruit and supply the soldiers. The merchants on the continent thus accumulated credits in London. These paper fortunes could be realised only by finding ways to move credit in the reverse direction, by using the German-owned bills in London to purchase British goods to supply to Europe. (The bills themselves did not move: rather, to balance the credits and debts, and to profit from their creation, goods had to move back to Germany.) The main vehicle used for this complementary movement of credit and the realisation of profit was the transfer of textiles. While the existing trade in English woollens and linens could provide some of the machinery of credit transfer, the principal means of unblocking and expanding finance was to put the bills to work in developing the production in England of a different fabric, cotton, which until then had been largely imported from the East. The London bills used to fund the war on the continent required a new clothing to return. As a London financier remarked in 1811, the blocked credit of wartime bills amounted to 'so much foreign capital invested in British industry'.[68]

The process was aided by the relocation of merchants from Paris, Amsterdam, Hamburg, and other trading centres on the continent to London. Fleeing the revolutionary forces and establishing trading houses outside their reach, the merchants profited from the wartime opportunities to finance the movement of goods. After Napoleon was defeated in Egypt, France imposed the Continental System in an attempt to block commerce with Britain, allowing additional profits for those with means to smuggle goods through the blockade.[69] These movements of credit driven by war, and of the goods required to create and mobilise credit, 'explain succinctly', according to the leading history of the rise of finance capitalism, 'why the British industrial revolution took place during that period'.[70]

The career of Nathan Rothschild, one of the best known of these displaced cloth traders and financiers, illustrates this machinic diversion at work.[71] His father was a successful Frankfurt merchant who served as a financial agent for the ruler of Hesse-Kassel, the main supplier of mercenary troops to the British. The father built a fortune dealing in the ruler's English bills of exchange, and later in financing the supply of war matériel to the Austrian imperial court. His son Nathan was sent to Manchester around 1799, while still in his early twenties, to begin trading in cotton goods, shipping cloth back to Frankfurt and other cities in Europe. He was one of at least eight German traders to settle in Manchester around then, a number that continued to increase.[72] Nathan could obtain bills from the merchant houses in London that handled the British side of his father's bill dealing – using, in other words, the surplus IOUs his father had built up in London as ready money. This access to plentiful credit money allowed him to undercut other Manchester cotton traders, who typically purchased cloth from local weavers with bills payable in two to six months. By paying upfront 'on present bill terms', Nathan could find weavers 'in want of money' and willing to lower their prices by 15 or 20 per cent.[73]

Rothschild's success in buying via the unblocking of German bills in need of London counterparts meant he soon struggled to find sufficient supplies. Within about a year, he arranged with the firm of Boulton & Watt to purchase a copping machine, a wooden contraption that allowed the yarn for the weavers' shuttles to be 'copped', or wound, onto several spindles at once. By supplying weavers with yarn already wound on the spindle, Rothschild increased the rate at which the weavers could supply him with cloth. While serving as a device to speed up the production of cloth, the copping machine was something more – a mechanism to improve the circulation of credit money. The bills of exchange accumulating in the hands of his father in Frankfurt could now be put to work.

Within a few years, Nathan Rothschild had moved to London, to focus on the financing of trade, the discounting of bills, and acting as investment broker to rulers. The detour to Manchester, having unblocked the flow of credit, could now be left to others. The city became the centre of merchant-financiers, who worked with and invested in the textile mills in the surrounding countryside but 'above all . . . financed the trade, serving as paymasters to their manufacturing suppliers and extending long credits to their overseas customers'.[74] A dominant role in the overseas trade was played by foreigners settled in Manchester, especially German merchants, such as the firm of Ermen & Engels, where Friedrich Engels, the son of one of the founders, managed the firm's commercial dealing, accounting, and intelligence gathering.[75] This intelligence helped inform the writing of his friend Karl Marx, shaping a Manchester-centred account of the capitalisation process.

The growth of German merchant houses was later matched by the increasing number of traders from the Ottoman world – Greeks,

Armenians, Arabs, and Turks, all referred to in Britain as 'Greeks', who financed trade with cities like Izmir, Aleppo, Cairo, and Alexandria, and later with West Africa and most other regions. In many cases, they established their own branches in overseas cities. The Manchester firm of Cavafy and Co., for example, founded by a Greek merchant family from Istanbul, opened an Alexandria office in the 1840s, run by one of the founder's sons, Petros. (The family returned to Manchester after the death of Petros and the financial crisis of the 1870s had forced the family to close the Alexandria office. A grandson moved back to Alexandria and made money on the side as a broker on the stock exchange, but, as a sign of the changing times, supported himself under the British by working for thirty years, appropriately enough, at the Ministry of Irrigation. From this corner of the Greek merchant world, standing 'at a slight angle to the universe', he became the noted poet C. P. Cavafy.)[76] The merchants from the Ottoman world continued to grow in number, and, by 1870, surpassed the Germans as the largest group among the 400 resident foreign merchants who dominated Manchester's long-distance trade.[77] Manchester, according to a contemporary report, 'came to be colonised' by these foreign merchants. With their credit networks, trading experience, understanding of overseas commercial law, and knowledge of multiple languages, the Ottoman and other merchant communities were the 'prime movers' in the development of cotton manufacturing.[78]

Provincialising Manchester

If the foreign merchants of Manchester financed the trade that helped drive the detour into cotton manufacturing, the 'Greeks' among them connected that detour to transformations under way in the Ottoman world, especially Egypt and Arabia. Before leaving the European-Mediterranean world, let us briefly follow this connection. Napoleon Bonaparte's short-lived occupation of Egypt (1798–1801) had been intended to make the country into a French colony, to compensate for the recent loss of colonies in India and North America and the threat of further losses from uprisings by enslaved people in the Caribbean, notably in France's most valuable overseas possession, and possibly the most productive colonial territory anywhere in the world, the sugar plantation colony of Saint-Domingue (Haiti).[79] French shipping interests in Marseilles were keen to support the Egyptian venture. Their profits lay in financing trade with the Eastern Mediterranean, a commerce they had briefly dominated in the first half of the eighteenth century (although they remained peripheral to the richer internal Ottoman trade), but were now losing to the Greek, Armenian, and other Ottoman merchant houses.[80] The result, however, was something very different: the defeat of the French plan for a Haiti-on-the-Nile and the strengthening of Egypt as a distinct centre of power, increasingly autonomous from the Ottoman authorities in Istanbul.

Cairo's autonomy had already been growing in the later eighteenth century, benefiting from an Ottoman policy of increasingly decentralised governance.[81] Local rulers like Ali Bey al-Kabir began to dominate rival governing households, embark on military expansion, and encourage and profit from increasing commerce, a development noted by the Egyptian historian al-Jabarti and marked by the building of large commercial complexes at sites like Bulaq, the river port serving Cairo, and Tanta, whose great seasonal fair has already been mentioned.[82] Another commercial and political challenge to Ottoman rule emerged in Arabia, where the Emirate of Dir'iya, under Muhammad bin Sa'ud, in alliance with wealthy long-distance Arab merchants and with the reformist religious movement of Muhammad ibn Abd al-Wahhab, extended its power across the entire peninsula from the Persian Gulf to the Red Sea, taking control of the holy city of Mecca, whose annual cloth fair was among the largest commercial events of the Indian Ocean world.[83] After the interruption by Napoleon, Cairo continued its expansion under Muhammad Ali Pasha, who ruled from 1805 to 1848. He built a large army and navy, crushed the Wahhabi movement in Arabia, and further expanded his rule, first over the kingdoms of Sudan and then to Palestine and Syria. To acquire the new empire, he sought arms and naval vessels from Europe, giving British, Greek, and other merchants, drawn to Alexandria from Aleppo and other Ottoman trading centres, the opportunity to finance the supply of the main imports, namely 'warlike stores, timber, and other commodities' required for the expanding Egyptian powers of war-making.[84]

Financing the expansion of a new power on the Nile created another force needing a diversion through the cotton mills of Lancashire. Muhammad Ali funded the purchase of arms and other warlike materials on credit, which was advanced by these merchants, especially the English firm of Briggs & Co. To obtain the credit, the Pasha offered credit payments of his own, in the form of contracts for the future delivery of crops.[85] But which crops? To profit from financing military supplies, Briggs and other merchant bankers needed a commodity they might ship back to Europe. At first, they could rely on Egypt's main export crop, wheat, which was in demand in Malta and the Iberian Peninsula to supply British forces during and after the Napoleonic wars.[86] By the 1820s, however, they needed a different material through which to sustain the movement of merchant finance. Alternatives included indigo, opium, silk, linseed oil (a by-product of the linen industry), and items brought from Sudan and further south, such as gum arabic and ivory. None of these, however, was sufficient to finance the expanding trade in arms.

The solution was provided by cotton. Every history of Egypt and its cotton industry repeats the story of a French textile engineer from Lyon, Louis Jumel, arriving in Egypt after a stay in New York, who happened to 'discover' a new strain of cotton with a long staple, or length of fibre, growing in the garden of a nobleman in Cairo, and who then persuaded Muhammad Ali to plant the crop widely and develop an export industry to serve the needs of European textile manufacturing. Almost every element of this story is a myth, elaborated by nineteenth-century French historians and American diplomats anxious to give France or the United States the credit for the transformation of Egypt under Muhammad Ali. Jumel was not from Lyon, had never visited the United States, and accepted employment from the Egyptian government to develop a woollen factory, not a cotton industry. He became ill and died in Cairo within five years of his arrival.[87] In exaggerating his role, the historians overlooked the more likely reasons for Egypt's attempted transformation from an exporter of wheat to an exporter of cotton. Those reasons lay in Ottoman-Egyptian expansion, in particular the colonisation of Sudan.

If conquering Sudan was the source of a need for arms, it also offered a solution to pay for them. The 'nobleman' cultivating long-staple cotton in his 'garden' near Cairo – no doubt a walled orchard on the outskirts of the city, a place where the wealthy of every large town used irrigation from wells to grow tree crops protected from the Nile flood – was Mahu Bey al-Urfali, a Kurdish Mamluk who had sided with Muhammad Ali and helped carry out the conquest of Sudan, serving as a senior cavalry officer in the Pasha's army and then as governor of Berber Province in northern Sudan.[88] Through Mahu Bey, it was quite probably Sudan, not a Frenchman, that provided the model, the motive, and the cotton variety for the development of the crop in Egypt.

Two kinds of cotton were found on the Nile, a perennial grown as a small tree, whose short staple was suitable for coarse yarns and as stuffing for mattresses and divans, and a smaller variety, grown as a bush and replanted each year from seed, whose long-staple fibres produced fine threads for luxury cloths. In Egypt, cotton was only sparsely cultivated, since, as a summer crop, it was incompatible, as we saw in chapter 2, with the summer inundation of the fields. In the south it was confined to Nile berms and elsewhere to walled orchards, where irrigation wells made water available in summer.[89] In Sudan, on the other hand, especially on the Dongola and Shendi reaches of the Nile, where most of the fields lay above the flood level and were irrigated with waterwheels, summer cultivation was widely practised.[90] The region had specialised in cotton production for centuries, using waterwheels both at the river's edge and, at least before the devastation of recent decades including the destructive Ottoman-Egyptian conquest, on wells accessing groundwater replenished by the Nile. The cotton manufactories of Sudan, it was reported, 'furnish the greater part of north-eastern Africa with articles of dress'.[91] The waterwheel, or *saqiya*, was used so widely that government revenue was based on the taxation of wheels, not land, and the term *saqiya* referred not just to the wheel, as we noted in chapter 2, but to the entire community organised around the households, animals, and fields required to keep each wheel running.[92]

In colonising Sudan, Muhammad Ali's initial aim had been to take over these rich agricultural regions and develop cotton, indigo, and sugar production using slave labour – the method that was proving successful in the American South. However, in 1822 the Nubian and Arabic-speaking communities of northern Sudan rebelled against the occupation and killed the commander of the invading forces, Muhammad Ali's son Isma'il Pasha. The Turco-Egyptian forces encountered similar opposition to their attempts to control Upper Egypt.[93] While the occupiers eventually suppressed the resistance and made efforts to develop a cotton and indigo industry, their main efforts moved back to the north. In the same year as the death of his son, Muhammad Ali ordered the digging of thousands of wells and the construction of waterwheels in the Nile Delta north of Cairo, presumably following the Sudanese model, in order to expand cultivation from irrigated gardens into large fields and allow widespread summer cultivation.[94] Another local merchant firm, the Jibara brothers, rivals of Briggs & Co, supplied credit for a new plantation in the Delta.[95] The introduction of mechanised spinning in Lancashire had made British mills dependent on these luxury varieties, not to produce finer cloth but because threads spun from the longer fibres were less liable to pull apart in the rough handling of the machine process. For this reason, in Liverpool long-staple cotton achieved a much higher price than the 'Levant' or short-staple variety.

The cotton could now finance the supply weapons to Egypt. Mahu Bey used the cotton as a means of acquiring 'weapons of superior quality'.[96] Samuel Briggs, who had already supplied vessels for the Egyptian navy, sent a frigate to England on behalf of Muhammad Ali, to be coppered and fitted with guns and other equipment. When the expense far exceeded the value of the cargo of linseed oil the ship had carried there, Briggs had the work and the weaponry supplied on credit, then sent a shipment of 270 bales of long-staple cotton to Liverpool, as a first instalment of the repayment of the debt.[97] This shipment inaugurated a lucrative trade in Egyptian cotton with Lancashire. Building and arming a powerful new state in the Eastern Mediterranean provided the merchant bankers of Alexandria with a new opportunity to reroute credit creation via the cotton mills of Manchester.

Packaging Tribute

The revolutionary wars in Europe and the transformations in the Ottoman Empire offer one explanation for the sudden diversion of credit creation into the difficult detour of factory production. The detour was driven not by a simple demand for goods, but by the need for a new mechanism of credit, when warfare broke up old routes of commerce and created an urgent need for mechanisms to move financial paper. However, the European experience of revolution and war, although more disruptive than any other events in the continent's recent past, and the building of a new regional power on the Nile were only part of a much wider interruption to the flow of credit, and the profits that depended on it. The broader interruption centred not on Europe and the Ottoman world, but on the transformation of colonial regimes in Asia and the system of slave-based production in the Americas.

In India, the rise of powerful merchants and credit networks in Bengal, a development similar to the changes under way in Cairo, had weakened Mughal rule and drawn agents of the British East India Company into the opportunities to profit from these commercial networks and to take sides in the power struggle.[98] As a result, the Mughal Empire lost control of its richest territories to the Company. In 1765, it had been forced to grant the Company the right to the tax revenues of Bengal and neighbouring provinces. The Company was transformed from a militarised trading corporation into a territorial power. It extended that power over the following years with further defeats of both European and Indian rivals. Consequently, as Edmund Burke reported, 'a very great revolution took place in commerce as well as in dominion'.[99] While the transformation secured the riches of the Company in India, it presented the novel problem of how to ship those revenues back to Britain. As a trading firm, its shareholders and company servants had grown rich from financing the movement of goods between Europe and India: mostly fine textiles, spices, and other luxuries to Europe. To profit from the financing of trade, through the discounting of bills of exchange, the Company needed something to ship in the other direction, from Europe back to India. Since Europe in the seventeenth and eighteenth centuries produced almost nothing worth carrying back to Asia, the trade had depended on shipping a commodity sourced in another place of colonial rule, silver from the mines of Mexico and Peru. By 1750, silver was becoming scarce and was supplemented with the shipment of military and naval supplies for the expanding colonial warfare against both Asian and European rivals, along with modest supplies of woollen goods, usually shipped freight-free, and copper and other metals.[100] By the 1780s, ships were sometimes travelling empty to India, returning with goods whose purpose was not reciprocal trade but a means of packaging tribute, especially the part acquired as private wealth by company servants. 'The idea of remitting tribute in goods naturally produced an indifference to their price and quality, – the goods themselves appearing little else than a sort of package to the tribute,' Burke explained. 'Merchandise taken as tribute, or bought in lieu of it, can never long be of a kind or of a price fitted to a market which stands solely on its commercial reputation.'[101] The result of this one-way transfer of tribute was the impoverishment of Bengal. But it also caused the collapse of the paper system in London built on earlier reciprocal commerce. This second and earlier disruption was a further source of the detour into manufacturing.[102]

The third source of disruption, and, arguably, the most important for the diversion of credit creation into manufacturing, was the interruption to profits from slave trading and slave plantations, caused by the American Revolutionary War of 1775–83 and by wider threats to those profits from the Atlantic slave war and the pressures to end the trade in humans.[103] The outbreak of the American Revolution was partly influenced by events in India. The British government had tried to address the crisis in the India trade by awarding the East India Company a monopoly on shipping tea to its North American colonies, causing Boston tea merchants to initiate their rebellion. The North American colonists sought to emulate the freedom enjoyed by British merchant-settlers in India, 'emphasizing constitutional similarities among the different parts of the empire, especially Bengal and the thirteen colonies'.[104] The Americans even appear to have copied the thirteen red-and-white stripes of the East India Company ensign for the flag of the new republic.[105]

But this rebellion was part of a wider crisis. The enslavement of Africans to work in the production of sugar in the West Indies and tobacco in the North American colonies had given movement to a vast circulation of credit. Enslaved humans generated a return to their owners over years or decades, so their enslavement produced not just a trade in human lives but an expanding system of credit. The price of a slave represented the capitalised value of those prospective years of labour.[106] It was typically paid not in cash at the moment of purchase, but as a debt repaid over the years to the Liverpool merchants and London commission agents who financed the supply of enslaved labour on credit, 'mostly through the mechanism of bill discounting'.[107] The discounting of bills was also a main source of profit in the wider triangular trade built upon enslavement. In the case of the Africa trade, for example, while an income could be generated from merchandise sales, the primary business and source of profit for those who dominated the trade was the income generated from the organising of credit through the discounting of bills.[108] The importance of the Atlantic slave trade to the entire system of credit positioned it 'at the very center of English political and economic life'.[109]

The interruption to this system of slave finance from the American Revolution was a spur to the development of cotton manufacturing.[110] The American war bankrupted three-quarters of the London merchant houses. In Manchester, four out of five manufacturers, most of them merchants as much as manufacturers, disappeared during the two decades of war.[111] Quarry Bank Mill, already mentioned as one of the first large cotton mills, was set up after disruptions to the slave trade and interruptions to the North American trade in which the proprietors, the Greg family, owners of slave plantations and merchant shipping, were mainly involved.[112] Like the German merchants disrupted by the French wars, those active in the Atlantic trade accumulated large numbers of bills of exchange, as well as the bonds acquired in the credit sale of enslaved people or in the acquisition of Caribbean slave plantations from owners forced into bankruptcy by the war. The close connections between Lancashire and Caribbean slavery made that region of England unique in monetary terms. As an element of the Atlantic system, it was the one part of the country in which bills of exchange, rather than specie, local banknotes, or other forms of credit, functioned as the main source of money.[113] Forming a dense network of credit money, the Atlantic system was thrown into disarray by the wartime blockages and interruption. Cotton production provided a means to unblock these flows, not least because it facilitated the development of 'the second slavery', the vast redeployment of enslaved people to the new plantation zones of the American South, where genocide and ecocide cleared the path of expansion.[114] As the harvest of another hemisphere, cotton was subject to the uncertainty of long-distance shipping and the unpredictability of a distant climate, creating additional opportunities for profit from both speculative trading and insurance contracts.[115] Cotton also had the unusual character of a commodity that could be shipped in both directions, first as raw cotton to Lancashire and then as manufactured products crossing back to the Americas. In other times, the producers of the raw material would have avoided this circuitous process by bringing the manufacturing process to where the cotton was first produced and later sold as a finished product. But, in this case, the producers, being enslaved, could be prevented from making their own cloth by the owners of the plantations, whose profit required the detour of shipping.[116] Profiting from bills of exchange depended on this ability to organise a detour, creating reciprocal movements of credit money.

To approach the rise of modern industrial life as a historical detour has several advantages. First, it helps avoid the issue of technical determinism, a problem already encountered in the history of colonial irrigation, the story of Uber, and many other areas, where technology is taken to be a self-moving force of improvement. Second, it avoids turning an 'exceptional discontinuity' in the history of a small corner of northwest Europe, and of an even smaller region within that territory, Lancashire, into a paradigm that can explain the success and failure of industrialisation or 'development' in every corner of the world.[117] Third, it replaces a productivist understanding of capitalism, which inevitably centres on the story of industrial production, with an approach that understands capital in terms of modes of capture of the future, or capitalisation.

This approach does not require minimising the subsequent significance of what happened in Britain. But it can frame that significance in terms of tipping points. Detours and disruptions can accidentally set in motion new processes, with wide repercussions.[118] One of those is the transformation in credit production enabled by wage labour, to which we now turn.

Evil Is the Root of Small Money

We can now consider the second modification to productivist, technology-driven, European-centred arguments about the history of capital, or capitalisation. The industrial revolution was a detour. It offered a new circuit, a way of bending onto a different path the opportunities for profiting from the creation of debt, amplifying the apparatuses of capture. But it did this not only through establishing new routes and goods to be financed. It incidentally created an extensive new machinery of credit: large bodies of workers, organised into factories and coal mines, who were to be paid a monetary wage.

Wage labour is often seen as a defining feature of modern, industrial capitalism, especially the disciplined, time-managed labour of factory production. The wage separates human agents into those who provide labour and those who own capital, dividing the product of manufacturing and other production into wages paid to employees and the interest and profit claimed by owners of capital. The contrast between an hourly or monthly wage, on the one hand, and interest and profit, on the other, appears to normalise the extraordinary difference between the earned income of the worker and the unearned income of capital. For Moishe Postone – among the most influential reinterpreters of Marxian theory of the last generation – the wage is central to the effect of 'value' through which social domination under capitalism is organised. Even if one acknowledges, unlike older Marxian scholarship, that not all of the surplus generated by processes of production is a product of labour, indeed especially with that acknowledgement, Postone argues, the wage still performs a powerful function. It makes all labour comparable, by measuring its value in uniform increments of time. This process of *valuation* transforms all productive work into units of the same seemingly abstract activity. The homogeneous, empty uniformity of hours and minutes of labour, paid at a given rate and appearing as the 'price' of that labour, operates as a governing abstraction, creating the mode of domination that allocates a larger and increasing share of earnings to those who do not work for them.[119]

This analysis, for all its persuasiveness, overlooks something. The wage operates as an abstraction not only through the abstractness of the hours and minutes in which, for many workers, its 'value' is presented. It abstracts in another, perhaps more effective way. Like the device of calculating value in hours and minutes, this effect rests on the deployment of time. However, it is not the meticulous 'time discipline' of factory employment, significant as that may be. It is the inexact, unmeasured, age-old mechanism of postponement and delay – practised on a new scale. As with older forms of profiting from productive labour, the delay takes the form of the gap in time between the performance of work and its payment, and, even more so, between the paying of the wage and its spending on consumption. To see the wage in this way is not to shift focus from the sphere of production to that of exchange and consumption – distinctions that make less and less sense today – but, rather, to understand the significance of the wage as a device for the production of credit.[120] The wage was paid in credit money, as we will see. The profit to be made from it was an unanticipated consequence of the *detour* into manufacturing. It was a consequence that would outperform and outlast the advantages of manufacturing itself.

What was a wage, and how was it paid? The factory was unusual in bringing under the control of one firm dozens and, in some cases, hundreds of workers. This produced a problem: how to find the money to pay them, in communities organised largely on the use of book credit, in a world where domestic servants and many other kinds of staff were paid mostly in food and clothing.[121] Merchants accustomed to using coinage only in exceptional circumstances, or to settle yearly accounts when the books could not be 'reckoned', or reconciled, with further credits, suddenly needed a means of making large numbers of small payments at the end of every week or month. Many studies of monetary history portray this as a crisis, leading to repeated efforts to find a means of introducing some widely used form of 'small money'.[122] However, we can see this not as an obstacle to the expansion of industrial production but as the opportunity it created. In fact, it may be a key to understanding that expansion. An influential argument in contemporary economic theory proposes that the need for money arises from the lack of trust between those involved in an act of exchange, and thus that 'evil is the root of all money'.[123] It may be better to see evil, in the shape of the exploitative conditions of factory labour, as the root of *small* money: the coins, chits, and banknotes now suddenly required for paying wages. These slight, token things – money in the form of *objects* – would soon come to be mistaken for the nature of money in general.

Previously, large merchants had often been involved not only in procuring goods but in financing their production, through the credit operation known as the putting-out system, where merchants loaned materials to a farmer, weaver, or clothier.[124] Factory production, based on the payment of a wage, intensified this control, but also represented a new method to profit from credit creation. In putting out, both in farming and in handicraft, the merchant typically loaned the seed, wool, flax, or other requirements, while the cultivators or weavers loaned their labour. The merchant paid a discounted price to 'purchase' in advance the finished product, while the producer repaid the advance in full with the product. The resulting product yielded a profit on these loans. The share of profit going to merchant and worker reflected their unequal strength. With factory production, the workers now advanced credit to the mill owner. They worked an entire day or week, loaning their labour to the employer. This loan yielded a surplus, the difference between the wage paid for the labour and the value of the product of that labour, a surplus that was acquired by the employer.

How does the large-scale organisation of labour in the production of goods and services become another apparatus of capture?[125] Modern commercial banking developed its powers of money creation initially through the financing of foreign trade, warfare, and enslavement. It was through those exceptional apparatuses that money was 'materialised', transformed from a mere unit of account into a concrete claim on future payments – into 'real' money. However, when industrial production generalised a domestic regime of wage payment, money could be materialised in a second way, through the labour process at home. Even the daily process of making goods could become an apparatus for the capture of the future. Cotton manufacturing was particularly suited to this role: as a 'freak' of world commerce, based on the capitalisation of labour both on slave plantations and in factories, it doubled the opportunities for the capture of surplus through credit.

The organisation of large-scale employment is usually understood as the 'purchase' of labour, via the payment of a wage. But that image of a market exchange is a fiction – although not in the way the fiction is usually explained. The employer or firm does not typically pay workers with funds accumulated from the past. The workers are paid in credit money: their wage is a loan. Initially, the working of the credit system was quite visible. In England in the 1820s, for example, the Yorkshire textile-weaving firm of John Foster and Sons 'paid' its workers one month in arrears, and not in cash but by assigning a credit to each weaver's account. Meanwhile, the firm supplied its employees with goods, including 'coal, candles, cloth, flour, malt, milk, sugar, treacle and cheese', leaving them continuously in debt. The wage was simply a deduction from these standing debts. Other firms paid their workers directly in 'truck', or material goods. The term trucking originally implied giving or trafficking one item in exchange for another, but payments in truck were not barter arrangements. The wage was recorded as book credit, advanced to the worker, to be spent on goods supplied by the employer. The goods were typically overpriced, and sometimes consisted of just unsold textiles or other produce of the business. Or the workers were paid in vouchers, tokens, or notes that were redeemable only at the company-owned shop.[126]

In 1806, Samuel Hibbert, a Manchester merchant, published anonymously a pamphlet on the evil of the use of 'small notes'. Their use had arisen from the practice of manufacturers issuing promissory notes

> to pay persons in their service, for whom a quantity of cash must be procured each week; and if they can be satisfied with such convenient paper, the wages due to them, which in the course of time, must amount to a very considerable sum, can be used by their employers in trade. Such a circulation is still farther encouraged, by the reflection, that for credit thus obtained, they pay no interest.[127]

In other words, just as colonial merchants made their profit not so much from the supply of goods, but from the creation of credit for which plantation slavery and the supply of goods was the mechanism, in the same way the emergent large textile manufacturers, most of whom began as long-distance merchants, discovered that the surplus from manufacturing did not have to arise from the increasing efficiency of mechanised production. Technical efficiency could never be a long-term source of differential profit, even once the breakdowns, setbacks, and accidents experienced by men like Jean-Baptiste Say or his brother's sugar company were overcome, as rival firms would soon adopt the same efficiencies.[128] Surplus could arise more reliably from the scale of the new system of credit, in which the wages of workers were paid not in cash, but in the employer's promissory notes or book credit.

Parliament attempted repeatedly to outlaw the practice of truck payment, banning it in many trades in the Truck Act of 1831, and extending the ban more widely in 1887. The 1831 legislation was a precursor to the Factory Acts, which represented a wider attempt to regulate wages and their relation to hours of work, helping to establish the impression that the wage was a direct exchange of time for labour rather than the aspect of a credit system in which the employer in effect captured the interest on a loan. But even before truck payments were outlawed, the creation of credit money to pay wages had become so general that it no longer appeared as a loan, and no longer depended on individual employers issuing their own notes or tokens.

In England, hundreds of small banks emerged with the development of manufacturing in the late eighteenth and early nineteenth century, typically formed by local merchants or merchant manufacturers.[129] Their power of credit creation, initially developed through financing trade, could be expanded in the new field of wage labour, requiring the issuing of paper notes by the banks in smaller denominations, but in larger quantities. 'In the provinces small notes became the common method of paying wages, and by 1825 one-half or more of the issues of the bankers in many districts were notes of less than £5.'[130] As the wider system of merchant credit and banking that spawned modern industry developed, merchants or banks would advance the money for wages, by the creation of credit money. Workers spent these credits on the goods that they and others produced. These purchases were the means of returning the funds to the banking system, in repayment of the loan through which the credit had been created. The purchases represented the entire value of what the workers produced, for credit money can be created only in relation to the future products and services through which it will be recaptured.[131] As with the system of truck, profit was 'captured' through this credit apparatus, by the manufacturer and retailer charging more for the goods than the cost of their production.

To generalise this system, the creation of credit came to be managed by a distinct entity, the banking house, which gradually separated itself from the merchant houses that had originally combined trade, credit creation, and the arranging of production. The credit money issued by banks was standardised as a means of payment of a worker's 'wage'. The issuing of notes by banks was gradually made uniform. Eventually, only one note, that of the central bank or treasury – in this case, the Bank of England – was used. In place of multiple kinds of money issued by merchants, employers, and local banks, money took on the appearance of a uniform, national institution.[132] However, while governments came to monopolise the printing of paper money, just as they had earlier monopolised the minting of coins, private commercial banks retained their power to create money. Almost all money continued to be privately created, through banks opening accounts for business and individual customers, who wrote cheques or arranged payments that transferred this credit money between firms or from employers to workers. Large enterprises, especially in the United States, also played a role in money creation, not only through issuing shares and bonds, but through the creation of commercial paper, widely used by large firms in the United States to pay wages.

This arrangement of creating credit money to pay wages confirmed the effect of 'abstract labour', the impression that workers were simply involved in an act of exchange, trading the use of their labour for a standardised payment in 'money'. From this perspective, the industrial revolution represented an extension and reorganisation of credit relations, rather than a shift from commercial capital to industrial capital.

City Business

There is a further way in which we can see industrialisation as something of a detour. Two other mechanisms were soon to overtake and displace the factory as principal sites of organising the capitalisation of the future. Like the factory, each of these provided an apparatus of capture that transformed future revenue into assets that could be acquired at a discount in the present; each also provided an arrangement or site for organising knowledge about money and material life. One of these came to be known as 'business', the other as urban life or 'the city'. Neither business nor the city was a new object for making sense of material processes. But, in the later nineteenth and early twentieth century, both in Europe and in many parts of the colonised world, each emerged as a matter of concern, much like 'industry' in earlier decades, as each became the focus of expansive new powers of capture. Looking at their emergence will help us resituate the significance of industrial production – and to understand the later turn to the idea of 'the economy', a question to be explored in the chapters that follow.

Nineteenth-century capitalists seldom seemed to enjoy having to capture revenue through the industrial process. Manufacturing was laborious and often dangerous, even for the mill owners. Jean-Baptiste Say was not exceptional in abandoning manufacturing due to the difficulties caused by the breakdown of machinery. The finger lost in an industrial accident by Ernest Cronier, the man who then turned the Say family's sugar-refining empire into a speculative trading business, and bankrupted it, was another reminder.[133] William Brooke, a merchant-manufacturer in Yorkshire, handed the management of his yarn-making mill to his sons after becoming blind from being splashed with hot indigo, and invested instead in railway shares.[134] The Greg family, whose mills had become the largest textile manufacturer in Britain, found themselves making little or no profit, in part because the father, having passed the firm onto his four sons, continued to claim a rent on the initial cost of the buildings and machinery, never depreciating their value. The sons soon tired of the effort, even Robert, who ran the trading and financing operations rather than the mills themselves. 'I shall find I have given my time, attention and anxiety in buying cotton well, selling well and financing well, for nothing,' he complained to his father. Another son wondered how long 'to continue the dog's life I am leading here. I never open a book . . . rise at 5:30 a.m., go to bed at 10 p.m. and toil like a galley-slave all day'.[135] Like the father, and like most business people, even the sons were more interested in becoming rentiers than the unrewarding work of manufacturing.[136] The company expanded through acquiring other mills, partly in settlement of debts owed from the more lucrative cotton-trading operations, but also because they could borrow funds at 3.25 per cent mortgaged on the mills, then earn 5 per cent by speculating with the borrowed funds. Eventually, neither Quarry Bank nor any of their other mills, those imposing, iconic sites of the industrial revolution, was making money from manufacturing.[137]

The experience of the Greg family, the Say refinery company, and many others typifies a larger change. Within a few decades, manufacturing had been overtaken by another transformation, the rise of 'business'. This was the term that emerged in the later nineteenth century to name a new phenomenon, the simultaneous development of the modern joint-stock company and its new partner institution, the investment bank.

As noted in chapter 1, the half-century from the 1860s to 1914 was marked by two interconnected developments. Within the space of two generations before the First World War, a handful of powers in the northern hemisphere consolidated imperial rule over most countries of the Global South, as well as the surviving native peoples of North America.[138] At the same time, it was an era defined by terraforming projects on a new scale: long-distance railways, large ship canals and sea ports, trans-oceanic telegraphs, the development of municipal lighting and electric power, and the re-engineering of river systems. Equally, it was defined by the novel forms of debt creation that the new scale of imperial power and technical projects enabled.

Critics of empire described the finance-driven expansion of domination with the new term 'imperialism'.[139] Many of the critics came to see this expansion of finance and its export abroad as the flow of an 'excess' of capital, a surplus created by the 'congestion of capital in the manufacturing industries which have entered the machine economy' and the 'saving up' of surplus because of low wages.[140] This view reflects the same misunderstanding of money we have examined here. Finance capital, for the most part, does not come into being as the surplus or savings from existing business, redirected through banks or private investors into new ventures. It is credit money, manufactured by the industry of finance, as a means of capitalising in the present a future revenue. Its creation depends not on the availability of a surplus to absorb from the past, but on the construction of a future in the form of a reliable claim on a surplus to be acquired, in a manner that can be realised and traded in the present.

This later-nineteenth-century expansion of credit mechanisms should therefore be understood not as a 'financial' expansion, in contrast to the industrial expansion that preceded it, but as a technopolitical one. The creation of credit required not just the efforts of a proliferating class of financiers, investment bankers, and stock promoters. It required the technical equipment and terraforming through which future flows of revenue could be manufactured and controlled via a mechanism whose forms of reliability, property, imperial expansion, and violent enforcement made them available in the present. These arrangements supplemented the older apparatuses for governing future revenue, whether through colonial trade, slave plantations, and government war-making, or even the detour of coal mining and manufacturing, in most cases eventually overtaking them.

As we saw, the technical and scientific developments of the period from the 1860s to 1914 had a set of interrelated characteristics that facilitated this new capture of revenue and were simultaneously encouraged by it – the new forms of technical and material durability, accompanied by changes in law and economics, allowing new ways to build apparatuses for the capture of revenue. These also involved new methods of money creation.

As a machine for the intergenerational transfer of wealth, the joint-stock corporation needed the assistance of another organisation, investment banking. Investment banks differed from the older merchant banks, which financed colonial trade and chattel slavery and created the government debt required for the new scale of warfare on which those depended. They also differed from the provincial and local banks that accompanied the development of manufacturing and the payment of money wages. Investment banking served a different purpose, providing a new form of money creation. They emerged to serve neither trade nor industry, but to underwrite and profit from the creation of joint-stock companies, principally for the building of railways and other infrastructure, often on an imperial scale. They managed the issuing of corporate bonds and shares, handled the reorganisation of failed ventures, and arranged mergers and acquisitions. Unlike older banks, which were usually run as partnerships, they were frequently themselves joint-stock companies, selling claims on the future profits to be made from the business of marketing shares.[141] They acquired an astonishing power to create money. The durability of the future revenue promised by new, large-scale infrastructure and the growing scale of imperial power allowed a vastly expanded apparatus for the creation of credit money.

Alongside the new sphere of business, there was a second object that emerged as a mechanism for constructing this expanded relationship to the future: the modern city. From the late nineteenth century, the metropolis was undergoing a similar set of technical changes, made possible by building with iron and steel and concrete, by the development of coal-fuelled electric power stations, by the invention of the electric safety elevator, the electric-powered tram, and electric light railways, and by employing legal arrangements along with race discrimination to plan and regulate the use of space.[142] Developers of urban property could now do the same thing as railway entrepreneurs (they were frequently the same persons): erect a structure whose value, from the day it was built, had little to do with the cost of its construction. The price of real estate, in the carbon-fuelled modernising city, corresponded to the capitalised value at which the speculator could sell the right to live for the next several decades on that spot of land.

Most writers failed to understand the city as an apparatus, like the large business corporation, for capturing payments from the future. When political economists and developers had to explain to people why the price of housing could become so much higher than the cost of erecting the building, they did not explain the difference in terms of the expanded ability of buildings to work as machines that extract revenue from the future; they disguised the mechanism at work by attributing the surplus to the changing value of nature. Urban development, they said, transformed natural or 'unimproved' land into real estate. The difference between the cost of construction and the price at which a building could be sold was ascribed to the rising 'value' of the 'land'. In 1900, as we noted in the introduction, a New York architect described a rapidly built commercial building as simply 'a machine that makes the land pay'.[143] But the payment came not from the land, but from the future renters or purchasers of the space. Land was merely the alibi. Writers like Henry George and other mostly forgotten political economists fought against this simplistic way of explaining the strange ability of speculators to capture future revenue through the device of property development.[144] However, the new theorists of the price system explained value as the product, not of an apparatus of capture, but of the simple mechanism pictured as an interaction of supply and demand. The technical novelty of the durable apparatus, in which the price mechanism played a minor part, soon disappeared from view.

One can contrast the economists with the founders of a new science of the city, the field of urban planning. The Scottish planner and social theorist Patrick Geddes, influenced by Thorstein Veblen and, like Veblen, by Henry George, was among the writers who understood that the unearned value of urban property now represented not the working of market forces but this claim on the future, and thus on the labour of those who would repay it through the increased cost of their housing. The wealth accumulated in the 'mean streets' and 'mean houses' of the modern city, Geddes wrote, consists chiefly in 'documentary claims upon other people's mean streets elsewhere, and upon their labour *in the future*'.[145] The remedy that he and others proposed was the new practice of town planning, to balance the accumulation of private claims upon future labour with the creation and distribution of civic wealth, in particular through the building of garden cities – a hybrid form in which the 'mean streets' of the industrial city would be replaced with 'a healthy, natural, and economic combination of town and country life'.[146] The pioneer of the garden city, Ebenezer Howard, explained that the purpose of this new urban-rural form was to capture the unearned surplus created by the development of property by collecting all ground rent and holding it in a community trust. The trust, in turn, would pay for public works, parks, and other amenities and for the provision of retirement pensions and insurance against accident and sickness.[147] Influenced by Howard, the planners and architects of Red Vienna, the city that made public housing the basis of a social democratic vision for urban life in interwar Europe, produced the most dramatic example of the shared capture of the value that urban living creates.[148] In practice, however, it was in Europe's overseas colonies that these eutopic schemes were most easily promoted, with the development of districts such as Heliopolis, Maadi, and Garden City around Cairo; the building of New Delhi as a new capital for colonial India; and the plan for Greater Jaffa, the conurbation of the Palestinian city of Jaffa and the expanding Zionist settlement on its outskirts, Tel Aviv, for which Patrick Geddes himself drew up the master plan.[149] In most places, however, planners lacked the power to impose such collective schemes, while the colonial versions, relying on racial exclusion to manage the use of space, implemented none of the plans for the collective capture and redistribution of surplus. Thanks to the new durability and scale of buildings and property rights, urban living became another apparatus, alongside the modern business corporation, in which entrepreneurs could realise revenues in the present captured from those who would come later. The colonisation of space relied upon the powers of colonising time.

The Financial and the Real

Let us return to the question of money and its manufacture as credit, and how this understanding of money was forgotten. As modern investment banking emerged alongside the modern business corporation and the modern city, those who wrote about how the new monetary system worked had no problem in explaining that money was something created by banks. A Scottish political economist, Henry Dunning Macleod, published the most influential account.[150] Mitchell-Innes, the Cairo-based financial expert with whom this chapter began, acknowledged his debt to Macleod's work. A bank, Macleod wrote, 'is not an office for "borrowing" and "lending" money, but it is a Manufactory of credit'.[151]

By the mid-twentieth century, this understanding had largely disappeared. The authors of a classic text published in 1960, *Money in a*

Theory of Finance, were aware that most money was created by private banks. But they treated this as a residual phenomenon. As the book's title indicated, their aim was not to explain how money works in banks – they made no study of actual banking – but how money can be made to work within 'a theory'. By theory, they meant the assumptions of neoclassical economics. Orthodox economic science assumes that all economic phenomena can be understood as if they were organised as a price mechanism, an arrangement of buyers and sellers trading goods and services. In this idealised world there are no impediments to trade. The supply of a good always self-adjusts, thanks to the operations of a frictionless financial balancing mechanism called 'the market', to equal demand. As a result, there is no need for money. In the classic Arrow–Debreu model that formed the scientific basis of modern neoclassical economics, money does not exist. So, the goal of a theory of money and banking was not to find out how money actually worked. It was to make an artificial place for it in this frictionless world.[152]

With the development of modern banking, the private power to create credit money became increasingly opaque. Everyday practices concealed the process of money creation. People became oblivious to the modes of extraction from the future. If industrialisation effected the separation of the 'bank' from the factory, and the moment of 'production' from the supply of credit, then the rise of the corporation, of the phenomenon that came to be called 'business', and of modern real estate, furthered this effect of separation.

As with the nature of money, the role of banking has been misunderstood. In the popular view, a bank is a place that stores and lends out money, not an industry that manufactures it. Banks appear to take in deposits from their customers, then loan out those same funds to other customers. Orthodox economic theory adopts and reinforces this common-sense view.[153] Banking provides a service that textbooks describe as 'financial intermediation', aligning those who purchase with those who sell, and those who borrow with those who save, and facilitating their interaction. Banks may have a further role, in this view, but only in their relations with one another, operating as a payments mechanism that carries out the nation's financial bookkeeping.

This popular understanding of how the monetary system works, reproduced in economic theory, has recently come to be understood as largely a fiction. When banks make funds available to borrowers, they are not lending out the money they have taken in as deposits. They are creating new money.[154] The manufacturing works as follows.

An individual customer or firm requests a loan from the bank, say in the amount of $100,000. If the bank agrees, it sets up an account for the borrower and records in its books that the account has a positive balance of that amount. This money is not transferred from some other fund, such as the savings of other customers. It is created from scratch, by the simple act of recording a $100,000 balance. The bank can create money out of nothing because the funds, although credited to an account in the customer's name, do not leave the bank. The borrower does not walk out of the bank's premises, or terminate an online transaction, with actual cash in a briefcase (except in rare cases, and only in small amounts). Rather, the borrower receives a paper statement or an electronic record indicating the availability of the new funds. When the customer decides to spend the funds, the money leaves the customer's account – but stays within the books of the banking system. If the money is spent as the deposit on the purchase of a property, for example, the bank marks the funds as debited from the account of the borrower, and credits them to the account of the person selling the property. If the seller is a customer of the same bank, the funds remain in the books of that bank. If the seller uses a different bank, the two institutions record a transfer from the ledger of one bank to the other, as part of the daily settlement of accounts between banks.

While the bank in this example has created new credit money, the funds have been brought into being through a corresponding future debt. The funds created within the banking system are a claim on the future; their creation depends on this reliable command of the future payment of the debt. Over the following months and years, the borrower will pay from future income the amount of the loan, plus the surplus charged by the bank as the cost of making that future debt payment available on demand to the seller today. This surcharge is described as the 'interest' on the loan. It is better to see it in reverse. The future debt will be repaid in full – its full value being the principal plus the 'interest'. The bank makes it available to the seller today at a discount – the discounted present value of the future payment to the bank. The discount (the lower value today) represents a premium the seller pays in exchange for immediate access to the funds, on demand. The bank is able to offer the future funds on demand today, not because it takes in loans from other customers, but because it continuously pools the debt repayments of other customers. The discounting of future debts is the very source of money.[155]

Depending on the duration of the loan and the discount rate (portrayed as the rate of interest) along with any associated fees, the money paid from future income may amount to double the original sum created, or more. In this way, the very creation of money by the banking system operates to transfer future income to the present, and to profit from this extraction from the future.

With modern banking, the mode of extraction from the future comes to govern not only the specific fields we have already considered – the conduct of long-distance trade, the payment of wages, the building of infrastructure, the provision of housing, or the creation of joint-stock companies. It governs all these and more, all of which enable the manufacture of money as a mode of indebting the future. All of them depend, in turn, on the role of the state, which operates not as the source of money, as is popularly imagined, but as the agency that recognises bank-created credit as money, in particular via its acceptance of this for the payment of taxes.[156]

We can now understand why we do not normally see how money works. There are three issues. First, what one could call the state effect: only a very small amount of money, estimated at around 3 per cent, takes the ready, visible form of coins and banknotes. That small, tangible part is issued by a public mint or monetary agency and printed with symbols of authority. We mistake this visible object for the whole and assume that all money originates in government and is produced through public means. Second, the effect of everyday experience: when individuals or firms do encounter the power of banks to create money – for example, when borrowing funds to purchase, say, a vehicle or a property, or when taking out a company loan – they are given an account in their own name listing the new funds. Since this is now the customer's money, it seems no longer to belong to the bank. Yet, as we now know, despite the name on the account, that money never leaves the bank.

Third, and most important, there is what we might call the economy effect. As we will see in chapter 5 and explore further in the final chapter, in the twentieth century, the construction of the economy as an object of politics and science allowed only a secondary place for finance. The economy was a way of measuring and picturing the movement of real goods and services, represented by prices. This movement happened largely between two sectors, business firms and households, represented as complementary parts of a whole. Firms produced goods and services, while households in the same temporal moment purchased those goods and services in exchange for their labour. Money was considered only a means of facilitating this instantaneous exchange, not as an apparatus of credit organised to extract a surplus from the temporal operation of this and other processes. Economic theory reflected this simple arrangement. This role of the economy in effecting the apparent separation of the material and the monetary, the real and the represented, the business and the bank, is a feature to be explored in the coming chapters.

We are preoccupied today with our relationship to the future, but largely oblivious of its most everyday mechanism. Money, which we think of as just a simple object of convenience or a system of representation for recording the values of actual things, turns out to be a ubiquitous mechanism for the extraction of future income. However, it cannot do this alone.

Most accounts of our relationship to the future are written backwards. The problem is a reversal not just of the temporal sequence, but of the explanation and the phenomenon to be explained. The habit is to begin from a simple materialism, where life is thought of as a set of future needs to be met. To satisfy those needs, goods must be produced and distributed. Their production requires land, equipment, the supply of energy, and an infrastructure of distribution. These, in turn, require the provision of credit, which can also finance the supply of raw materials and the purchase of finished goods. Simple materialism starts in a simple future, the future conjured by the decision to begin the explanation from the standpoint of prospective needs that must be satisfied. Those needs connect the future back to the present. The supply of credit is portrayed as a response to a set of requirements that appear to arise from the material conditions of life. Credit comes afterwards, in this back-to-front account, as a supplement. It supplies the present with a path back to that future.

We have now reversed this sequence, avoiding its simple materialism. We already know that, with the spread of large-scale commercial relations, credit was not something secondary, created to serve the needs of trade. If anything, trade was organised to serve the creation of credit. Likewise, banking was not just an 'intermediary' that served the needs of households and businesses, by matching the demands of borrowers to the availability of lenders. It became a 'Manufactory', a large workshop for the manufacturing of credit money.

This reversal does not require us to abandon the analysis of material processes in favour of studying the immaterial forms of finance. It requires a different materialism, one in which the financial is not the opposite of the real, but, rather, one in which the production of this opposition, in which finance appears to stand apart from the real, is an effect to be studied and explained. And it requires a different approach to the future, seeing it not as the imaginary world conjured by the projection of material needs, but as the concrete apparatus built for the appropriation of wealth and livelihoods from those who come later, and at the expense of the common future.

4

Reading the Book of the Future

We have come to inhabit a world in which extraordinary wealth is claimed from the future, then repaid from the livelihoods of those who come later. Meanwhile, the earnings of most people in the present carry the burden of repaying the cost of previous claims. This mode of capturing income across time is not part of the standard picture of our social and economic lives. That picture is organised around a different object. We imagine ourselves to be connected in the present by an arrangement that we name 'the economy'. The economy is not a mere fiction, for it has real effects on the way we live. In fact, the idea of the economy is a keystone of the alibi of capital, producing a different understanding of relations to the future. It obscures the mode of living at the future's expense.

The worlds we have been exploring in the last three chapters precede the emergence of 'the economy'. The contemporary idea of the economy dates only from the mid-twentieth century. While the term itself is much older, only in the years just before the Second World War did economists begin to refer to '*the* economy' as an object in itself. In the interwar period, statisticians had developed methods of estimating aggregate levels of prices, incomes, purchasing, and savings. Meanwhile, economists devised a schema that pictured these aggregates as parts of a whole. To make the numbers appear to form an object – a freestanding structure of interacting parts – the schema had to estimate various totals, for products, firms, households, incomes, and more. But it also had to determine, as we will see, how to disguise the cost of living at the expense of the future, providing an alibi for our modes of impoverishment. By the 1950s, this whole came to be described routinely as 'the economy'. Over the following decades, in the industrialised countries of the West but not necessarily elsewhere, the economy became the central object of politics.[1]

On learning how recently the economy emerged as an object in itself, most scholars ignore the significance of its invention and continue to project the concept of the economy onto earlier periods. The term is used ahistorically, especially among historians, as an abstraction that can apply to any period, as in discussions of the medieval economy or the economy of Ancient Egypt. Since older societies engaged in agriculture, manufactured goods, developed commercial relations, created credit, and levied taxes, surely, they can be said to have had an economy?[2]

One can take that approach to the idea, but there is no need to do so. The previous chapters have explored several centuries of the history of money, commerce, and material life without once referring to something called 'the economy'. The chapters avoided the term because its ahistorical use comes at a cost.

Climate breakdown and other threats to Earth systems and collective futures require us to loosen the hold of 'the economy' over the way we think and act. In recent decades, economic reasoning and the needs of the economy have provided the most powerful justifications for failing to adequately address those threats. To counter the force of arguments and assumptions about the economy, we must ask how this twentieth-century entity was organised, how it governs us in the present, and how it shapes our relationship to the future. Those questions will be taken up more fully in the next chapter and again at the end of the book. But answering them will be more difficult if we do not first ask, as this chapter sets out to do, how wealth, improvement, obligation to others, and well-being were governed and understood prior to the invention of the economy.

The language that any human collective employs to order its forms of life and establish duties and obligations is part of what constitutes those forms – even if language never fully apprehends the worlds it designates. Ignoring the key words in which a collective life is grasped and imposing terms from another age, treated as universal concepts, or universally modern concepts, fails to properly understand that life. While this is true of all language and all forms of human sociality, it acquires a special pertinence with the language of economics. The method of economic science, as several recent studies have shown, is performative. That is, it seeks to bring into being, and to optimise or make more effective, the schemas that it describes.[3] While almost any science can be shown to have this two-way relation with its object, seeking to make of the world a place where its models or explanations can thrive, economics plays a marked role in attempting to format its object, the economy. The reasons for this lie partly in the calculative nature of economics, as a science oriented towards equipping agents and agencies with the means to better determine courses of action. They also lie in the special role that economics plays, compared with other sciences, in justification – in explaining *why* the collective world exists as it does.

Justification operates not only by offering specific reasons for the way things are, but by establishing a natural order. In naturalising the economy as its object, and assuming that self-interest is innate and works through the economy as a fundamental principle of order, economics operates in a manner analogous to the early modern role of religious practice. Philosophers of religion use the term 'theodicy' to refer to the role of religious discourse in rationalising the presence of suffering and evil in the world, despite the existence of an all-powerful and benevolent deity. Economic discourse can be said to perform an analogous operation, positing the existence of a natural order, the economy, whose processes of exchange tend towards harmonisation. Drawing on the Greek term *oikos*, from which the word economy (*oikonomia*) was originally derived, Joseph Vogl suggests we refer to this operation of economics as 'oikodicy'.[4] The analogy is not intended to imply here that economics operates today as a replacement for religion. In fact, the concept of 'religion' emerged in the West in the same period as the ideas about 'economy' to be explored in this chapter, and, indeed, as part of the same reordering of modes of justification.[5] Rather, it is to suggest that the coproduction of 'religion' and 'economy' facilitated the unacknowledged role of economics in the justification of suffering.

There is another cost to using the concept of the economy to understand worlds prior to the invention of the term in the mid-twentieth century. As we will see, the language of economics appears to address the present in relation to the future. It diagnoses current conditions to offer predictions about where things are headed. Leaving aside the issue of how successfully it does this, and, indeed, how it causes us to misapprehend our relation to the future, the predictive, future-oriented, ahistorical mode of economic reasoning conceals another consequence of this mode of thought: it helps produce the past. That is, it frames the present as the outcome of general principles, such as material needs, self-interest, the development of technology, and the dynamic of growth, that are always determining.

Taking a concept like the economy as applicable to all times, or even just to the era known as capitalist modernity, creates a false continuity. One of the surprising things about the history of ideas about opulence and suffering, improvement and decline, virtue and self-interest, nature and society, and law and government is the frequency of sudden breaks, even over just the last two centuries, in which earlier forms of understanding were abandoned.

This discontinuity has a special significance. The sudden shifts and interruptions, in which a new object of knowledge and language of justification emerge, tell us something about the process of living at the expense of the future. Capturing the future is always open to uncertainty and moments of dislocation and crisis. These dislocations can be tracked, in part, in the abruptly changing forms of knowledge through which collective life is comprehended, organised, and justified. By adopting over the last century a uniform conceptual language of 'the economy' and 'the economic', and of its government and its 'growth', projected back onto the nineteenth century and earlier, the very political uncertainty and instability of extracting from the future is papered over in a continuity of conceptions.

More than that, the writing of political economy in the nineteenth century, and of economics in the twentieth century, played a decisive role in attempting to manage this discontinuity. Starting in 1824 with J. R. McCulloch's *Discourse on the Rise, Progress, Peculiar Objects and Importance of Political Economy*, every significant text on questions of wealth, improvement, and material progress presented itself as the development, clarification, and critique of earlier writings. Earlier writings were portrayed as simply immature and imperfect versions of the same form of knowledge about the world. The teaching and writing of political economy and economics played a role in occluding the discontinuities in thought and dislocations in forms of life.

There are four questions to consider to bring these discontinuities into view and connect them with wider dislocations explored in other chapters of this book. First, how was the term economy used before it referred to *the* economy and how did those uses repeatedly change? The answer will rely on not confusing a process with a thing. Second, even if it can be shown that the economy was invented only in the mid-twentieth century, why not see its invention as mainly a change in vocabulary? Surely there had already emerged in the nineteenth century, if not earlier, a distinct sphere of 'the economic'? Was this new sphere not the object of a new science, political economy? In fact, some would claim that the very separation of 'the economic' from 'the political' is what distinguishes the organisation of modern capitalist societies from earlier modes of acquiring wealth and ordering collective life – that 'the economic' designated a distinctively capitalist mode of exercising power. The answer here will depend on not confusing an incomplete process of restructuring the world with an accomplished structure – in fact on understanding a structuring effect as a process defined by the impossibility of its completion; and, indeed, by the incompleteness of any description of it. Third, if we acknowledge that the economy may have been invented only in the mid-twentieth century, what led to its invention? What upheavals in the late nineteenth and early twentieth century preceded its creation? Since that period coincides with the formation of the modern academic discipline of economics, what was the object of that discipline and why by the interwar period did it appear to have failed? Finally, moving into the interwar years, to what immediate problems did the concept of the economy first emerge as an answer?

Political Economy Lies Dead

How was the word 'economy' used prior to the invention of *the* economy? To what did the older science of political economy refer? And when the study of economics emerged as an academic field, in the later nineteenth century, what was its object?

Before the mid-twentieth century, the term economy referred to a process, not a thing. Earlier uses of the word in English connoted husbandry, household management, or the administering of resources. It could be used in reference to any entity with affairs to be managed, from the human body to the household and the large agrarian estate. As Keith Tribe notes, the word suggested stewardship and preservation rather than acquisition and trade. Cognate terms were found in other languages. In the Ottoman world, for example, a discourse concerned with household management or the stewardship of an estate (*'ilm tadbir al-manzil*, or the science of managing the household) developed from the sixteenth century in both Turkish and Arabic. Drawing on classical Arabic authors, the science was a branch of practical philosophy offering 'guidelines for the sustenance of the upper and lower classes'.[6] By the mid-eighteenth century, as centralising states like France, England, and

Prussia developed enhanced means of violence that allowed them to displace the older empires of the Spanish, Portuguese, Dutch, and Ottomans, the term economy was repurposed. French, Scottish, and English writers used it to refer to arguments for and against expanding the role of the sovereign in the improvement of trade, population, and welfare. Retaining, at first, some of the older sense of stewardship, the new discourse of *political* economy addressed the duty of the sovereign to pursue the 'economy' or good government of the people and of other productive resources of the country. Much of the debate concerned the question of whether sovereign rights, or 'reasons of state', should prevail, for example in the government of commerce, or should be limited by principles of natural justice. In a Europe of competing commercial states attempting to justify or expand state-sponsored colonial trading monopolies, the term political economy was adopted to refer to an older set of texts from the seventeenth and eighteenth centuries on the topic of commerce. Their main concern was the role of government in the protection and improvement of colonial trade and the fiscal regime at home on which this expanded, militarised power depended.[7]

This deployment of 'economy' could draw upon not only earlier notions of husbandry and proper management, but also medieval and early modern uses of the concept in theological debate. In Christian theology, the 'divine economy' referred to the role of providence in regulating earthly affairs. This theological heritage resurfaces, Giorgio Agamben suggests, in the invocation of a providential 'hidden hand' in the regulation of human commerce, or in the more general appeal to laws of nature.[8] One school of French writers, the group known as *la secte économiste*, insisted that nature or natural law was the source of the principles of good government and advocated 'physiocracy'. The term invoked the rule of this providential nature.

The most influential eighteenth-century writer on these questions, Adam Smith, was an opponent of political economy. In *The Wealth of Nations* (1776), he attacked the very idea of political economy, or the view that the political body needs some economy or 'regimen' for its improvement. Like the French Physiocrats, he argued that natural law rather than reasons of state should guide legislators and sovereigns. Unlike them, he rejected the argument that legislators needed a 'system' of political economy, meaning a set of prescriptions to follow, to compensate for the failure of natural laws, or the principles of natural jurisprudence, to correct for the ill-effects of bad government. Just as the human body 'contains in itself some unknown principle of preservation', he wrote, so in the political body 'the natural effort which every man is continually making to better his own condition, is a principle of preservation capable of preventing and correcting, in many respects, the bad effects of a political economy'.[9]

In the same year that *The Wealth of Nations* appeared, a group of Britain's North American colonies declared a revolt against the system of European commercial monopolies and colonial rule that had been the focus of debates for and against political economy. By the early nineteenth century, as we saw in the last chapter, revolutionary change in the Atlantic world, in South Asia, and in the Ottoman Empire and Europe had forced Britain, France, and other powers to begin supplementing capitalisation through imperial commerce with expanded domestic modes of capturing the future. To address this changed world, and the new modes of justification it demanded, the meaning of the term political economy was reinvented. Political economy became a discourse for establishing not the providentially ordained principles of proper government but the secular laws of human society.

A secular science of society appeared to require a break with the idea of a 'divine economy' regulating human affairs. In England, the Anglican establishment, entrenched at Oxford and other centres of learning, opposed the teaching of doctrines that derived rules of human conduct not from scripture but from science – especially a science that turned the vice of self-interest into a public virtue, and the Christian virtue of poverty into a collective vice. To overcome these objections, in 1825, Henry Drummond, a wealthy London banker and Protestant evangelist, endowed the first chair in political economy in England, at Oxford.[10] A fellow evangelical, Richard Whately, elected to the new chair in 1831, proposed abandoning the 'most unfortunately chosen' term political economy, with its ungodly preoccupation with the acquisition of wealth, replacing it with the new name 'catallactics', or the 'science of exchanges'.[11]

However, the leading figures in the field, the French writer Jean-Baptiste Say and his English follower John Stuart Mill, decided to revive the term political economy as the name of the new science, understood as a subfield of the science of social economy. (In Prussia, Say's German translator coined the parallel term *Nationalökonomie*, which began to replace the eighteenth-century German science known as cameralism, which had guided the public administration of finance and welfare.[12]) These writers agreed with Adam Smith's rejection of the old political economy, but also largely rejected Adam Smith. They repurposed the name political economy to describe not the duties of the sovereign, nor the principles of natural law, but certain 'laws of society'. The term designated the laws, as Mill wrote, that concern man 'as a being who desires to possess wealth, and who is capable of judging of the comparative efficacy of means for obtaining that end'.[13] These social laws, Mill argued, should take the place of sovereign powers. Political economy was redefined to refer to governing not through the sovereign or legislature but through the unchanging rules that regulated what Mill called 'the enrichment of society'.[14]

The insistence that the rules of political economy now formed a secular science did not, in fact, remove the question of a providential order. The assumption that man acted on a principle of self-interest in the pursuit of wealth, a scientific principle whose operation was the source of social order and collective enrichment, may have eliminated the reference to a divine hand; but the assumption still operated as a transcendental truth – perhaps more powerfully so with the reference to the deity elided. The justification of suffering and evil still rested on the work of an unseen principle, albeit one now expressed in the language of secular science rather than sacred scripture.[15]

At the same time, this 'formation of the secular' was not itself simply secularising.[16] Establishing the laws of 'economy' was equally the establishing of a new Christian concept of 'religion', as the container for what lay beyond the realm of political and social economy – of rational, self-interested enrichment. The terms economy and religion were codependent, each acquiring a novel meaning and import through reference to one another. And the 'beyond' to be grasped through the lens of religion was now simultaneously a private and a geographical-colonial space. In Europe, religion was coming to designate a residual realm of private belief. Beyond Europe, as its merchants and missionaries turned their attention from the lost colonies of North America to the new opportunities for enrichment in North Africa, Egypt, and the Levant, the concept of 'religion' took on an expanded, colonising force. In the lands of Islam, where rational self-interest, it was thought, did not yet govern, religion took the place of a secular science of society. Indeed, the science of Orientalism elaborated the idea of religion into a master concept, as central to understanding the Islamic world as economy was becoming for the world of Europe.

At the centre of this Christian preoccupation with the Orient, as Gil Anidjar notes, there stood 'one particular Oriental city, one particular Oriental land, and one (or two) particular religion(s)'.[17] It is useful in this regard to note that the banker and evangelist Henry Drummond, while endowing the teaching of the new field of political economy at Oxford, simultaneously founded a prophetic movement warning of the imminent Second Coming of Christ. He funded the Swiss Orientalist Johann Burckhardt's travels to Egypt and the Levant in the hope of converting Muslims to Christianity. The proselytizing effort failed – Burckhardt converted to Islam and was buried at his own request as a Muslim.[18] More promisingly, Drummond's movement advocated the settlement of European Jews in Palestine, in preparation for the apocalypse.[19]

Like its eighteenth-century predecessor, the nineteenth-century science of political economy had a short life – for reasons connected in part with its unresolved relation to the Orient. 'Our Political Economy', wrote the English essayist and banker Walter Bagehot in 1876, 'lies rather dead in the public mind.'[20] Its critics complained that the laws of economy were hypothetical, based on assuming humans to act only on selfish instincts, ignoring 'man as a responsible moral being'. The science described a society governed by self-interest, a paramount force 'only in certain parts of Europe in very recent times', and not the only human motive even there.[21] So it dealt only with 'a very limited and peculiar world'.[22] Even in Europe, the science could not adequately explain the rise of collective arrangements such as trade unionism and co-operativism, and the increasingly powerful institutions of private land ownership and large business. 'Wages, profits, accumulation, consumption, population, poor-laws, land-systems, partnership, tenancy, trade-unions, co-operation, these are things which involve the great human instincts, wants, and institutions; and they are for the most part beyond the reach of a mere economist.'[23]

The critics acknowledged that there could be exceptional circumstances where the laws of economy might apply. These exceptions are interesting to note. They were situations that tended to be 'independent of general institutions and habits', resting on 'temporary' human relations or efforts and on 'special and highly artificial processes'. One writer gave the following list of examples: 'Prices in market overt, currency, bills of exchange, monetary practices, insurance, restrictions on trade, [and] taxation'.[24] So the laws might apply in financial markets and such related areas as currency exchange, taxation, and insurance, and in 'market overt', or open market. The last of these was a legal term for an authorised public market, an exceptional space in which prices for goods are openly negotiated and thus made visible.[25] In other words, the rules of economy appeared to have little relevance to the majority of human material life – to crafts, the professions, and most other employment relations (many of which were governed by new master and servant laws, which criminalised the breaking of employment contracts and other worker 'disobedience'), to land ownership and tenancy, to the supply of goods among businesses and their customers, to almost every kind of material relationship.[26] But, in the specialised and expanding world of finance, along with certain other uncommon sites of exchange, the principles of economy might apply.

Why did these exceptional sites and practices seem to lend themselves to the theorising of the economists?[27] Bagehot, the leading English writer on the money markets, explained it as follows: they are exceptional because they trade a standardised good that generates knowable prices. The stock exchange, for example, in one key respect, 'is the simplest of markets', because 'there is no question in it of the physical quality of commodities: one Turkish bond of 1858 is as good or bad as another; one ordinary share in a railway exactly the same as any other ordinary share'. In other situations, by contrast, 'each sample differs in quality, and it is a learning in each market to judge of qualities, so many are they, and so fine their gradations'. Economists had tried to solve this difficulty in determining 'the price' of an ordinary commodity, such as corn (wheat), by compiling tables of statistics. But tables cannot tell you the price: 'in a hundred cases you may see "prices" compared as if they were prices of the same thing, when in fact they are prices of different things. The *Gazette* average of corn is thus compared incessantly, yet it is hardly the price of the same exact quality of corn in any two years.'[28]

This is the clue to the way out of the failure of political economy that began in the 1870s, leading to the birth of a new discipline, which came to be called 'economics'. The discipline did not define its object with the broad term 'the economy'. Rather, it came into being as a field that abandoned the study of man 'as a social being who desires to possess wealth' and took as its focus these 'special and highly artificial processes'. A critical phenomenon, as we noted in chapter 1 and as Bagehot confirms, was the development of one particular process, the stock market.

The Birth of Economics

While the unusual role of stock markets was new, this was not the first time that a shift in the object of knowledge had followed the emergence of a novel, circumscribed, and closely documented site. Political economists and their critics had for a long time relied upon 'limited and peculiar worlds' as the places where they might describe 'artificial' processes and translate them into models for explaining arrangements of human inequality and well-being and the laws of nature or society that governed and justified them.[29] Adam Smith had proposed that the increase in a nation's wealth arose from the increasing division of labour. To make his argument, he famously offered the unusual example of the multiplication of discrete operations in a workshop for the manufacture of dressmakers' pins. He borrowed the illustration from a series of French accounts, commissioned mainly by the French Academy of Sciences and reproduced in Diderot's *Encyclopédie*, each text based largely on earlier versions, originating in reports of local trade inspectors and responding to the desire of the French monarchy to document and regulate specialised craft knowledge, for which pin making served as a paradigmatic example.[30] Smith's argument about the multiplicity of operations rested in part on an accumulation of misreadings in successive reports. However, the more interesting fact is that it relied on the availability of an exemplary case. The closely documented making of the pin formed an artificial microprocess in which Smith could read the entire history of human material progress.

Less famous than the pin factory, but more important in shaping the argument of *The Wealth of Nations*, was Smith's experience in advising his young benefactor, the Duke of Buccleuch, one of the largest landowners in Britain, on measures to 'improve' the 434 farms on his Scottish estates. Improvement took the form of enclosure, road building, the draining of wetlands, and the clearing of upland heaths, in order to lease the farms commercially by auction. Such cases of 'improvement' demonstrated Smith's point, more critical to the arguments of classical political economy than the division of labour, that human improvement depended on the increase in relations of commerce, or peaceful exchange, and that 'land should be as much in commerce as any other goods'.[31] In a similar way, to demonstrate the laws governing and justifying the distribution of gains between labour, invested capital, and landowners' rent, David Ricardo drew on contemporary accounts of practical experiments in farm improvement – crop rotation, drainage, manuring, animal husbandry – published in pamphlets and journals, many of which were collected in a House of Lords report in 1814 on the Corn Laws. A response to the food crisis brought on by the Napoleonic wars, these reports on experimental farming, as Mary Morgan suggests, formed a laboratory generating numbers and examples from which Ricardo produced a model explaining, through this emergent 'oikodicy', the unequal distribution of wealth among the different classes of society.[32]

The work of Marx provides a third example of this relationship between the circulation of reports and experiments on exemplary sites, often in connection with official investigations into current threats and crises, and the production of the changing paradigms of political economy. In Volume 1 of *Capital*, Marx famously made a distinction between two forms of labour, the distinction that he claimed to be the key difference between his own understanding of value and the writing of earlier political economists. Although his predecessors understood labour in general to be a source of surplus value, Marx argued that labour had now been transformed into two kinds, the 'concrete' labour of particular productive tasks and 'abstract' labour. Industrial machinery, driven by steam power, had taken control of human bodies, allowing factory owners to control minutely the time of each operation and to continuously speed up the making of goods. The control of accelerating production time increased the relative share of the surplus going to employers, as labour's share was limited to the payment of a wage, determined by the cost of sustaining and reproducing the labourer. More importantly, however, that surplus was now measured not in different kinds of concrete labour, but in homogeneous units of labour time. These units, emptied of the particularity of actual human labour, appeared not as the result of the productive activity of specific individuals but as an abstraction – a structure of universal units of 'value'. Marx named this abstract structure, or 'form' of social life, the 'value-form'. Under the capitalist organisation of production, the domination of capital over labour operated through the alien power of this abstraction.

Marx's account of how the 'value' of labour was now measured in homogeneous units of time, appearing as an abstract mediation or structure, was not based on his own observations of what happened in factories. He drew it from reading the proliferating reports of factory inspectors and medical examiners, whose work was the result of liberal reforms intended to ameliorate the abuse of industrial labour. The reformers were keen to emphasise the systematic and relentless increase in the control of time and labour. So, the 'abstractness' was partly a result of the regulating, recording, representing, and replicating accounts of factory life. Only a very small part of the population worked in factories. Not all followed the logic described by the reports and by other sources on which Marx relied, such as Charles Babbage's study of the organisation of mechanised production, *On the Economy of Machinery and Manufactures* (1832), and Andrew Ure's *The Philosophy of Manufactures* (subtitled *An Exposition of the Scientific, Moral and Commercial Economy of the Factory System*).[33] We know from contemporary accounts that factory labour was often very different from these descriptions of the 'economy of manufactures' or the 'economy of the factory system' (again, 'economy' refers here to efficient organisation).[34] But the work of measuring, recording, reporting, and referencing contributed to the production of abstractness. Like Smith's model pin workshop and his and Ricardo's experimental farms, the factory provided an exceptional and unusual site from which a science could be developed – in this case, one in which abstracting itself became not just a mental exercise but a real process, part of the daily mechanism of social domination.

The subsequent demise of political economy and the birth of economics can also be situated in this sequence of shifting locations of experimentation and abstraction. As we will see shortly, the shift caused Marx himself to largely abandon the project of writing a critique of political economy, after publishing just the first volume of *Capital*, with its focus on factory production. In the last third of the nineteenth century, another exemplary site became the focus of observation and concern, and the source of a new paradigm of explanation. The stock exchanges and other money markets, where investors and speculators traded in shares, currencies, and bills of exchange at an increasing volume and frequency, operated in a similar yet more focused way, creating another form of abstraction – and, we might add, another mode of domination. They too allowed the possibility of a different science: the marginal utility theory of the emergent academic discipline of economics. Abandoning the study of the great human 'wants and institutions', the new discipline substituted the study of 'prices', taking them to represent the scale of human wants and the volume of commodities supplied in response. A financial market was arranged to produce, as Bagehot noted, the uniform 'quality' of the commodity, a synchronised movement of knowable prices, and the simple identification of buyers and sellers, each one equivalent to all the others. It was the stock market, rather than the regulated labour process of the factory, that generated the simplified, numerically organised world on which economics could now be built.[35]

The new microsite of scientific knowledge also created an increased need to establish such knowledge. The need arose from the fact that stock and commodity exchanges produced not just a measurable price system, as a 'real abstraction', but a pattern of repeated crisis. The crises were connected with the process on which the price system was constructed – Europe's intensifying attempt to profit from and reorganise the non-European world. That world began to appear as a spectre: a figure just beyond the grasp of the new science, the unknown that both demanded and escaped its understanding.

Half-Finished and Half-Civilised

There was thus a further reason why the stock markets suddenly seemed an ideal place to understand, not just the working of mechanisms of enrichment, through the formation of prices, but the crises affecting this process. The money markets were the site that connected the phenomenon of prices to the circumstances that seemed most responsible for their instability: the new powers, as noted in previous chapters, of 'imperialism'.

By the mid-1870s, the money markets were in turmoil. The Panic of 1873 was the worst in a series of financial crises that seemed to occur almost every decade.[36] Triggered by the failure of North American railway ventures and the collapse of speculative urban development in Chicago, Berlin, and Vienna (where the absence of anticipated crowds at the 1873 World Exhibition caused property values to crash), the crisis destroyed trade and businesses across North America and Europe, caused the value of the Indian rupee to collapse, and led to the bankruptcy of the Ottoman governments in Istanbul and Cairo (destroying the value of Bagehot's 'Turkish bond of 1858' and all the subsequent loans created in its support) and to the British conquest and occupation of Egypt.

The connection between economics and imperial crisis can be seen in the writings of Walter Bagehot. There was a need, Bagehot wrote, to better understand not just 'the facts of commerce', but those of 'the great commerce', especially in relation to the 'new problems' of the time. He focused on two such problems. One was how to understand the repeated, worldwide financial downturns – the panics and crises that seemed to afflict the money markets in cyclical fashion. Some writers had proposed a meteorological explanation for the cycles. It appeared that 'changes in the state of the sun' followed regular cycles that could be linked to 'the general productiveness of the earth'. What effect, Bagehot asked, would the resulting 'cyclical variation in the efficiency of industry' have upon commerce? Was it therefore true, as seemed likely, that the solar variations, affecting the whole Earth, 'have much to do with the regular recurrence of difficult times in the money market'?[37] The other problem was how to understand the effect of the large loans England had been making for the first time to governments outside Europe. At the moment when the Ottoman and Egyptian governments were declared bankrupt before the European bankers who had extended credit to them, using the language of racial superiority on which such imperial finance relied, he noted how 'we press upon half-finished and half-civilised communities incalculable sums'. These communities had been offered credit money in almost any amount. 'No incipient and no arrested civilisations', he added, 'ever had this facility before.'[38]

Both examples of the new problems for which economic science was needed arose from financial markets.[39] And both concerned the new financial relations between Europe and the world it was colonising. While this connection between the instability of finance and the instability of imperialism was clear in the case of bank loans extended to so-called 'half-civilised' communities, the connection could also, it now seemed, account for credit cycles. The most promising explanation for the cyclical abundance and scarcity of credit was based on the relationship between agriculture – the main source of wealth – and climate. The idea that meteorological cycles might explain the recurrence of financial crises had been discussed since the early nineteenth century. Variations in weather, perhaps associated with cycles of solar activity, it had been proposed, would cause fluctuations in the price of grain and other crops, leading to cycles in the availability and cost of credit. But no conclusive relation had been found. In the 1870s, however, the economist William Stanley Jevons developed statistical evidence that might explain this relationship. It was true that there was no correlation between sunspots and the price of grain and other staple crops in Europe. This was because, Jevons suggested, 'the conditions of our climate are too complicated'.[40] But large disturbances in the credit market might be a problem caused by Europe's expanded trade with the tropics. In tropical countries the weather was less variable from season to season and the climate therefore less complicated. This stability on the one hand facilitated the creation of long-term credit. On the other, it made that credit vulnerable to long-term cyclical variations, such as those associated with sunspots. This was probably the case with the Indian trade, and 'it is quite possible that tropical Africa, America, the

West Indies, and even the Levant are affected by the same meteorological influence'. Jevons claimed that sunspots caused decennial famines in the tropics, so 'it is the nations which trade most largely to those parts of the world, *and which give long credits to their customers*, which suffer most from these crises'.[41]

If the stock markets connected financial instability to imperialism, there was a curious consequence. While, in Europe, political economy seemed to lie dead, in the non-European world, it was suddenly brought back to life and repurposed. The repurposing served the creation of another of the alibis of capital: the 'alibi of empire'.[42] If the new science of economics was born out of the model space of the money markets, the old science of political economy was reborn at the same time with a new purpose. It was becoming a means to govern the victims of these financial processes, those disparaged by Bagehot as half-finished and half-civilised. By the 1880s, the doctrines of political economy were being revived as an instrument of colonial government by men like Lord Cromer, the British viceroy in Egypt, the country Britain had occupied after the crisis in the financial markets pushed its ruler into bankruptcy. What followed might be labelled 'colonial materialism'.[43] The operative distinction in colonial rule was not between the economic and the political, but between the material and the moral, words that refer not to discrete domains of the social world but to the government of human action. The Egyptian peasant, in Cromer's view, like all human beings, was naturally self-interested, and therefore capable of governing his own *material* improvement. But in his *moral* life he was backwards, incapable of working towards a higher, common good. This distinction between individual self-interest and moral backwardness provided the justification, the oikodicy, for the great injustice of colonial domination. In the two-volume account of his colonial administration of Egypt, Cromer used the word economy only with the conventional meaning of parsimony. In the early years of the British occupation of the country, he wrote, 'the utmost economy had to be practised'.[44] He used the word 'economic' only a handful of times, and in reference not to a realm of the social world but to rules and questions of government. His 1,200-page text touched on every aspect of the material and moral improvement that colonial rule claimed to bring but used the word economy with no meaning other than frugality, and the word economic to characterise not material life but the moral laws or principles of policy. Colonising Egypt was never concerned with an object called 'the economy' or a realm of 'the economic'. That particular language of justification would emerge only when the colonial order was collapsing.

In a subsequent chapter examining neoliberal economic thought a century later, we will see how the doctrines of neoliberalism acquired a second, spectral life in the Global South, lingering as a kind of zombie thought long after they had been largely discredited in the north. So too the old, classical political economy acquired a ghostly afterlife – in Egypt, to govern natives. The colonies were not in this case a laboratory for a new science, but a spectral home for a dying science.

Meanwhile, apart from its use in the title of the increasingly irrelevant science of political economy, the word 'economy' continued to be used in English mainly in a more restricted sense. Still associated with the original meaning of husbandry or householding, but narrower in scope, the word denoted the prudent use of resources, or the efficient management of an undertaking. One of the rare uses of the term 'economy' in Marx's *Capital* occurs in a passage on the growing of timber, quoting from a German handbook on land economy. Forestry, the author of the handbook notes, 'requires for its regular economy a larger area than grain culture'.[45] More frequently, the word referred not just to efficient administration but to cost-saving or frugality, as when Marx remarked in the manuscript that later became volume three of *Capital* 'how much *economy* on space and therefore on buildings, etc., results in crowding workers together in cramped conditions'.[46]

Just as the original meaning of the word economy in the sense of husbandry had previously been borrowed to denote the wider idea of good government, economy in the sense of frugality could be used not just for specific acts of prudence and efficiency but to describe government policy in general. It designated a programme of more efficient and prudent use of resources.

When advocated at home as an alternative to imperialism, such a policy might now be called '*national* economy'. The leading English critic of imperialists like Cromer, the socialist writer J. A. Hobson, used the phrase in this way to promote a programme of domestic improvement. In his classic essay *Imperialism*, published in 1902, Hobson argued that the external drive by imperial powers to acquire territory and resources overseas, usually involving warfare, could be replaced with the more efficient use of resources at home, by 'substituting an intensive for an extensive economy of national resources'. He concluded that 'such a national economy would not only destroy the chief motives of war, it would profoundly modify the industrial struggle in which governments engage'.[47] The word economy, and the phrase national economy, refer in such writings to an efficient administering of material resources, not to the object being administered.[48] This usage should not be confused with the later deployment of the phrase '*the* national economy', from the mid-twentieth century, to denote such an object.[49] The wording refers to a process, not a thing.

Economisation

Such arguments overlook the significance, to be explored in subsequent chapters, of the twentieth-century invention of the economy.[50] The economy, as we will see, was not 'just' a representation, but a complex device for governing a population through its future (chapter 5), and for managing the alibis under which the appropriation of that future proceeded (chapter 8). However, they also misunderstand what came before. Underlying such arguments is a larger view of the way a capitalist, or capitalising, society works. As we noted in chapter 3, most historical accounts of capitalism are evolutionary: they attribute its emergence to the gradual development of commercial relations, methods of production, military power, or human acquisitiveness. But a different line of thought, drawn from Marx, locates the birth of capitalism in a more disjunctive change: a process of dispossession, first seen in early modern Britain, that created a large landless and propertyless population dependent on selling its labour to those who now controlled the means they required to survive. Once the majority of people depended for their existence on the sale of their labour to others, in this analysis, a new political order emerged, based on a different arrangement of coercion. The majority had no choice but to sell their labour to those with the means to buy it, and the latter acquired not only the use and control of the labour but ownership of what it produced. This reorganisation created a different method of governing populations. Previously, the powerful acquired control of surplus wealth only *after* goods had been produced. Its acquisition relied upon creating relations of personal dependence that required social justifications, religious and judicial prerogatives, and coercive powers. Now they acquired the surplus through the very way goods were produced. The apparatus of production, whether a coalmine, a corn field, or a cotton mill, became the privatised site where the powerful organised control and captured the surplus.

Let us leave aside for now the fact that older forms of commercial society often allowed merchants and rulers, through the instruments of debt and tax obligation, to arrange and govern production.[51] Let us also leave aside for now the fact that most people today are not employed in material production, and that forms of consumption, debt, policing, and political coercion are significant aspects of the apparatuses for the capture of surplus (as Marx himself was aware). Let us hold off, until the next chapter, an explanation of how the economy and its growth became among the most powerful abstractions through which many populations are governed. Staying, instead, with the forms of dispossession and coercion associated with the rise of industrial production, a simple way this change is sometimes described is to say that the apparatus of capture had been relocated from the political realm to the economic. That formulation often implies that the political and the economic are trans-historical categories describing the functional divisions of all societies. A better approach might be to say that this method of arranging the capture of surplus took the historically specific form of a separation of the economic from the political.[52] This is what is meant by the argument that, in capitalist modernity, the economic becomes a *real* abstraction. It is not just an emergent category of thought, but an abstraction that is realised in the organising of the collective world. How is it realised? The key element is that human labour becomes a commodity that can be bought and sold. Labour under capitalism, in the words of Georg Lukács, 'becomes a category of society influencing decisively the objective form of things and people in the society thus emerging'.[53] The goods produced by that labour are not the property of the producer, but commodities owned by whoever purchases the use of the labour. A series of private legal arrangements and market mechanisms – contracts to buy and sell labour and rules of ownership that award an individual absolute control over land, labour power, or goods irrespective of who cultivates, nurtures, or creates them – render the capture of surplus a private arrangement governed by laws of property and rules of market exchange.

However, to describe this process simply as the creation of 'the economic' (and of 'labour') as a real abstraction seems to simplify what occurs. Such abstracting does not happen easily, instantly, or in isolation from other processes on which it depends.[54] We can refer again to the work of Moishe Postone (see chapter 3). Moving beyond Lukács, Postone develops a powerful analysis of real abstraction. If factory production, based on time discipline, turns labour into a real abstraction, its significance lies in producing the effect of 'value', counted in uniform units of labour time. 'Value', Postone argues, becomes the abstract measure that appears to determine how wealth is distributed. Causing us to misapprehend the source of inequality, this measure, this abstraction, operates as a structure of domination.[55] Unlike the overt institutions of political domination, an 'economic' category, value, operates through the mundane organisation of material life. In this sense, capitalism depends upon the 'real' separation of the political from the economic.

We can follow this analysis but still be left with a disagreement. Postone describes an abstraction, but not the process of abstracting. Producing the abstraction is not so much a singular historical transformation as a daily reoccurrence. For a particular use of labour or item of merchandise to acquire 'value' and operate as a 'commercial' or 'market' phenomenon, it must be separated from other ties and entanglements, such as those of householding, mutual obligation, personal dependence, state power, and physical coercion. Furthermore, as Carolyn Hardin suggests, the apparatuses of capture producing governing abstractions operate through spheres other than the payment of the wage, for example through relations of debt and the compulsion to manage 'risks' to one's future. While such 'governing by debt' may appear more prevalent today, the force of indebtedness was, as we argued in chapter 3, internal to the operation of the wage.[56]

We can see this difficulty of grasping 'the economic' in the work of another scholar, one more widely associated with the idea of the emergence of the 'economic' as a separate sphere in the nineteenth century, Karl Polanyi. For a long while, Polanyi's classic work *The Great Transformation* (1944) was misread as an account of the nineteenth-century disembedding of 'the economy' from society. This was not what the book said; its argument was conceived in the 1930s, before the idea of 'the economy' had properly emerged. It perhaps contributed towards its emergence, especially after Polanyi and his followers, from the 1950s, restated its argument in the novel language of that decade as an account of the nineteenth-century disembedding of 'the economy'.[57] The book itself was an argument about 'market relations' or 'market society', describing how 'the market' in the nineteenth century was formulated as an institution disembedded from the ties of reciprocity, obligation, and moral argument that had previously contained it. (It was an argument against men like Friedrich Hayek and other Austrians, the pioneers of neoliberalism, to whom we will turn at the end of this chapter.) However, disembedding is a misleading term.

On the one hand, as Polanyi himself makes clear, 'market society' never successfully frees itself from those entanglements. The domestic household, political authority, physical compulsion, and social obligation are all reorganised to support, enforce, or compensate for the operation of new laws of property and contract. On the other, the idea of the disembedding of the economic assumes that the economic market exists as a kind of abstract space, already populated with its buyers, sellers, and commodities. As Michel Callon points out, the concept tends to overlook all the continuing work that must be done to 'embed' labour, capital, or merchandise in relations of market exchange. In fact, to constitute some agency as a buyer and another as a seller, to recognise a claim on future payments as 'capital', and to identify some arrangements as items of merchandise and others as economic agents, in each case depends upon rules for defining certain property as assets, mechanisms of price formation, negotiated descriptions and categorisations of goods, ways of arranging work processes, and techniques of assembling agency.[58] Every act of economic exchange occurs only by invoking, establishing, and re-enacting these rules, mechanisms, and categorisations. An abstraction such as 'the market' or 'the economic' designates not some freestanding sphere of social reality but an arrangement under continual construction and contestation.

Another way to say this is that the concept of 'the economic' that emerged, not in the eighteenth century but towards the end of the nineteenth, should be grasped not as the creation of a distinct sphere of collective life, the making of 'the economy' before the word itself was used in this way, but the designation of a process. The process of forming and designating what is economic – of rendering the world calculable in distinctive ways – can be described as 'economising'. The transformation characterised in schematic accounts of the development of capitalism as the creation of 'the economic' as a real abstraction is better understood as the continual, never-completed work of 'economisation'.[59]

While our focus in this book is on the process of assetisation or capitalisation, by which future goods are converted into assets that can be controlled, discounted, and traded in the present, capitalisation always depends on acts of economisation.

We will come back to economisation later in the chapter, to understand how the analysis of the process of economising in the work of Max Weber, motivated, in part, by the challenges of capitalisation, inadvertently contributed to its transposition into an object, 'the economy'. For now, there are two features of economisation to note. First, a significant aspect of the process of rendering things calculable was the writing of texts on economics. In establishing political economy in the nineteenth century as a scientific field, such works were not so much describing 'the economic' as attempting to secure the acquisition and protection of wealth as a distinct realm of social life, subject to its own calculations and laws. Take the most influential work in English, John Stuart Mill's *Principles of Political Economy.* Its subject is not 'the economic' (a term Mill does not use as a noun) but wealth, and more specifically 'mankind as occupied . . . in acquiring and conserving wealth.'[60] In his influential essay 'On the Definition of Political Economy; and on the Method of Investigation Proper to It', Mill insists that political economy is a branch of the wider field of 'social economy' (or the 'science of politics'), being the part concerned with man 'solely as a being who desires to possess wealth.'[61] This mode of being, man as desirer of wealth, is formulated and facilitated in the *rules* of economic science. (The term 'economic' in nineteenth-century political economy is not only an adjective, rather than a noun; it is also one that never describes an aspect of the material world but rather, as we noted just now in discussing Lord Cromer's account of governing Egypt, an aspect of the moral world – a set of desired rules, principles, social laws, or mental habits.) Thus, political economy performs part of the work of abstracting, of economisation. Writings on the subject operate continually to separate its topic, distinguish those material processes to which it applies, and construct the principles on which its agents should act. The 'real world' to which a real abstraction is said to correspond is a world that includes the work of political economy.

Second, this process of differentiation and specification, whether through the work of political economy, law, government, or productive organisation, is continually undermined. It is a work of difference that can operate only by repetition, by repeatedly excluding what does not belong to the calculation of 'the economic'. But, in the process, what is excluded continually exerts pressure upon, shaping by its exclusion, the very category from which it is excluded.[62] Thus categories such as 'the political', 'the domestic', and the external or 'colonial' all operate as the internalised outside. The incompleteness of the separation is not just a symptom of a historical process, remedied over time as the process tends towards completion. It is a necessary and never-ending feature of the way an abstract effect like 'the economic' is produced.

Spokesmen of Credit

One nineteenth-century thinker understood the economising work of political economy, and then learned how it was undermined from within, was altered, and created new objects. Karl Marx began his study of what he called 'the classical works' of Adam Smith, Jean-Baptiste Say, and David Ricardo in 1844, when still in his mid-twenties. Three years later, he remarked that these writings had a particular relationship to the truth. The laws they developed were true only under the supposition that

> competition be perfectly free, not only within a single country, but upon the whole face of the earth. These laws, which A. Smith, Say, and Ricardo, have developed, the laws under which wealth is produced and distributed – these laws grow more true, more exact, then cease to be mere abstractions, in the same measure in which Free Trade is carried out.

As a consequence, he concluded, the economists 'know more about society as it will be, than about society as it is. If you wish to read in the book of the future, open Smith, Say, Ricardo.'[63] Political economy was a science that would become more accurate with time, as social and material relations were gradually organised – or economised – according to its principles.

Yet the future did not unfold that way. The process of economising, by which the laws of economics 'grow more true', Marx was to discover, was always incomplete. Periodically, the process was forced onto a different path. He had set to work on a 'critique' of political economy, a project whose writing and rewriting consumed the following two decades. In 1867, he finally published the result, volume one of *Capital*, which, as we noted, dealt mainly with the English factory system of the mid-nineteenth century. But, by then, as Walter Bagehot was soon to point out, political economy was already losing its hold on the public mind. Marx himself had evidently given up on the project of its critique. He had originally planned several further volumes of *Capital* but never completed them. Two further volumes were pieced together posthumously by Friedrich Engels from Marx's earlier notebooks and fragments.[64] The abandonment seems to reflect the fact that the 'commercial society' or 'bourgeois society' built on free competition that the works of political economy envisioned had not come about. That vision of the future had succumbed, however, not to the 'critique' exposing its internal contradictions that Marx had attempted to develop, nor to the revolutionary self-destruction predicted to follow from the spiralling expansion of those contradictions. It was giving way to new definitions of property ownership, new collective experiments, and new modes of capturing the future.[65]

Some of these alternatives to the future that political economy had once anticipated were being consolidated elsewhere, such as forms of communal ownership re-emerging in the village communities of Russia following the abolition of serfdom, in which Marx developed an interest.[66] Others, however, grew out of the English factory system itself. One example was the cooperative owning of enterprises, which Marx celebrated as 'not only a great practical success', but a 'victory of the political economy of labour over the political economy of property'.[67] A more significant development was the modern joint-stock company, together with the expanded powers of credit creation that enabled it. Marx famously saw the transition from the private ownership of production to its ownership by shareholders as 'the abolition of the capitalist mode of production within the capitalist mode of production itself' – not least because capital could now be created 'fictitiously', as he put it, as a claim on future production.[68] He borrowed the notion of fictitious capital from the classical political economists, who had used the phrase to describe bills of accommodation (mentioned in the previous chapter), meaning bills of exchange created as credit money where there was no underlying sale of goods.[69] By the time Marx published volume one of *Capital*, however, so-called 'fictitious' credit was not just a device used by merchant bankers to create bills. In the proliferation of joint-stock companies promoting the construction of railways, ports, and urban property, most capital was coming into being through the 'fiction' of creating and selling shares in a revenue to be paid from the future. The fiction was now quite real.

Marx realised that it was no longer the classical political economists whose work shaped and predicted the future course of society. It was the pioneers of expanded credit creation. He offered two examples. One was John Law, the infamous eighteenth-century Scottish French financier who as Controller General of Finances in France took over a failing colonisation project, the Mississippi Company, and granted it a monopoly of all French trade in North America and the West Indies. Combined with establishing a private national bank with the power to issue paper credit, the scheme created a speculative bubble that burst in 1720. The other example was a contemporary of Marx, Isaac Pereire, a French railway entrepreneur who with his brother Émile founded the finance company Crédit Mobilier. Pioneering a novel model of joint-stock finance, the Company replaced private, gentlemanly merchant banking with the new business of 'investment banking', selling shares in financial ventures, and in the bank itself, to the public.[70] The venture created hundreds of millions of francs in credit money by marketing railway shares, government war loans, property development, and other debt, including the supply of loans to the Ottoman government and the development of port and rail connections at Marseilles, a city that was expected to rival Liverpool as a node of international shipping following the building of the Suez Canal.[71] In September 1867, the same month in which volume one of *Capital* had finally appeared, the bank and its property empire collapsed in debt. The causes included the failure of Marseilles to become a new Liverpool and the weakening of Paris property speculations when the Exposition Universelle that year failed to attract the millions of expected visitors – a precursor to the failure of the Vienna world exhibition six years later that was to trigger the global financial crisis of the 1870s.[72] John Law, Isaac Pereire, and other such 'spokesmen of credit', Marx later observed, were endowed 'with the pleasant character mixture of swindler and prophet'.[73]

Marx fell back on the idea that capital created by such swindlers was 'fictitious', on the grounds that the future surplus they were selling in the form of company shares and other assets had not yet been created. This view followed from his earlier understanding of capital as a fund accumulated from the past and then transformed through the labour of material production into a larger amount. But, as we have seen, capital itself comes into being as an effective claim on the future. Industrial labour and its wage, advanced as credit against future consumption, is only one aspect of the proliferating methods of claiming from the future. The future is not a fiction conjured up only by swindlers and prophets, but an elaborate apparatus of capture organised to allow those who control it to realise future revenues at a discount in the present and then impose repayment as a burden on the future.

The work of economising, it follows, is never as simple as constructing a divide between the economic and the political. It requires the proliferation of new forms of property, in particular those forms of property that exist as claims on the future, such as the company share. And it is open to the instability of all relations to the future.

The Business Problem

The stock market helped create the phenomenon of the singularised commodity that summons into being a buyer, a seller, and a knowable price, in association with which the new and narrower discipline of economics could be organised. Yet the science of this price mechanism struggled to establish itself. In these early years, during the late nineteenth century, its practitioners still felt the need to speak about something more than the system of prices; several decades passed before most economists agreed to narrow their field of knowledge to the study of the price mechanism. That narrowing would be accompanied by the emergence of other concepts, to reference the material processes outside the price system that economists by then had largely excluded from their study. The most significant of these, as we have suggested in earlier chapters, was to be the word 'technology', which eventually operated as a catchall term – and indeed an alibi – to describe so-called non-economic forces that repeatedly disturbed the stabilising pendulum of price formation.[74] But, in the late nineteenth century, many economists still felt the need to incorporate into their explanations something of the technopolitical world beyond the price mechanism.

The work of Alfred Marshall, the leading Cambridge economist, can illustrate these difficulties. At Cambridge, Marshall removed the study of political economy from its existing home as part of the field of moral sciences and established a separate faculty of economics.[75] His *Principles of Economics*, whose first edition appeared in 1890 and was initially intended as one-half of a two-volume textbook, helped define the scope and method of the new discipline for more than four decades. Marshall popularised a diagram representing a commodity or good as a mechanism, in the form of a graph with the quantity of the good increasing along the x-axis and its price rising up the y-axis. A curve ascending towards the right depicted 'supply', assumed to grow as the price rose, while one descending towards the right pictured the 'demand' for it, assumed to decline as its price increased. This created an abstract representation of a competitive system in which supply and demand formed counteracting 'forces' whose intersection was a point of balance or 'equilibrium'. The second volume of the textbook was intended to complement this abstract portrayal of the market as an automatic mechanism with a 'realistic' account of other processes, beyond those of the price system. Marshall eventually abandoned the attempt. 'I had laid my plan on too large a scale', he noted, 'and its scope widened, especially on the realistic side, with every pulse of that Industrial Revolution of the present generation, which has far outdone the changes of a century ago, in both rapidity and breadth of movement.' He suggested that 'four thick volumes would be needed for the task'.[76]

Marshall had hoped to make sense of the wider movements at work by means of a novel idea, the process of *organisation*. Leaving aside the physics of the price mechanism, he turned to metaphors from biology. He proposed that the concept of organisation expressed a historical process of increasing complexity, characterised by interdependence and the interaction of forces. The concern focused on the 'industrial struggle', to which J. A. Hobson's *Imperialism* referred, and on what Marshall and others began to call the 'industrial *system*'. The concepts of system and organisation referred to the interlocking world of railway networks, iron- and steelmaking, coal and oil production, electrical power generation, industrial chemistry, and large-scale machine processes, together with the sizeable new joint-stock companies and financial firms that grew to control these processes and made them what Hobson termed 'a single organic whole'.[77]

While stock markets, commodity exchanges, currency dealers, and other financial mechanisms manufactured a world of prices that could be recorded and charted, they were the visible, simplified part of this much larger reorganisation. The wider arrangements that made possible the production of such prices included, as we have seen, the ecological remaking of regions to supply bulk commodities, from the prairies of the American Midwest homogenised into wheat fields, to the flood basins of the Nile dried out to create cotton and sugar plantations; the railway schemes that moved armies, pathogens, financial paper, and crops and accounted for the majority of firms whose shares could be traded; the banks and insurance companies that arranged and guaranteed the supply of credit; and many other interlocking processes. Terms like 'system' and 'organisation' attempted to distinguish these forms of interdependence and instability from the discrete forms of 'industry' that earlier political economists had studied. But as Marshall's failed attempt to produce the subsequent volumes of his textbook testified, the abstract science of prices proved impossible to reconcile with understanding this wide space of ecological, imperial, and industrial processes and the struggles unfolding there.

The solution that emerged to this impasse, and that was to prepare the ground for the birth of the economy, can be grasped, again, as a process of economisation. The object of study itself produced the modes of reason, rationality, and calculability that the science in turn helped to abstract, formalise, and calculate upon. We can examine how this happened by considering the development of another idea that came into common use in English, in the decades prior to the invention of 'the economy' – the object whose emergence we noted at the end of the previous chapter, the realm of 'business'. Economists referred to a world populated by 'men of business', or 'city men'. They even acknowledged that it was to the calculations of these men that they thought their work mainly applied, and whose reckonings it would improve.[78] Among American economists a wider term for these processes was 'business enterprise', as portrayed in books such as Thorstein Veblen's critical 1905 work, *The Theory of Business Enterprise*.[79] Increasingly the phrase was shortened to just 'business'. The fluctuating fortunes of this organic system became an important subject of investigation. In particular, its repeated episodes of expansion and decline were now studied as the problem of 'business cycles'. In 1913, Veblen's most famous student, Wesley Clair Mitchell, published a 600-page study, *Business Cycles*, arguing that episodes of depression and financial crisis were caused not by 'abnormal phenomena' such as war, crop failure, or technical inventions, but by the very 'interdependence of business enterprises', as a consequence of which they now 'constitute a system'.[80] One way to think of the subsequent birth of the economy, as we will see, is that the latter term provided a translation of the idea of 'business' as an organic system into an expanded concept, equipped with a more regulated temporality.

The railroad companies, especially those of the United States, were the main example of this expanding apparatus of capture – a mode of taxing the future, as it appeared even at the time.

A sudden wave of mergers and acquisitions took place at the end of the nineteenth century – the 'great merger movement' of the decade after 1894, in which companies in food manufacturing, petroleum, coal, metals, and transport merged or were taken over to form many of the largest corporations of the twentieth century. By the mid-twentieth century, indices of the stock market value of these companies – notably the S&P 500, representing the largest firms selected to represent every sector of industry – became, alongside the gross domestic product, or GDP, one of the principal measures of the health of the economy and the success and failure of government.

Here, one can see succinctly the processes of economisation at work. There is a standard story in business studies about the rise of the large business firm. It was laid out in Alfred D. Chandler's classic work of 1977, *The Visible Hand.* Countering the views of Veblen, who had explained business enterprise as a mode of capturing surplus based on the ability to disrupt the industrial system, Chandler argued that the large, integrated firm was a natural response to the problem of coordination. As nation-states expanded in size, as their populations grew, and as transportation infrastructures spread, market relations were no longer the most efficient way to coordinate exchanges among suppliers, manufacturers, distributors, and consumers. The coordination of functions and allocation of materials was more efficiently handled by professional managers. These were located within the different divisions of the large corporation. The growth of the managerial firm was a response to the growing technical complexity of modern industrial life.

Chandler's account was based largely on the history of railroad firms. This is no surprise. In the 1880s, some ninety of the hundred-plus firms traded on the New York stock exchange were railway companies, and most of the rest were related businesses – telegraph companies, express mail firms, and similar enterprises. Chandler argued that the great challenges and opportunities presented by the running of railways – setting timetables, calculating freight prices, coordinating the movement of locomotives and rail cars, and so on – enabled and spurred the development of the new managerial skills out of which the large corporation was formed.

Let us consider this story differently, in a way that is not driven by a simple logic of increasing technical complexity – by the force of 'technology' – but in the way it exemplifies economisation at work. The building and running of railways required new kinds of engineering and calculative knowledge. The *American Railroad Journal* was a publication that had been launched as early as 1832 to educate young engineers by gathering know-how about the engineering of bridges, turntables, locomotives, and other machinery and technical structures. But railways were not only, or perhaps even primarily, a technical system for improving the speed and scale of transport and communication. As we know, they were also a means of attracting shareholder investment and profiting from speculation in land prices and other opportunities to which railway construction gave rise. So, the railways required and encouraged the growth not only of engineering know-how, but of new forms of financial information and calculation. Their own published freight rates, ticket prices, track mileage, share prices, and other operational data were the building blocks of this financial world. They were a vital part of the process of economisation.

This relation between an industrial apparatus and the formatting of economic and financial information can be seen in what happened to the *American Railroad Journal*. In 1849, a New England railway magnate, John Poor, purchased the journal and gave its editorship to his brother, Henry Varnum Poor. Alongside engineering knowledge, Henry Poor began to publish financial information, as railways were incorporated and issued bonds and stocks. The journal developed ways of representing and circulating the financial data that railroads were generating and depending upon. From 1868, Varnum Poor tabulated this information systematically each year in a new form called the *Manual of the Railroads of the United States*, detailing for each railroad their bonds, stocks, mileage, earnings, and expenses, along with a history of each corporation. Eventually the *Manual* was publishing annual data on 128 railroad companies.

This economising work was driven by the difficulties of capitalisation. The introduction to the first volume of the *Manual* drew attention to the 'great evil' of the 'constant increase in the nominal *capitals* of railroads, without any addition to their capacity to earn'. Varnum Poor gave the example of a railroad built at a cost of $4,868,428, which had made successive issues of stocks and bonds to create a capital account of $11,250,000. Instead of reducing rates, 'the commerce and travel of the route is taxed to the utmost extent', to pay the interest and dividends on this capital. 'A traffic which should have paid on a cost of $4,868,428, is made to pay on two and a half times this sum.' This was one of many such violations, Poor wrote, 'not only of the true principles of railway economy but of popular right'.[81] The *Manual* was intended as a check on these 'violations', hoping to monitor and regulate the very process of marketing such claims to the future and their practice of thereby taxing future commerce and travel 'to the utmost extent'. The very practice of capitalising the future began to generate in reaction the forms of statistical knowledge through which 'the economic' might be known and regulated.

In 1916, Poor's publishing firm began to compile and sell ratings of the creditworthiness of the debt issued by the firms on which it collected information.[82] Another company, the Standard Statistics Bureau, founded a decade earlier to compile the same information for non-railroad companies (published in the form of 5 x 7–inch index cards that could be easily updated), began to sell its own ratings in 1922. In 1941, the two firms merged to form Standard and Poor's, the credit rating and financial information agency that established the world's first value-weighted Stock Market Index, and what is today known as the S&P 500.[83]

This illustrates how economisation works: not as the mapping of a pre-existing terrain, but as the laying out of tracks and connections that are simultaneously a means of moving goods and people, of creating the capital that 'taxes' future commerce and travel, and of constituting and circulating quantities, prices, indices, and futures that enables this taxation of the future to proceed. It was out of these same interconnected processes that the economy would emerge, as an object of knowledge and regulation.

Economisation also works through the production of academic knowledge. In fact, through the same connections and circulations, a *history* of the phenomenon of 'business' was generated, one that could serve to justify and become a further alibi for this mode of capture of the future. Henry Varnum Poor's personal papers were kept by his daughter, Lucy Poor. They were later discovered in a storeroom of her house in Brookline, near Cambridge, Massachusetts, by her great nephew, who moved into the house after her death. The nephew, a great grandson of Varnum Poor, used the papers to write his PhD thesis at Harvard, comprised of a study of Varnum Poor, the history of the American railways, and the development of corporate finance and management.[84] The author of the PhD was Alfred DuPont Chandler. As a graduate student at Harvard, he worked under the influence of Talcott Parsons and the Parsonian interpretation of Weberian organisation theory, to which we will shortly turn, and indirectly under the influence of the late Joseph Schumpeter.[85] He went on to write *The Visible Hand*, the book that established him as the dean of American business historians. Poring over the statistical and organisational data that his great-grandfather had compiled, Chandler did for his own generation 'just what Poor had done', deriving from the account books and organisational charts of the railways an understanding of American capitalism – indeed a theory of its distinctive form, the large investor-owned corporation.[86]

Chandler's thesis helped overturn Thorstein Veblen's account of the rise of big business, in which the pecuniary advantage pursued by businessmen led to inefficiency and sabotage.[87] *The Visible Hand* argued instead that in the period from the 1880s to the 1920s, after pioneering by American railroads, corporate *organisation* – the organisational chart – took the place of market mechanisms, and did so as a more efficient way to manage large, technical processes.[88] The world of charts, account books, business records, and projections through which economisation proceeds helped generate the justification – the oikodicy – on which the world of business increasingly depended.

Economy Shall (No Longer) Mean Economising

Let us turn to a final question about the history of 'economy' before the birth of 'the economy'. What dislocations or disruptions led to the change from understanding economy as process to economy as object? In what way was the shift connected to the rise of 'business', and to urgent attempts to understand the credit crises and 'business cycles' to which the world of large corporate firms seemed to belong? The answer is that by the interwar period these crises were beginning to threaten the very future on whose production the capitalising process depended. Weakened by the crises of capitalisation, capitalism itself, as the system would now be named, appeared about to collapse. The response was the invention of *the* economy.

The shift appears to have occurred among the works of German writers before it appeared in English. This reflects perhaps the concern of German scholars to define their field of research as 'social economy' (*Sozialökonomik*). While John Stuart Mill, echoing Jean-Baptiste Say, had situated political economy as a branch of this wider field, men like Alfred Marshall, as we just saw, struggled to expand their analysis beyond the narrow topic of the price mechanism. It was in Germany in the late nineteenth and early twentieth century that leading scholars such as Max Weber insisted on studying economy and society as connected processes.[89] It was there that modern conceptions of the management of economic processes were most clearly developed. The crises of German politics, including the country's defeat in the First World War by the imperial powers of the North Atlantic world, and the possibility of socialist revolution in the wake of that defeat, called for new conceptions.[90]

Anglo-American thought was slower to adopt these ideas, which often arrived in English through the translation of the work of German and Austrian social theorists. The difficulty in translating key German terms into English, starting with the word *Wirtschaft*, or economy, helps illuminate these challenges. We can close this chapter by looking in detail at the case of the social economist Max Weber. His masterwork, *Economy and Society*, published in German in 1922, was mistranslated when it appeared in English twenty-five years later.[91] The mistranslations reveal something. The problem begins with the two words of the book's title. In the period after the Second World War, when the English version finally appeared, economy and society were coming to be understood as objects in themselves, each forming a stable domain composed of relations that can be identified and measured. Weber, on the other hand, as Keith Tribe points out in his landmark retranslation of the text, was still interested less in structures than in processes. For the term society, he focused not on the word used in the title, *Gesellschaft*, but on the process of forming society – and other associations – or *Vergesellschaftung*, meaning literally 'societisation'.[92] But what about the other term, 'economy'? Given the weight of the new meaning the word was to acquire in English by the mid-twentieth century, it is more difficult to translate in a way that avoids the later sense of economy as an object. In chapter 2 of his text, where he lays out the categories for understanding economic life, Weber makes clear that the term refers to a process: 'in practice economy [*Wirtschaft*] means the careful choice between ends', taking into account the scarcity of available means.[93] In the formal definition of the concept with which the same chapter opens, he confirms this meaning by defining his use of the word in terms of its own verbal noun: economy, he writes, shall mean '*Wirtschaften*' – literally, economising. The latter term refers to 'a peaceful exercise of a power of disposition' over resources, a power that is 'planfully' (rationally) oriented to economic ends.[94]

As we have seen, the word 'economising' needs to be used in an unusual sense to capture the meaning of *Wirtschaften*. While the English term economy was used in Weber's time to indicate the efficient disposition of resources, economising usually conveyed the narrower idea of acting with frugality. The word had lost its wider meaning referring to the processes and reasonings that render certain goods or actions economic. William Smart, who translated key texts from German into English around the turn of the century, tried using the term in this broader sense in translating *The Positive Theory of Capital* by Eugen von Böhm-Bawerk, a leading member of the Austrian school of economics:

> Of the vast natural endowment which serves as foundation for man's productive combinations, one portion particularly claims the interest of economics, and that is, those useful things offered by nature only in

> limited amount . . . These limited gifts and energies of the natural world obtain for us a peculiar *economic* importance. It would be foolish not to economise them.[95]

However, even then, and more so today, 'economising' suggests thrift or parsimony, rather than rendering things economic.

One route around the difficulty might be to translate *Wirtschaften* as 'economisation', the term, as noted above, that Michel Callon proposes for the process of qualifying specific arrangements and actions as economic, in which both economic agents and theorists engage. Using a contemporary vocabulary, this meaning seems to echo aspects of Weber's approach.[96] In the sense of economisation, the term 'economising' connects Weber to the Austrian school of economics, with which, as we will see, he was closely associated.[97] The connection allows us to understand how Weber's work contributed, perversely, to the future disappearance of the understanding of economy as economising and its transformation into *the* economy as an object.

For Weber, as for other writers of the nineteenth and early twentieth century, the process of economy could be identified at different scales. It did not presuppose '*national* economy', which became the default scale only as part of the way in which 'economy' was made into an object in itself.[98] One could refer to the processes of household economy, village economy, urban economy, and national economy (*Volkswirtschaft*).[99] Contemporaries of Weber read this difference in scale as a historical progression, from simpler to more complex forms. In *Economy and Society*, however, Weber was not interested in writing a developmentalist account. His goal was to clarify concepts for understanding the present.

The interesting point for our purpose, in connecting economisation to capitalisation, is not that Weber, like his contemporaries, understands economy as a process – economisation – but that, in defining the term, he sets up a distinction between two kinds of economising. The distinction points towards understanding the problem of business or business enterprise and its role in capitalisation. Weber introduces the distinction to establish the very definition of economy: '"Wirtschaft" soll ein autokephal, "Wirtschaftsbetrieb" ein betriebsmäßig geordnetes kontinuierliches Wirtschaften heißen.[100] ("Economy" shall mean an autocephalous, "enterprise-economy" a continuously ordered economising.)'

Weber defines economy not in isolation, but via a contrast between

Wirtschaft (economy) and *Wirtschaftsbetrieb* – literally, enterprise-economy. The first term, he says, shall mean an 'autocephalous' (self-governing) economisation, whereas the second shall mean an enterprise-scale 'continuously ordered' economising.

The term *Betrieb*, used to form the compound word *Wirtschaftsbetrieb*, can be translated as 'enterprise'.[101] So the wording might suggest the distinction between autocephalous economy, such as household-based economisation, and enterprise-based economy. However, 'enterprise' here is misleading, as it usually designates a discrete business unit – an individual firm or factory. Weber makes clear that his meaning refers not to a unit but, once again, to a process. In fact, the purpose of economisation at this larger scale is precisely that it is the calculative activity, as we will see, that allows separate economic units and processes to be connected. It enables the arranging and disposing of operations across differences in product, location, or, critically, period of time. Such calculability allows the transformation of self-governing forms of economic life into parts of an enduring, interconnected, continuously ordered process.[102]

Economisation, then, takes two forms, discontinuous (or autonomous) versus continuously interconnected. Why define the very concept of economy in terms of the difference not between household and business enterprise, but between the discontinuous and the 'continuously ordered'? The reason is that, for Weber, durability and continuity – extension in both space and *time* – are made possible by the economising power of a specific form of calculation, one that is both replicable and continuously replicating. The text's explanation of contemporary capitalist life rests on this difference. It is an explanation that places at the centre of economisation the forms of *calculability of the future* on which capitalisation depends.

Weber distinguishes between two forms of economic calculation, accounting in kind and accounting in money.[103] Put simply, accounting in kind, or *in natura*, compares materials or processes that are similar in nature, but differ in quantity or in output. Accounting in money, on the other hand, refers to a form of calculation in which entirely different processes or materials can be made comparable, by expressing all of them in a common unit of account, typically a monetary unit.

The distinction between the two forms of accounting relates directly to capitalising the future. Any business at any scale can employ accounting *in natura* to calculate the costs or profitability of a process, by estimating the cost of one item or process in terms of others. For example, a machine of a given type can produce a certain product using various combinations of energy, labour, and materials. However, as Weber explains, to scale up and replicate such calculations using evaluations *in natura* soon becomes unwieldy. If the firm diversifies into another kind of product, it becomes difficult to calculate cost and income across different production processes, and even more so if it considers adding an operation at a second location or wishes to compare present costs with those at a subsequent moment in time, *for the purpose of discounting future revenue*. Such business expansion across differences in process, location, and, above all, duration of time depends upon the second form of calculation, money accounting.

In describing 'accounting in money' as the basis of one of the two modes of economisation, Weber is not concerned simply with a particular technique for calculating profitability, such as double-entry bookkeeping (although that may play a part). Rather, he is laying out the specific role that economisation plays in building a world of durable, future-oriented, large-scale enterprise: the conversion of difference into a single standard of comparison, based on money as a unit of account. In other words, economisation of this kind is required for the expansion of capitalisation. The conversion does not necessarily require the actual use of monetary payments, or market transactions. The money prices can be deployed to create 'the fiction' of exchange processes or to determine the opportunities for future gain.[104] In fact, their main usefulness may lie within the firm, whose opportunities for expansion and for 'continuous' economisation across time, including the projection of future costs, rest on the use of money accounting. Of course, such accounting will also extend to transactions outside the business and can inform decisions about whether to obtain a certain product or service in-house or from an outside vendor. Such commercial interaction with other firms, or with consumers, establishes and generalises the process of monetary accounting. Although economisation is largely organised within business enterprise, without the possibility of commercial exchange the expansion and coordination of business activity would fail. This expanding network of calculability, and centrally the calculability of the future, made possible by money accounting, is what Weber has in mind in distinguishing 'continuously ordered' economisation from mere self-governing (autocephalous) 'economy'.

Without Economic Calculation There Is No Economy

Why does Weber place at the centre of his understanding of contemporary capitalist economisation this process of accounting – rather than, say, the expansion of markets, the exploitation of labour, the spread of monopoly power, or other historical developments that also figure in his account? Weber's interest in accounting, it turns out, arises not from a concern with understanding the historical development of capitalism, but from something more urgent: the imminent possibility of its demise.

As Weber was writing his text, socialist parties were coming to power, or threatening to do so, in Russia, Germany, the Bavarian Republic, Austria, and other parts of Europe. As Weber saw it, the problem with the state or community taking over the ownership and management of industry and finance lay not simply in the elimination of property rights, the abolition of markets, or the removal of competition. It lay in undoing the coordination mechanism based on money accounting, the principle of calculation that allows extended, continuous, and expanding economisation – and thus capitalisation. Under socialist 'planned economy', where centralised decisions about resources replace commercial exchange, the process of economisation fails to generate a set of prices. Such prices may arise from exchange, but also from cartel arrangements, labour agreements, internal business accounting, or other sources. What disappears is not simply 'the market', but the entire process of 'effective' prices, however they arise, through which large enterprises allocate resources, continuously expand, and calculate and discount future revenue.[105]

Weber develops the distinction between calculation *in natura* and money accounting not just to understand capitalist forms of life, but to argue for the technical impossibility of large-scale 'planned economy'. As a result, much of the text resolves into a contrast between just two modes of economisation, 'planned economy' and 'money economy'. In making this contrast, the term economy begins to lose its association with calculative methods, which can occur at multiple scales, and appears to designate a way of grasping the totality of economic processes as an object in itself. To demonstrate the impossibility of the socialist project, which aims to govern economisation as a whole, Weber has inadvertently contributed to the possibility of thinking of that whole as an object.

Thus, Weber's preference for the study of processes rather than structures can be seen as not merely methodological. It registers the political context in which he is writing. Central economic planning presents the problem of how the entirety of economic life can be grasped, both by economic planners and, as a consequence, by social-economic explanation. Weber's answer is that it can be grasped only as the outcome of a dispersed, but interconnected, process. However, in the years after Weber's work, thanks to the appearance of a fundamentally different mode of governing collective life, in socialism, planned economy and money economy will come to be read as two modes of governing a common object: *the* economy.

Weber is known today as perhaps the most influential twentieth-century social theorist of economic life. But he owes his influence not to his writings on the economy, but to his efforts to oppose the idea. He died in June 1920. Meanwhile, in postwar Vienna, the group of intellectuals known as the Mises Circle, who were associated with and influenced by Weber, had begun developing similar arguments about economic calculation. The arguments were to become the core of the intellectual movement later named neoliberalism (a movement to whose later significance we will turn in chapter 6).[106] An essay by Ludwig von Mises, 'Economic Calculation in the Socialist Commonwealth' (1920), came to Weber's attention while his own work was in proof. Weber died shortly thereafter, just as his publisher was typesetting part one of *Wirtschaft und Gesellschaft*. The other parts of the book were assembled later from drafts of Weber's earlier writings, creating an unwieldy, disconnected, and difficult tome. Thanks to the Mises Circle, however, the book was rescued from relative obscurity and eventually made available in English. Begun a decade and a half later, the rescuing was part of a renewed effort, now centred in the English-speaking world, to oppose the emergence of *the* economy as an object.

In 1935, Friedrich Hayek, a member of the Mises Circle and lynchpin of the emerging neoliberal movement, published an English translation of the Mises essay on economic calculation. He included it in a collection intended to alert English readers to the argument against the feasibility of economic planning, which was now emerging as a threat even in Britain and the United States.[107] In his introduction to the book, Hayek names Weber alongside Mises as those who had first developed this argument.

Hayek himself, however, no longer used 'economy' in the old way. For Mises, as for Weber, economy was a process of calculation. Indeed, the elimination of money accounting under socialism signalled for Mises the very absence of economy: 'Ohne Wirtschaftsrechnung keine

Wirtschaft', he wrote (without economic calculation there is no economy).[108] Introducing this essay to English readers fifteen years later, in a new political context, Hayek reverses the use of the word.

In summarising the argument against planning, he uses 'economy' only as a negative term, to designate what was now emerging as a Keynesian or New Deal alternative to the crisis-ridden capitalism of the interwar period. He employs the word in phrases such as 'a planned economy', 'a centrally directed economy', or simply 'a socialist economy'. To describe its opposite, a mode of life governed by the calculation of individual or corporate self-interest, he never once uses the term economy, preferring phrases such as 'the institutions of a free society', 'a system based on exchange', or a 'purely competitive society'.[109] Four decades later, Hayek was still objecting to the use of the term 'economy', proposing instead the word 'catallaxy' (borrowed – through Mises – from Richard Whately's Oxford lectures of 1831, mentioned above).[110] By then, however, as we will see in chapter 6, Hayek's followers were replacing phrases like 'the institutions of a free society' with a simpler abstraction, later to be widely adopted as an alternative to the idea of the economy: 'the market'. Even in the 1930s, however, for Hayek the totality to which 'the economy' now referred designated only the impossible object of centralised planning, the antithesis of 'a capitalistic society'.[111]

Around the same time, as Tribe tells us, Hayek helped recruit a young Scottish economist to start translating part one of Weber's *Wirtschaft und Gesellschaft* into English. The translation was taken over and reworked by the American sociologist Talcott Parsons (the teacher of Alfred Chandler).[112] Delayed by the war, it appeared only in 1947. By then, as we will see in the next chapter, economy had become an object, and indeed in the hands of Parsons, 'a system'. Weber's definition of the term – 'economy shall mean . . . economisation' – was now rewritten accordingly: 'An "economic *system*" (*Wirtschaft*)', Parsons makes Weber say in his mistranslation of the text, is a '*system* of economic action'.[113] The word 'system' occurs nowhere in Weber's text. Nor does the original contain the indefinite article 'an', which in the English translation operates to transform economy as process into economy as an object. Twenty years later, a revised English translation of Weber's text was published. This version removed the word 'system', while retaining the indefinite article.[114] By the 1960s the concept of 'an economy' signified by itself the structural system that Parsons had tried to identify.

Keith Tribe's retranslation of *Economy and Society* allows us to read the two key terms of its title as sociation and economisation: that is, as processes, not objects. But it also allows us to read, in the text itself, the way Weber's urgent intellectual response to the politics of his times contributed to the developments through which economy would no longer be understood as economisation.

Paying attention to Weber's use of the term economy enables us to recover a way of thinking about economy before the invention of 'the economy'. This is not only a matter of historical interest. Over the last two decades, some of the most important writing addressing how to think about the economic has proposed that we replace the term economy with the concept of economisation. The concept can be used to denote the arranging, calculating, qualifying, and reasoning through which processes and materials are rendered economic.[115] The term brings into view the realisation that 'the economy' is not a universal realm of human social life but the product of particular calculations, devices, and actions. This draws attention to the issue explored in this chapter, obscured by our everyday use of the word economy: the continuous academic and practical work that must be done, in service of the process of capitalisation, always incomplete and open to disruption, to render things economic.

5

Economentality: How the Future Entered Government

In 1947, an Egyptian entrepreneur named Adriano Daninos published a proposal in a scientific journal in Cairo to build a new dam across the Nile. Placed upstream of the smaller masonry barrage built fifty years earlier by the British at Aswan, discussed in the opening chapters of this book, the new rock-filled structure would be so large that the reservoir it created would stretch more than 500 kilometres to the south.[1] 'Daninos is a man with a mission', reported an official at the World Bank in Washington, where he later went to pitch his plan. The official noted that the scheme concerned not just building the dam but 'land reclamation and irrigation connected therewith, production of power, and construction of plants to use that power in mining, making fertilizers, other manufacturing, and possibly iron and steel'.[2] Before publishing his plan, Daninos had visited the Tennessee Valley Authority in the United States and similar integrated hydro-electric, river control, irrigation, and industrialisation projects in the Limousin in southwest France. These gargantuan schemes for reorganising forces of nature, systems of agriculture, flows of energy, powers of labour, and material production were known as 'total development'.[3]

Despite the scale of the venture, Daninos was not asking the World Bank for money. He merely sought help coordinating the involvement of US and European engineering firms. The capital would come from London or Paris – either from the large credits in sterling built up by the Egyptian government at the Bank of England during the war or from the Anglo-French-owned Suez Canal Company, which was due to revert to Egyptian ownership soon after the building of the dam. The company's income and profits were surging with the postwar growth of oil shipments through the Canal.[4] As a director of the Suez Canal Company pointed out in private to officials at the World Bank, the future canal tolls paid by shipping companies could cover the cost of repaying the construction loans.[5] The dam was a means of capitalising the wartime debts and postwar oil revenues that the prospective petroleum boom would bring.

But the Bank was unwilling to play only a marginal role. Excluded from postwar reconstruction in Europe, it was keen to reinvent itself as the conduit for the supply of economic advice and the selling of Wall Street loans to countries of the South.[6] So it put itself forward to coordinate government and private financing of the dam. The Bank's involvement did not make the project any cheaper. On the contrary, European engineering firms had estimated the cost of the dam at about £200 million sterling, but when the Bank finished its calculations the estimate had more than doubled, to £500 million.[7] The Bank had added to the scheme the cost of digging irrigation canals that would allow Egyptian farmers to replace fast-growing food crops with more hectares of the slower-growing industrial crops – cotton and sugar cane – needed to pay off the bonds the Bank would raise on Wall Street. But the Bank had added something more: a new set of calculations about the future. The novel way to represent this calculated future was to refer to it as 'the economy'.

Misplaced Definiteness

Around 1948, it became common in American political debate to talk about the economy. References to this object in government and newspapers were starting to appear in a routine, repetitive way that made the economy appear for the first time as a matter of fact. It was no longer always necessary to explain what the term meant or to qualify it in some way.

The idea of 'the economy' itself had been around for at least a dozen years. But in the period before the Second World War the word still often carried echoes of the older sense explored in the previous chapter, when 'economy' described a process, not a thing.[8] In everyday usage, the term still referred to the act of economising, or making prudent use of limited resources. The US Economy Act of 1933, for example, was not a law to regulate the economy. It was an act of economy, or economising, intended to cut the federal deficit by reducing the pay of government workers. In a wider sense, as we saw, the word economy had often meant modes of government itself. As in the older phrase 'political economy', it still referred to forms of administration and know-how concerned with the efficient management of human lives and material resources. Even in the late 1930s, as a new meaning of the term emerged, this older sense of governing was usually still present. The word occurred frequently in newspaper discussion of topics such as planned economy versus market economy, the war economy in Japan, and the German experiment in controlled economy.[9]

Occasionally, in the years before the war, the new meaning of the term appeared explicitly, referring to *the* economy, as a distinct space or object. In January 1938, the *New York Times* reported on a monograph published by two economists at the New School for Social Research.[10] Using data from the new Roosevelt income taxes, the study argued that taxing the better-off had created a scarcity of investment capital 'in the American economy'.[11] It was perhaps no coincidence that the authors were, as the *New York Times* noted, 'members of the school of economic thought that holds governmental intervention in the economy to be a necessary datum of modern crisis policy, of which John Maynard Keynes of England is a prominent spokesman'. Equally significant was the fact that the revenue acts of the New Deal had turned a large number of citizens, through their tax returns, into producers of the data that made it possible to discuss 'the economy'.[12] Lehman Brothers had provided funds for the two economists to hire a team of statisticians to transcribe figures from the tax returns and translate them into measurements of the economy.[13]

Although useful to a bank interested in the evidence that income taxes reduced investment, this kind of calculation alarmed many economists. One reviewer reported that the book had 'grave defects', in particular its 'misplaced definiteness' and its use of 'highly tentative savings statistics without adequate warnings to the reader'. 'In view of these defects', the review concluded, 'it is to be hoped that the book will not have the popular following which its attractive form and lucid style and the manifest competence and honesty of its authors would otherwise deserve.'[14] The review's author, Albert Gailord Hart, was a convert to Keynes's ideas but was working on a study of uncertainty.

Unpopular with academic economists, this kind of 'misplaced definiteness' found a more effective way to shape opinion. The following year, in 1939, the US Bureau of the Budget was moved from the

Commerce Department to the new Executive Office of the President. The lead author of the 'defective' book, Gerhard Colm, was appointed as a chief economist in the Bureau's new Fiscal Division. There, he introduced into federal budget making the novel methods of estimating national income accounts devised by the statistician Simon Kuznets. After the war, Colm joined the staff of President Truman's new Council of Economic Advisers. At the Council, he transferred the accounting methods used for the federal budget to the estimating of something called 'the nation's economic budget'. In the annual reports of the Council, intended to educate the president and the public in economic forecasting and calculation, the work of formatting the economy proceeded. The 'misplaced definiteness' of the economy acquired a larger definition.[15]

A decade after the introduction of national income accounting into the calculations of the federal budget, it was becoming more commonplace to refer to 'the economy'. How should we understand the emergence of this new way of referring to collective life? Was the economy a new object, or just an old name for things that already existed? As we saw in the previous chapter, neither option is satisfactory.

We can show how the economy appeared to be a new object: a thing made up of aggregate price levels, wages, consumer spending, money supply, purchasing power, and savings, all of which could now be estimated, thanks to new business procedures, banking reports, household tax returns, and other forms of accounting. But the economy was more than just a set of aggregates. These accounts could be used to divide the collective world into a series of units: firms, households, and government departments. The monetary interactions among these accounting units could be estimated and the estimates then adjusted, so that additions to one account equalled the subtractions from another. Instead of attempting to estimate only annual additions and subtractions, the accounts could be broken down into monthly reports, so that the movement of funds between units could be depicted as a dynamic process.[16] These movements, from unit to unit, month by month, now appeared not just as aggregates but as parts of an interacting system.

The accounting procedures that made this system visible and measurable were more than just forms of representation. They were new ways of raising revenue from a population, managing national finance, marketing and consuming goods, and earning livings. The economy provided not just a new object of government policy, in the way that governments had also become concerned with, for example, public health, or urban renewal, or social welfare. The economy provided a more pervasive effect, one that has since then escaped attention: a way to *bring the future into government*.

The making of the economy established a new temporal scheme in which past, present, and future were relocated. We can study this shift as a new prognostic structure, one in which the future was mobilised as a mode of adjudicating and managing claims in the present, and stabilised – after a period of crisis – for the purposes of capitalisation.

As we noted briefly at the end of the previous chapter, the interwar period produced a deep crisis in the processes of capitalisation. The Great War, the coming to power of socialist governments, the Crash of 1929, the Great Depression that followed, and the rise of fascism and total war marked profound setbacks not only for the system that had recently come to be known as 'capitalism', but more specifically for the sense of a stable, predictable future on which the capitalising process depended. This was also an era of crisis in the order of imperialism that had emerged as part of the same process, nowhere more than in the Middle East. Following the collapse of the Ottoman Empire after the First World War, Britain and France had attempted to control the entire region from Egypt to Iraq through a system of protectorates and mandates, and largely failed. The greatest failure was in Palestine, where the Great Revolt of 1936–9, the Palestinian uprising against Britain's occupation of the country and the Zionist colonisation it promoted, required 100,000 British troops for its violent suppression.[17] We will return below to the aftermath of this violence, in Palestine and other parts of the region, to connect it to the making of the economy and the kinds of future, through schemes of postwar 'development', that the discovery of 'the economy' appeared to promise. At the same time, this chapter aims to connect the birth of development, for which the Middle East formed a prime testing ground, with a parallel production of the future in the politics of the emergent imperial centre, the United States.

The government of the present, as it was imagined through new forms of the future, would come to operate within a new metric of temporal change, the measurement of 'growth'. This metric can be traced in novel methods of economic calculation. However, establishing the process of measurement involved more than just devising calculative techniques. It required securing and stabilising a number of sites that would support the new metrology. This chapter illustrates how this happened by tracing one aspect of the process that was increasingly vital to the work of capitalisation: the flow and regulation of energy supplies.

1948: An Infrastructure of Democracy

It is sometimes argued that the new energy regime of the modern era, based on fossil fuels, which allowed the quantity of energy available in industrialising societies to increase exponentially, was an important source of the sudden acceleration of time and the unpredictability of futures. While the consumption of energy, in particular from oil, climbed rapidly from around the middle of the twentieth century, this consumption, paradoxically, contributed to an era in which the future became, for a while, part of a stable instrument for governing populations.

Power in the modern era is said to work through forms of reason, modes of calculation, and tactics of control that we sometimes refer to as governmentality: methods of power that take the population as their object and govern through the improvement of its health, wealth, and well-being. Michel Foucault argued that the spread of these methods leads to the 'the governmentalization of the state'.[18] If new forms of political reason and calculative practice emerging in the mid-twentieth century formed the economy as their object and introduced the future into government, perhaps we can refer to this mode of governmentality as 'economentality'.

Economentality draws upon those processes that, in the previous chapter, we identified as 'economisation': the assembling of actions, devices, and fields that are characterised and formatted by economists and others as being economic.[19] The aim is not to shift attention from the economy to economisation but to understand the making of the economy as an effect in which economisation plays a part. Economics never merely represents the economic, but we can understand the work of economics only in its relation to other modes of government that draw upon the possibilities of economentality.[20]

Economentality, and thus the economy, emerged as part of a struggle over capitalisation. The crises of the interwar period represented the near collapse, around the world, of the entire system of capturing income from the future through the capitalising of corporate profit. After the war, in North America, Europe, and many other parts of the world, organised workers and socialist governments made efforts to develop alternatives to corporate capitalisation, based on the state ownership of industry, the strengthening of labour unions, worker rights in the allocation of corporate profits, and – in Egypt and many parts of the South – limits on the foreign ownership of businesses. These efforts provoked a period of unprecedented labour struggle – unprecedented in the number of workers involved and the range of sites around the world where the battles occurred.

One can trace the way in which governments responded to these struggles by following in a single year of strife the making of new connections among America, Europe, and the Middle East. On 10 May 1948, President Truman ordered the US army to seize control of the country's railways and place them under military operation. At the same time, a federal district judge issued an injunction against the leaders of the railroad unions. The seizure and injunction prevented a national railway strike planned for the following morning. A month before, the courts had been used in a similar way to end a four-week strike by 400,000 coal miners seeking company retirement pensions.[21] The Taft–Hartley Act of the previous year gave courts a new power to prevent strikes, undoing some of the labour rights won during the New Deal. The act was a response to the largest wave of industrial action in US history. Starting in the country's oil refineries, which had been placed under army control, the 1945–6 strikes had spread to railways, electrical workers, power companies, the steel industry, and automobile manufacturing. The federal government had deployed the armed forces to seize control of the oil refineries, then the coal mines (twice), and, on at least one previous occasion, the railways.[22]

The new Taft–Hartley powers could not be used against rail unions, which were governed under a separate legal regime, the Railway Labor Act. To get around this problem and prevent the new strike in May 1948, the court claimed a general power to prevent rail strikes. It justified this expanded power on the grounds that a nationwide rail stoppage would 'imperil national health and safety' and obstruct 'vital and necessary Government functions'. The court was empowered to act, it argued, to prevent this 'irreparable injury' to the country.[23] Justifying the peacetime use of military power by invoking the threat to national safety linked the government's domestic fight against the claims of organised workers to a new idiom of national security. The National Security Act of July 1947 inaugurated the remilitarisation of government, setting up the Department of Defense, the Central Intelligence Agency, and the National Security Council. It seems only appropriate that President Truman signed the act into law aboard his presidential aircraft, a Douglas C-54 Skymaster nicknamed *The Sacred Cow*. Peacetime militarism was to be America's sacred cow, acquiring an increasing immunity from criticism. Peacetime militarism was now required, it was claimed, for the protection of Europe and, increasingly, to defend the oil fields of the Middle East. The novel idea that the energy resources of the Persian Gulf were important to the 'security' of the United States was connected to the labour struggles at home.

In response to the postwar coal and railway strikes, electric power companies and railroad firms began to strengthen themselves against labour demands, thus weakening the expanded claims of popular democracy by developing an alternative source of energy, switching from using coal to burning oil. Partly as a result, in 1948 the United States became, for the first time in its history, a net importer of oil, even though it was then producing 60 per cent of the world's supply.[24]

The United States hoped to engineer a similar weakening of the labour movements in Europe. On the same day that the national railroad strike was defeated, across the Atlantic in Bordeaux, the first ship carrying Marshall Plan aid began unloading its cargo, 8,800 tons of American wheat.[25] US aid was intended not just to rebuild Europe but to strengthen its business class against miners, railway workers, and other unionised forces even more powerful than those in the United States. In Britain, France, and other parts of the continent, strikes led by coal miners and their allies in other unions were demanding an extension of the social and democratic rights won before and since the First World War, including a collective right to a role in the management or ownership of industry. For the American businessmen who ran the European Recovery Plan, as the Marshall Plan was formally known, US aid was to be used to weaken this left-wing threat to the corporate ownership and control of industry. Partly for this reason, the largest expenditure of Marshall Plan funds was not on US grain but on oil from the Middle East. The United States subsidised the conversion of Europe's principal source of fossil fuel from coal to oil. The conversion was intended, in part, to weaken the coal miners of Europe and thus undermine the power of the left.[26]

Oil had a number of advantages as a source of energy that made it less vulnerable to the demands of labour and thus to threats against capitalisation. Simply by supplementing coal supplies, it weakened the ability of coal miners, in alliance with railwaymen and dockworkers, to shut down a country's carbon energy system. The weapon of the general strike – assembled with the building of coal, rail, and electric networks in the 1880s, and used rapidly thereafter across the industrialising world to win battles for more democratic and egalitarian forms of collective life – was now dismantled. Oil could be raised to the surface and distributed using pumps and pipelines, so its production and distribution required fewer workers and was more difficult to interrupt with strike actions. Unlike coal, it was light enough to be moved easily across oceans. So, regions that had previously industrialised using coal, building industries close to the source of energy and thus vulnerable to the interruption of local energy supplies, could now begin to outsource the supply of energy.

The postwar development of Middle Eastern oil can be viewed as an outsourcing of European industrial production. The outsourcing of manufacturing did not become significant for another two or three decades, following the development of containerised shipping, which allowed the advanced industrialised countries to begin importing large quantities of manufactured goods from East Asia. Energy, on the other hand, had already been containerised, as it were, ever since the development of the ocean-going oil tanker and the long-distance pipeline.

In 1947, workers in Saudi Arabia had begun constructing the Trans-Arabian Pipeline, the world's largest pipeline at that time. The US companies that had purchased rights to produce oil in Saudi Arabia planned to pump the oil overland from the eastern shore of the country to an oil depot at the port of Haifa in Palestine, the terminus of an existing British-built pipeline from Iraq. Such pipelines and the associated tanker routes provided the infrastructure for moving the West's energy production offshore. In the process, they were also building a new infrastructure of democracy, in which struggles for a less precarious collective life would be waged over a more extended network.

In the same week of May 1948 that Truman was declaring military control of American railways, another military confrontation was culminating in a defeat at the Iraqi town of Falluja, on the Euphrates River, forty miles west of Baghdad.

The Falluja defeat was the last step in a struggle that had been building for two years. In June 1946, a strike at the Anglo-American-owned Iraq Petroleum Company fields in Kirkuk, the site from which oil was piped to Haifa – by oil workers demanding the right to a union, sickness and disability insurance, and a pension – had been crushed.[27] A force of mounted policemen charged the meeting of several hundred workers gathered in the Gawupaghi Gardens outside the town, killing ten workers and injuring another twenty-seven. In early 1948 a series of student and worker strikes led to nationwide protests against the British-backed monarchy and the British-run oil company. This time, the oil workers focused their action not on the oil fields but at 'the point of bifurcation of the Kirkūk–Haifa and the Kirkūk–Tripoli pipelines, the K3 pumping station near Hadīthah', which they shut down.[28] The stoppage lasted two weeks, until the Company surrounded the site with machine guns and armoured cars and cut off supplies of food. Unable to risk a repeat of the massacre at Gawupaghi, the strikers decided to march on Baghdad, more than 150 miles away. They set out on 12 May, the day after the defeat of the railroad strike in the United States. After three days of marching, with increasing support along the way, they entered Falluja and 'fell into . . . a police trap'.[29] The oil workers were sent back to K3, and the strike leaders to prison.

Defeat for Iraqi oil workers was repeated in the breaking of oil strikes in Iran in the same period, and later in Bahrain and Saudi Arabia. The Saudi oil workers demanded better pay and living conditions from the American oil company Aramco, the right to form a union, a political constitution from the government, and the removal of extra-territorial privileges for the American oil company. In Egypt, as we will shortly see, a wave of strikes was brought to an end in the same week as the oil workers' protests in Iraq.[30]

The year 1948 was marked by two setbacks that would shape the postwar Middle East. One was the defeat of the Iraqi oil workers at Falluja, a reversal for worker movements based in oil extraction – movements whose dispersion across the region and distance from Europe deprived them of the possibilities for interconnected, disruptive, democratic collective action that coal workers in Europe, tied closely to the sites of energy use, had enjoyed. The second was a reconfiguring of colonial power, centred in the calamity that same year in Palestine. Before the war, colonial powers had typically justified their presence in the region by claiming the right to 'protect' certain populations, specifically colonies of European settlers and communities of religious minorities. As the United States began to replace Britain and France as the principal imperial power in the region, it started to recast the justification for its power around a new doctrine, the doctrine of 'security' – first of oil and later of the new settler state that had just adopted the name 'Israel'.[31] The first of these setbacks, as already mentioned, took place in Falluja, Iraq; the second, in Palestine, happened to culminate in defeat at a place that shared the same name.[32]

The Birth of the Economy

In 1948, the new Council of Economic Advisors (CEA) in Washington published its second 'Annual Report of the President to Congress'. Its inaugural report the previous year had contained the first official statistical representations of the American economy. But that report did not name its object as the economy and contained few uses of the new term, preferring to discuss 'the nation's economic budget'. But, only a year later, in the 1948 report, one can see the familiarisation of the idea of the economy, which begins to be used with increasing frequency. Five years later, in 1953, the report dropped the term 'nation's economic budget'. The introduction to the annual accounts explains that they illuminate 'the process of change and adjustment within the economy'. The report warned, at the same time, that annual accounts could not capture the dynamic movement of this object. The accounts were more like 'snapshots taken at intervals than like a moving picture which shows the process of change'.[33]

Historians have emphasised two points about the setting up of the CEA. First, its creation marks an unprecedented influence of academic expertise within executive office. For economists, the CEA was 'the institutional zenith' for their service to the state.[34] Second, this expertise represents the emergence and triumph of 'economic growthmanship'.[35] We can propose two modifications to this story for understanding the emergence of new modes of securing capitalisation through governing the future.

First, the CEA was set up not to allow economists into government but to keep them out.[36] More precisely, it was intended to effect a separation of the powers, creating an arrangement analogous to the quasi-independence of the Federal Reserve, the Supreme Court, or the new Joint Chiefs of Staff, or analogous in another sense to the new realm of 'national security' for which the Middle East served as an important reference point. The CEA would produce public economic knowledge outside the day-to-day contestations of political debate, contributing to the parallel effect of the economy as an object separate from the state that operated as a visible mechanism independent of the process of government.

Congress created the CEA in reaction against postwar proposals for a Full Employment Act, which would have established full employment as a collective right. Opponents of the right to employment inserted the plan for the Council into the Employment Bill, to weaken the influence of wartime economists in government and the mechanisms they had developed during the war for maximising employment and controlling prices. The opponents hoped to undermine these government economists by giving a voice in Washington to academic economists: the CEA would express the views of specialists from outside the decision-making process. In contrast to economists hidden from view inside government departments, these academics-on-loan would be forced to make public their advice and the evidence on which it was based, through an annual report to a new joint committee of Congress. By making economic calculation public, right-wing forces hoped to be able to expose its uncertainties and limit its influence. The CEA was to be a device for this making of things public, for an 'economic education' (starting with the education of the president). Its regular reports would help bring the economy into effect.

Second, economic expertise had often been concerned with problems of growth: the increase in population, the expansion of trade, a surfeit or shortage of natural resources, or inflation in the supply of money.[37] What was new in this historical juncture was not growthmanship but the object that would grow: not population, trade, resources, or wealth, but something less material and therefore more effective: the economy. The growth of the economy was not a question of the government of resources or the management of national finance. It was a means of bringing the future into government – of governing populations through their futures.

What was the effect of the economy good for? Part of the answer is that it allowed a mode of government that both represented opposing social forces and indicated the means of bringing these opposites into balance. This was the chief method and purpose of the CEA reports: to represent and balance the claims of 'business' and 'labor'.[38] Exposed to the changing metrics of the economy, businessmen would be pressured to consider the impact of excessive price increases, while workers would be forced to moderate wage demands. A third party, the bankers, would see the effect of directing credit creation towards consumption versus investment.

The economy was a device for achieving this balance by deploying the future, calculated in terms of the changing movements of funds among economic units. Rival political forces would now see how different demands and choices affected the future balance. The economy would embed people's political lives in the future, by leading them to calculate according to its representation. It would locate them in relation to a future formed in a particular way, as a balance, or trade-off, between forces now inscribed as equivalents in the structure of national accounts, with wage earners/consumers on one side, and business and banking on the other.

Until this point, economists had typically been suspicious of the work of forecasting. It was an unpredictable and therefore dangerous art, whose excesses and failures were said to be visible in the example of Soviet-style planning. Writing about 'the crises of confidence which afflict the economic life of the modern world', John Maynard Keynes had pointed out that 'our basis of knowledge for estimating the yield ten years hence of a railway, a copper mine, a textile factory, the goodwill of a patent medicine, an Atlantic liner, a building in the City of London amounts to little and sometimes to nothing'. He did not believe that economic calculation could overcome 'the dark forces of time and ignorance which envelop our future'.[39] This fear of unknowable temporal forces could now give way to a calculated future, projected with increasing confidence onto the politics of the present.

To govern through the future was not just a question of calculating the new circulations of funds that were to be known as the gross national product (GNP) (and later the gross domestic product, or GDP) or discussing the economy as a political object. We can trace two more specific innovations which illustrate the kind of work required to produce and command the future.

The first example is the price deflator, or the method of adjusting figures to account for inflation so that comparisons could be made over a wide span of years. The GNP was an annual figure. Like a company's books, it required an arbitrary starting and ending point, creating a unit of time within which everything had to balance, generating a total – a fixed quantity manufactured out of the endless circulation of payments between economic entities. The total produced from within one unit of time could then be compared to the next, just as a firm compares its quarterly or annual profits. But, unlike those reporting company profits, those calculating GDP wished to make it possible to compare the total from one year not just to the preceding or following year but across several years or even decades. This extended temporal sequence required a method of adjusting the numbers to allow for the declining purchasing power of money or the phenomenon of price inflation.

Inflation was one of those words borrowed by economists from nineteenth-century physics.[40] It was previously used to refer to the mechanical act of pumping additional supplies of credit money into the currency system. It was now a term that could refer to the new national-level aggregates whose measurement was taking shape as the economy. The staff of the government budget office initially refused the request to calculate the rate of inflation in prices, arguing that gathering representative data on the prices of numerous commodities at regular intervals and manipulating the figures to calculate an aggregate annual inflation rate was far too complex and costly an operation. Eventually, however, the usefulness of the deflator in constructing the economy overcame these objections.

The deflator made it possible to measure the change in GDP not just from one year to the next but over a sequence of years or decades. This extended measurement made visible a dynamic process – the growth of the economy. With adjustments for inflation, it was now possible to estimate totals for the gross national product of the United States for every year back to 1875. The estimates suggested that the American economy had been growing, with ups and downs, at a rate of 3.75 per cent per year.

A second innovation came in the way economists pictured growth. In its annual reports, the CEA began portraying changes in the nation's 'economic budget', and later its economy, with a chart. The vertical axis indicated the level of GDP, while the horizontal axis marked the succession of years. The phenomenon of growth, portrayed as the increase in GDP, could now form a line ascending from left to right across the surface of the chart. The line started in the past, moved through the present, and was projected forwards into the future. The visualisation was intended to educate the public, encouraging workers, employers, and financiers to act in relation to this projected future. Unreasonable political demands, it could be shown, especially the demand of workers for higher wages based on an increased share of revenues, rather than on increases in their own productivity, would threaten to throw the line off its path. Similarly, excessive creation of credit by the financial industry would risk the stability of the future.

However, with production growing at close to 4 per cent a year, this line that was to rule people's lives was not straight. Drawn on a chart, it formed an exponential curve. It started as a gently rising slope but gathered height rapidly and before long would reach an angle that was almost vertical, with values approaching infinity.

The problem of the exponential curve was twofold. First, with the line moving close to vertical, the beneficial effect of minor adjustments in the demands of different political forces could not easily be made visible on the graph. Second, if the future was an accelerating curve – a line of exponential growth – how could it be made an instrument of discipline and control? It appeared out of control.

Until now, this had been the problem not just with the drawing of graphs but with the future itself. The modern experience of growth had not previously offered a point of stability or a focus of control. Geometric rates of change always appeared unsustainable. Thomas Malthus's *Essay on the Principle of Population*, first published in 1798, was the best-known expression of this problem. But many other nineteenth-century political economists carried out similar calculations with the same resulting alarm about the future. William Jevons, whom we met in chapter 4 as one of the architects of marginal utility theory in the 1870s, had abandoned the question of growth after completing a report on Britain's coal reserves, published in 1865. The study predicted that the exponential growth in Britain's mining and consumption of coal would lead to a peak in coal production within about fifty years, and then an accelerating decline. The prediction proved accurate, with the peak arriving in 1913. By the early twentieth century, there was a widespread preoccupation among economists with the consequences of geometric growth, including the exhaustion of natural resources.

This concern was found not just among political economists. Reinhart Koselleck argues that in the modern era geometric rates of change defined the very experience of modernity. Modernity was encountered, he suggests, in terms of 'a future that transcended the hitherto predictable' because of the phenomenon of the ever-increasing speed of change.[41] The future, in other words, was experienced in the present as this lived sensation of acceleration. The experience created a rupture with the past, which now had to be refigured as an inaccessible 'tradition' because earlier history 'might not count against the possible otherness of the future'.[42] The new horizons of expectation caused a repositioning, as it were, of the relative locations of past and future.

In the early decades of the twentieth century, writers continued to associate the experience of modernity with accelerating change and thus unpredictable futures. Very often the experience was related to the 'natural energies' that humans could now exploit. 'One phase of the workings of a technological age, with its unprecedented command of natural energies, . . . needs explicit attention', John Dewey wrote in *The Public and Its Problems* (1927). 'The older publics, in being local communities, largely homogeneous with one another, were also, as the phrase goes, static.' Dewey argued that 'the mania for motion and speed is a symptom of the restless instability of social life, and it operates to intensify the causes from which it springs. Steel replaces wood and masonry for buildings; ferroconcrete modifies steel, and some invention may work a further revolution.'[43]

The later, postwar visions of development presented by men like Adriano Daninos, the promoter of the scheme to build the Aswan Dam, had not yet taken shape. On the contrary, in the interwar period mammoth infrastructure projects could exemplify the speed of change that made reliable calculations of the future impossible to carry out. The US Congress spent more than a dozen years debating what to do with the stalled project to build a giant dam on the Tennessee River at Muscle Shoals, Alabama, before devising the scheme for the Tennessee Valley Authority that was to inspire Daninos and others as a model for post-war visions of integrated development. For more than a decade, the problem of the dam occupied more time in the US Congress than any other issue.[44] For Dewey, the failure at Muscle Shoals, with its associated nitrate plants built to produce munitions and fertiliser, epitomised not a mastery of the future as a programme of economic and material growth but the way in which proliferating sources of energy made new projects almost immediately redundant. 'New methods', he wrote, 'have already made antiquated the supposed need of great accumulation of water power.'[45]

For other writers, the motor of acceleration was identified as money. Writing earlier in the century about the mental experience of what he called 'money economy', Georg Simmel explored 'the accelerating effects of an increase in the supply of money on the development of the economic–psychic process'.[46] No wonder there was a suspicion of financial forecasting as unreliable and dangerous. The new powers of credit creation, to which Simmel was responding, generated the instability of the future.

The example of Soviet central planning might have offered an alternative way to control the future, but it appeared no better. As a centre-piece of the 1929 'Five Year Plan for the Development of the National Economy', the Soviet government hired the Chicago engineering firm of Henry Freyn and Company to design an integrated iron and steel complex at Magnitogorsk, the largest single venture in Soviet planning.[47] The design reproduced on a gargantuan scale the plant that Freyn's engineers had designed for the US Steel Company at Gary, Indiana. This was crash mobilisation, on the German military model, creating 'a rallying point for a technologically perfected future'.[48]

Planning meant the chaotic multiplication of factories, repeated improvisation, a future of proliferating numerical quantities and impossible annual targets. Soviet central planning inspired governments around the world with its ambition and speed, but its deployment of the future calculated as the rapid multiplication of physical plant and material output rendered the present more rather than less chaotic.

Keynes, it is worth remembering – one of those who followed the Soviet experiment – was not a theorist of economic growth. He imagined the period of expansion required to escape from the downward spiral of economic depression as a limited period of temporary acceleration, leading to a future steady state. For Keynes, growth was not a metric of success but a potentially destabilising force.[49]

Before the late 1940s, economists discussed growth as a question of periodic physical expansion. In the context of the New Deal, the war, and postwar restructuring of production, they saw growth as a question of putting idle plants back into service and creating jobs for those out of work. Such 'expansion' was envisioned as part of a cyclical process of the growth and contraction of productive activity – the business cycle. Expansion was not an end in itself. Instead, the goal was to smooth out peaks and troughs, restricting expansions where necessary to decrease the severity of downturns, and attempting to reduce bottlenecks in the supply of goods or labour that might hamper recovery from stagnation.[50]

By the 1930s, earlier prospects of a long-term expansion of material life had given way to arguments about a stagnation brought on by reaching the limits of physical growth. In a presidential campaign speech in 1932, Franklin D. Roosevelt warned:

> Our industrial plant is built; the problem just now is whether under existing conditions it is not overbuilt. Our last frontier has long since been reached, and there is practically no more free land . . . We are not able to invite the immigration from Europe to share our endless plenty. We are now providing a drab living for our own people . . . Clearly, all this calls for a re-appraisal of values. A mere builder of more industrial plants, a creator of more railroad systems, an organizer of more corporations, is as likely to be a danger as a help . . . Our task now is not discovery or exploitation of natural resources, or necessarily producing more goods. It is the soberer, less dramatic business of administering resources and plants already in hand . . . of adapting existing economic organizations to the service of the people.[51]

In 1938, the same views were expressed by Alvin Hansen in his address as President of the American Economic Association, as the doctrine of 'secular stagnation'. A decade later, the postwar effect of the economy and its growth replaced this process of secular expansion and stagnation of commercial life with a politics oriented around a new object that was somehow capable, it seemed, of perpetual growth.

Imponderable Resources in Technical Knowledge

How, given all this previous experience, was it possible to turn a pattern of exponential growth into a stabilising effect? One small but significant solution to the problem was a new method of representation, the log plot. In drawing charts of the growth of this new object, the economy, one could plot the growth of GDP over time not against a linear vertical scale but with the scale adjusted to show the rate of change; in other words, by using a logarithmic scale, in which the intervals on the vertical axis represent ratios. This trick turns an exponential curve into a straight line. An economy growing at 3.75 per cent per year can then be represented as a steady line, rising gently and uniformly across the page. What is steady is not the process of increase, but its rate. The effect of small changes in political demands or economic decisions could now be clearly projected on paper – and the future no longer appeared as an unstoppable acceleration.

Through such material inscriptions, the economy emerged as a non-material object, a set of calculations that converted the accelerating growth of modernity into an apparently stable future. In ways like this and many others, the economy, perhaps just for two or three decades, made the future an instrument of government.

The log plot, although useful, was far from the only device to make the future appear governable. The calculation of GDP – and what that excluded – was itself important. But so too were those workers in Iraq and other countries of the Middle East. They helped in two ways.

First, the outsourcing of energy production to the Middle East helped break the postwar wave of labour protests, in North America and even more so in Europe, in which organised workers were demanding a more far-reaching reconfiguration of claims to collective wealth. Coal miners had played a critical role, in alliance with dockers and railway workers, in bringing popular democracy and more egalitarian forms of life to the West. Working against the forces of capitalisation, they had engineered an unprecedented redistribution of wealth in the interwar and wartime decades.[52] From now on, however, increases in the income of workers were to come not from a redistribution of profits but from the steadily rising line. Pipelines and graph lines would work together. If the rate of growth levelled off or dropped back, of course, the income of most people would 'naturally' stagnate or decline.

Second, productivity could grow, and thus incomes increase, thanks to oil, which for a quarter century after 1945 increased in supply and gradually declined in price. Oil companies tried hard to restrict supply and fix the price of Middle Eastern oil supplied to Europe at the much higher level of oil produced in the United States – a price set by state-level production quotas designed to maintain the scarcity of oil and thus extraordinary levels of corporate profit. But these restrictions were never adequate, and there was a constant pressure from the governments in oil-producing countries for increased output or an increased share of revenues, a pressure the oil companies found it increasingly difficult to manage. As a result, the price of oil gradually declined. By 1970, oil's price in real terms had been declining in every decade since 1920. Thus, for several decades, the supply of energy seemed unlimited and its costs, for many purposes, negligible. If the economy appeared capable of unlimited growth, this was for practical reasons of the mid-twentieth-century energy regime.

One could also explore how the new calculation of growth in an economy solved the problem of peacetime military spending. Before the making of the economy, it was difficult to imagine how the decision to build a permanently militarised society after 1947–8 was not somehow at the expense of improvements in welfare and wages for the majority. Those points of vulnerability in the Middle East, where US 'national security' was soon deemed to be especially at risk, formed part of this new political-economic effect.

Governing people through their future also became a device for managing populations in the formerly colonised world. In 1940 the economic statistician Colin Clark, who had developed the methods used in Britain for the calculation of national income, published the first estimates of national incomes everywhere. His calculations of the degree of inequality between rich and poor countries caused a shock.[53] But this generalisation and repetition of the effect of the economy also raised questions. Simon Kuznets, for example, had not expected his methods of estimating GDP to be used to produce simple inter-country comparisons. The statistics implied that every country had an equivalent economy, a similar arrangement of monetarised relations, which could be compared and ranked. However, even if every country was said to have an economy, producing this effect in what were now to be called underdeveloped areas required different methods and offered different political possibilities. One of the most significant of these possibilities was that industrialised states could attempt to govern not only their own populations but also relations with the rest of the world through their futures.

In international relations, the future entered government unexpectedly. Harry Truman announced the arrival of the era of 'development' in his January 1949 inaugural presidential address.[54] The announcement was an afterthought. Asked to formulate a vision of the future to guide the president's second term, officials in the State Department came up with three points: support for the United Nations (UN), the Marshall Plan, and the North Atlantic Treaty Organization (NATO). But these were existing policies that the president had been discussing throughout the election campaign in the fall of 1948. An assistant for Truman told the State Department that he needed a fourth point, something 'that's just a bit original'. The men at State found it difficult to imagine the future. 'One heard a great deal of the need for long-range thinking and long-range planning in those days, and the State Department men tried to organise themselves for it', one of them later explained. But 'their efforts at planning generally bogged down in topical questions, such as what to do with the Italian cable under the Atlantic, or what price to pay for tin'.[55]

An official who had been helping with technical assistance programmes in Latin America suggested that such aid could also be offered to other parts of the world. The idea was hastily inserted in the speech. Truman announced what became known as Point Four (there had been no time to name the idea): 'A bold new program for making the benefits of our scientific advances and industrial progress available for the improvement and growth of underdeveloped areas'.[56] The post-war US politics of development emerged not as an imperial vision but as an afterthought.

As with the device of the national economy at home, the new politics of development represented a novel way of bringing the future into government.[57] The vision was not one of how America would share its wealth with the world or support the rights of oil workers in the Middle East trying to access a better share of oil company profits and create more equitable forms of life. In March 1948, the UN Conference on

Trade and Employment in Havana, Cuba, had adopted the charter of an International Trade Organization. Forming a third body alongside the International Monetary Fund and the World Bank, the organisation was intended to govern international trade, including trade in oil and other commodities. Under the agreement, oil company cartels and other private trade arrangements would be replaced with publicly regulated agreements. International investment and trade would be governed by principles of labour rights and the right to full employment. India and other countries from the South hoped to use the new organisation to redirect a greater share of the profits from production and trade in goods to the producer countries. There were concurrent demands for a Marshall Plan for such countries, to return some of those profits in the form of material assistance. The United States opposed such proposals. The agreement to create the International Trade Organization collapsed after the US Congress refused to ratify its charter.[58]

As a substitute for the international regulation of trade and the extension of labour rights to people like the oil workers of Iraq, Truman's speech mobilised the future, now understood as the development or growth of national economies. The contribution the United States would make was not material assistance, or support for labour rights, but something unlimited: technical know-how. 'The material resources which we can afford to use for assistance of other peoples are limited', Truman explained. The 'interest of the people whose resources and whose labor go into these developments', moreover, had to be 'balanced' by 'guarantees to the investor'. In contrast to these necessary limitations, 'our imponderable resources in technical knowledge are constantly growing and are inexhaustible'.[59] There was no capital to spare, and there would be no Marshall Plan for countries such as Iran, Egypt, or India, and no support for the right to unionise or act collectively to improve terms of employment and exchange. In place of material resources and the labour rights that had created widespread affluence in the West, the United States would offer the alibi of know-how and technical advice. The know-how would be offered in various fields, but, increasingly, in one main form: economic calculation. A know-how that, in contrast to limited material resources, was inexhaustible.

The Materially Minded Egyptian

Let us return to the Arab world in May 1948, where the protests of Iraqi oil workers had ended in defeat at Falluja. As it happens, the town gave its name to a second setback that summer, an event in Faluja, Palestine. A village in the District of Gaza, lying about halfway between the town of Gaza on the coast and the hill town of al-Khalil, or Hebron, Faluja was named after a shrine the villagers had built to a fourteenth-century Sufi master from Iraq, Shahab al-Din al-Faluji, and thus indirectly after the Iraqi town.[60] The same week of May 1948 was a fateful one for Palestine.

Britain had announced that Friday, 14 May, would be the final day of its occupation of Palestine. The UN had proposed a scheme to partition the country between the Zionist settlers, who were concentrated in half a dozen of the largest towns and a number of agricultural settlements, and the indigenous population, who inhabited every region of the country and outnumbered the immigrant Jewish population by about two to one. But no one had agreed to implement the partition or organise the removal of the Palestinian people from their towns and villages, as the plan for ethnic separation apparently required. While railway workers in the United States were forced back to work by the Truman government, and the mass oil strike in Iraq met defeat, Palestinians faced a larger calamity.

By the time the British departed, military actions and terror campaigns carried out by the well-organised Zionist forces had already driven as many as 400,000 Palestinians from their homes.[61] On the following day, 15 May, Egyptian soldiers entered Palestine from the south, while a rival army from Transjordan (following secret talks with the Zionist leadership) took control of the west bank of the Jordan River. The Egyptians initially held the southern parts of the country, but, when the Zionist forces later broke through their lines into the south, an Egyptian brigade was trapped in a pocket of territory at Faluja, the village linking what would become the West Bank and a remnant of Gaza District, later termed the Gaza Strip. In the armistice negotiations four months later, Egypt was forced to abandon the Faluja Pocket, as the area became known, although the partition proposal of 1947 and a revised UN plan in the summer of 1948 had called for retaining that area as part of the rump Palestinian state. In exchange for Egypt withdrawing its besieged force, the government of the new Zionist state, whose leaders – after some debate – had decided to call it the state of 'Israel' rather than Zion, agreed to leave the area's Palestinian population in place. Within weeks of the agreement, the villagers of Faluja were forced out, through a campaign of shootings, beatings, attempted rape, and other acts of terror, 'orchestrated' by the commander of the Zionist ethnic cleansing operations in southern Palestine, Yitzhak Rabin.[62] The population link and territorial contiguity between the coastal strip and the hills was severed, and the way cleared for the encirclement of the Gaza Strip and the Zionist settlement of southern Palestine.

The two Faluja defeats of 1948, in Iraq and Palestine, mark two sides of a connected process. The Egyptian defeat consolidated the Zionist control of southern Palestine and the confinement of most of the Palestinians who managed to remain in the country within two enclaves, the West Bank and the Gaza Strip. Britain had supported Zionist settlement in part because the creation of a European settler minority established a point of vulnerability that required an imperial presence for its protection. After the Second World War, Britain could no longer afford to run this old kind of protectorate, based on the need for armed protection of European settlers. And the settlers themselves, strengthened by their own wartime militarisation and by a competition between the United States and the USSR to support them, had helped force the British out. After 1948, imperial protection took new forms, based on new vulnerabilities. Expanding oil production enabled the rise of labour organising in Iraq, Iran, and later in Arabia, but it also allowed the United States into the region as the protector of regimes in countries such as Iran and Saudi Arabia that would police and prevent labour's organisation. The attempted removal of Palestinian society would create another vulnerability that Washington was later to take advantage of, as protector of the state of Israel.

In Egypt, a labour movement that was larger than those of the Gulf states and which had enjoyed a long history of political action had organised a wave of strikes and industrial protests in 1947–8. The strikes began in the large textile mills and other industrial enterprises, then spread to the Anglo-French company operating the Suez Canal, the Shell oil refinery at Suez, petroleum distributors, and many others. The emergency in Palestine in April and May 1948 allowed the government to impose martial law, break the strikes, and imprison thousands of leading opponents of the regime.[63]

The defeat of the Egyptian army in Palestine galvanised a group of junior army officers, who formed a secret organisation, the Free Officers, to combat the corruption of the country's military and political leadership and support a popular movement to end Britain's continuing imperial presence in Egypt. The leader of the group, Gamal Abd al-Nasser, was one of those who had been trapped for four months in the Faluja Pocket. After overthrowing the British-supported monarchy in July 1952, the Free Officers introduced a modest programme to redistribute the royal estates and other very large landholdings to the landless. These policies met resistance from the old order, with its base among large landowners, while leftist groups demanded more radical change. In March 1954, a showdown with opponents culminated in a general strike organised by pro-Nasser unions – transport workers, bank employees, and the workers at Suez. The show of strength enabled Nasser to defeat opposition from right and left. To extend its popular support, the regime declared a new era of revolution and adopted a grand plan to remake Egyptian material and collective life. The revolution, however, was to operate not through a further redistribution of land in the Nile Valley but through a scheme to expand the valley, by building the High Dam at Aswan.[64]

The story of the World Bank's offer to help fund the Aswan project is well known.[65] Having lost its mission of funding postwar reconstruction in Europe to the Marshall Plan, the Bank seized the opportunity to demonstrate its transformation from a financial institution into what could now be called a 'development' agency, by offering to finance the largest rock-fill dam in the world. The Bank even contemplated using the dam as an instrument for solving the Palestine question: bringing irrigation water under the Suez Canal to northern Sinai and settling Palestine refugees there might have allowed the UN Relief and Works Agency, the agency set up to provide relief for the refugees, to contribute to the cost of the dam.[66] But the Zionists had other plans for Sinai. In February 1955 Israel launched an attack on the Egyptian forces in Gaza, taking the first step to capture the remainder of southern Palestine and seize Sinai, occupying Egyptian territory as far as Suez.

Meanwhile, the British were developing plans for a military counter-coup in Egypt and the assassination of Nasser, to reinstall a neocolonial regime. The plotters within MI6 had difficulties with the design of the assassination devices – a Remington shaver fitted with explosives failed to operate properly – and, in any case, the plot was uncovered.[67]

The Americans decided that there was an easier way to deal with the threat of a nationalist government in Egypt. Rather than eliminating the president, they would kill the mechanism that was supposed to increase his power, the construction of the dam. In July 1956,

Washington cancelled its participation in the funding for the dam, surprising the World Bank and leaving it unable to proceed with the loan.[68] Like the Remington shaver, however, the device failed to destroy its target. Cancellation of the loan offered Nasser the opportunity to nationalise the Suez Canal Company, whose capitalised revenues had been originally discussed as the means to finance the dam.

Britain responded by putting aside its assassination plans and supporting the Israeli scheme, already backed by the French, to invade Egypt, take the canal, and overthrow Nasser's government by force. Yet the Anglo-French-Israeli attack, launched at the end of October, quickly exposed the limits of neocolonial violence. The invaders tried to portray the seizing of Sinai and the canal as a police action that secured a vital communication route, an imperial version of Truman's seizure of the US railways eight years earlier. But British war planners, aware of the limits of military power, also hoped to enlist an economic device to assist the military operation: a form of calculation operating among ordinary Egyptians.

The British would not attempt to invade Cairo by force. They planned to land troops at the canal and destabilise the country by destroying railways, power stations, and oil refineries. The economic consequences of this destruction of the country's energy infrastructure, the military planners argued, would create public support for a coup. Since 'the materially-minded Egyptian civilian is unlikely long to sustain the rigours of wartime or the actual experience of battle', they reasoned, 'it seems less likely to be a question of "if there is a collapse of public . . . support for the Regime" as "how quickly that will take place" '.[69] Backed by this military-engineered economic force, Nasser's generals would remove him from power and the British would occupy Cairo without a battle.

The British never had the chance to try provoking militarily these forces of local economic calculation. Anxious to protect the corporate-built 'petroleum order' it was constructing in the region, based on an alliance with Saudi Arabia, the United States refused to support the invasion. Saudi oil workers had staged another general strike that summer, so, while the United States was keen to weaken Nasser, it was unwilling to associate its imperial methods with colonial occupation.[70] Having decided that deploying economic calculation on a larger scale was more effective, the Americans thought depriving Egypt of development assistance was a more effective weapon. But whereas the new methods of economentality had offered an alternative to military action at home, the attempt in Egypt to impose the discipline of collective economic calculation abroad had failed to achieve results.

The failure of the US plan to discipline Egypt through the development of its economy began with the fact that Egypt did not need the World Bank loan. Initial plans for the dam had been backed by a consortium of German engineering firms. The Germans brought British firms on board, as a way to access credit money: Egypt had large foreign reserves, sterling credits built up in London from goods and services it had provided to Britain during the Second World War. British firms lobbied London to release the credits and tried to get the release conditional on building a dam through the Anglo-German consortium.[71] To engineering firms, desperate for business in a depressed postwar Europe, the dam was not a scheme to store up and divert the Nile so much as a scheme to divert the sterling credits in London, via a detour through southern Egypt, into company revenue. An alternative source of future hard-currency funds was the shipping tolls from traffic passing through the Suez Canal, which totalled 12 to 15 million sterling per year and were increasing rapidly as oil shipments grew. Two-thirds of this income went from the shipping countries directly into Paris and London. But by 1968, about the time the delayed interest on a loan for the dam might become due, the canal would revert to Egyptian ownership.[72]

In fact, Egypt did not even need the dam. This was probably the view of the Bank's leading development economist in the early 1950s, Paul Rosenstein-Rodan.[73] He had argued against developing 'backward regions' through investment in infrastructure such as dams, canals, and hydro-electric power stations, which required imports of construction machinery, turbines, and other expensive capital goods. Such countries, he believed, should focus on small-scale, labour-intensive manufacturing. This had to be done extensively enough to take advantage of the social profit that arises from the clustering of interrelated small firms, producing wage earners who would create a demand for goods produced by other wage earners – the kind of complementary effects at the level of the 'national economy' that individual investors, figuring future profit at the level of the enterprise, were unable to calculate.[74]

But the Bank could not fund industrialisation. It had to raise investment capital by selling bonds. To market these debts on Wall Street, it needed large, measurable, capitalisable structures.[75] Agriculture and small industry were too localised, too small, too short-term, and too dependent on forms of social profit for bankers to calculate the current value of future debt payments. Large infrastructure projects such as dams, roads, and power plants, as we will explore in chapter 7, produced future returns that could be calculated at the level of the economy, rather than the enterprise. And because this calculable future, the economy, was now the responsibility of government, a sovereign state could be made liable for the repayment of loans, reducing Wall Street's risks.

When the Bank became involved in the funding of the dam, it performed its own calculation of the costs of the project. By doing so, it introduced a different calculus. The construction costs, first of all, comprised not just the work of building the dam and supplying turbines. The project needed a nationwide electrical grid and water distribution system. But, more importantly, the costs had to calculate the future. These included interest payments on the loan, which were relatively simple to determine. But they also included the future effects of the loans spent on construction of the dam and its associated infrastructure. As wages for construction workers were turned into forms of consumption, demand would rise, prices would climb, and the currency would be devalued. The cost of imports, in turn, would increase, and the cost of international borrowing would go up. All these interconnected future costs could now be brought into view by the measurement and calculation of the economy.

Given these much higher costs, the Bank calculated that the dam would consume the entire disposable spending of the Egyptian government for the period of the loan. So, in exchange for the loan, the Bank proposed that the government would have to submit its budget to international inspection and control for the fifteen years or more of loan repayment. The scale of the dam created a project that could be monitored at the level of the calculation of the national economy. Building the dam required putting in place a future, an economy unfolding in time, through which Egypt would now be governed. In opting instead to nationalise the Canal, the Nasser regime refused to accept this mode of government.

In the years after the Second World War, a new future arrived. It was brought into being as a specific set of techniques for governing relations in the present in order to manage the economy, understood not just as a numerical totality but as a dynamic set of forces, caught in snapshot and under management. The economy was not simply a new object of government but rather an effect: the product of a series of iterated calculations.

Yet the effect depended on more than just new calculative techniques. It was constructed technically and materially. In the case examined here, we have seen how the economisation of domestic political life in industrialised countries depended upon the outsourcing of energy, a change brought about partly in response to the efforts of organised workers in the postwar years to claim a different collective future, one in which the profits of shareholders – the beneficiaries of capitalisation – would have to be adjusted to the claims of those in the present. In turn, the decreasing cost of the energy, in particular of oil from the Middle East, helped generate the effect of the economy as an object capable of unlimited growth. The growth of the economy, as we will see further in the final chapter, would provide the alibi for an expanded capitalisation. The calculative practices of the economy depended on maintaining forms of autocratic and inegalitarian rule in the Middle East. The postwar economentality of the West, as a mode of government organised around the economy and its growth, was an effect manufactured partly in the Middle East.

The economy worked effectively as a mode of government through the future for only two or three decades. By the late 1960s, the forms of productivity growth, energy use, cheap oil, and Middle Eastern politics on which it depended were all under pressure. We think of modernity in terms of accelerating time, of time experienced in the present in relation to a future that appears to be moving ahead at an exponential rate of change. From around the mid-century, for just a few decades, a different future had entered government.

6

The Properties of Markets

In the last quarter of the twentieth century, across the North Atlantic world and beyond, a new mode of government took hold, emerging alongside the government of 'the economy' and soon rivalling it in importance. People found their lives managed through a novel and more powerful abstraction, an idea that was to provide a new alibi for living at the expense of the future. The abstraction was called 'the market'.[1]

The idea of market exchange, of course, was nothing new. The classical political economists had written extensively on the topic of 'commercial society'. However, two changes transformed the market into a more effective mechanism of government. The first of these we have already noted. Towards the end of the nineteenth century, in the age of empire, the act of exchange came to be understood as a mathematical device. As we saw in chapter 4, the sudden growth of money markets helped create the new science of economics. Stock markets and other exchanges produced 'highly artificial processes', where a standardised good was traded at instantly knowable prices. Taking such sites as exemplary, economists formulated the laws of supply and demand, based for the first time on the principle of marginal utility. The value of human labour, goods, or money was now said to be determined not by their intrinsic qualities, but simply by their price. Pricing was an automatic process, in which small increments of supply and demand reached a point of equilibrium. It took some time before the tenets of what became known as 'neoclassical' economics were generally accepted. But, among the adherents of this view, the market could be considered a machine that efficiently determined the allocation of resources and rewards.

A second change occurred a century later. From the late 1970s, these 'artificial processes' came to be seen as something much more: a general mechanism for the government of relations in the common world. The proliferation of such mechanisms was now referred to in the singular, as simply 'the market'. This second change was given force by a well-organised political movement. Known as neoliberalism, the movement defined its project as curtailing the role of the state to allow free operation to the underlying laws of the market. In practice, however, the movement helped to produce the market as a general abstraction.[2]

The appearance of the market as a governing abstraction was a process that paralleled the emergence of the economy a generation earlier, as explored in the last chapter. In both cases, what was previously understood as a diverse set of processes was transformed into something like an object in itself. One could argue, in fact, that, initially, 'the economy' and 'the market' emerged side by side, as rival concepts for capturing the totality through which collective life should be governed. In the interwar years, economists who emphasised the role of government in the regulation and planning of economic relations began to refer to those relations, revealed through new forms of statistical knowledge, as 'the economy'. In the same period, opponents of government planning and regulation, as we saw earlier among those who took up the work of Max Weber, argued that statistical knowledge could never accurately measure the complexity of economic relations. As an alternative, they proposed that the dispersed, often tacit knowledge of individual economic actors is translated by market exchange into the prices at which they buy and sell. Taken as a whole, prices operate as a knowledge system, analogous but superior to the government's statistical knowledge of the economy.[3] Such arguments made it possible to imagine 'the market' as a double or doppelganger of the economy, mimicking its claims to form an object in itself. The ascendancy of wartime and Keynesian economic planning in the middle decades of the twentieth century pushed this rival conception into the background. A generation later, as the neoliberal movement emerged from the margins and began to determine the framework of economic discourse and policy, the market came to the forefront.

Critics of neoclassical economics often accuse it of misrepresentation. Economists reduce the complexities of social life to the outcomes of the calculations of rational agents responding to incentives, expressed as prices, in ways that maximise their interests. In doing so, the critics claim, they leave out of the picture any account of how agencies are constituted, interests are formed, incentives are managed, or resources are initially distributed. This criticism has plenty of force, but also a clear weakness: it assumes that the work of economics is to model a world external to itself.

What if we think of economics differently? Suppose it operates from within the technopolitical world, not from some place outside it. Suppose it provides a set of instruments of calculation and other technical devices, whose strength lies not in their representation of an external reality but in their usefulness for organising technical practices, understood as markets. This is the process of economisation.[4] In the words of Donald MacKenzie, the algorithms that produce market prices should be thought of as 'an engine, not a camera'.[5] The narrowness of neoclassical economics then serves a purpose. It helps perform the operations that Michel Callon calls 'framing', or the process of 'entangling' and 'disentangling'.[6] Economic theory must assume a world that already consists of buyers, sellers, and goods or services to be exchanged. However, an exchange cannot take place unless the world is simplified to create the buyer, the seller, and the good. Complex powers of attachment and relations of dependence among human and nonhuman forces must be disentangled to create something as simple as a rational economic agent, or an instantly available and measurable good or service. Market exchange would not work if arrangements were not in place to exclude some attachments, leave many costs or claims out of the calculation, render large parts of the common world unknowable, and deny responsibility for catastrophic consequences, such as climate collapse, species extinction, and genocide. Equally, as Callon shows, transactions could not take place if there were not some process of producing the re-entanglements that progressively create attachments between agents and goods.

Economic analysis does not describe these processes of exclusion and attachment. Rather, it helps to produce them. It provides ways to distinguish what can count in the act of exchange from what cannot, which arrangements constitute agents and which do not, and what must be accounted for and what should not. From this perspective, economics should be analysed not in terms of the reality it represents – or fails to represent – but in terms of the arrangements, entanglements, and exclusions it helps to produce.

Economics places out of sight the work that is performed to format the world it models, a world that seems already made up of buyers, sellers, and commodities. In doing so, it does something else. It places out of sight the work of assetising, or the process of turning future forms of life into streams of revenue and making that future revenue into a commodity that can be traded in the present. This is the very process on which extraction from the future depends. In this respect, the idea of 'the market' performs a double abstraction. It abstracts from all the framing and (dis)entangling that scholars like Callon describe, in which complex worlds are simplified into market relations. But it also abstracts from a temporal process, in which the command of the future is mistaken for the atemporal, instantaneous event of a market transaction. The market is an effective abstraction, not only because it abstracts from, or simplifies, the work of producing a world of agents and commodities, but because it abstracts from the work of producing a future open to capture in the present. In this way, the seemingly instantaneous temporality of a market transaction, as framed in the work of economics, contributes to the alibi of capital.

Economics does not work alone. It operates together with other techniques, sets of information, arrangements, and agencies, with different strengths and resources. Enframing a market and assetising the future do not happen only by employing the methods of classification or calculation that economics may provide. The exclusions on which assetisation and marketisation depend take a variety of forms and acquire different degrees of force and effectiveness. Another, interrelated set of arrangements is the law.[7] Contracts must be calculable, but also enforceable. Property must be transferable and open to being forfeited or reclaimed. Employees must be put to work, or they may be locked out or fired. Businesses may be taken over or shut down. Such processes involve a mixture of technologies, calculative devices, methods of control, and trials of strength.

Where do these arrangements acquire the forms of compulsion they require? In the first place, from what is called the rule of law. Markets depend upon a form of politics in which relations among agents, and across time, are governed by rules of property and contract. These are among the distinctive technologies of power and obligation in market societies, generating a variety of micro-sovereignties, disciplinary regimes, and coercive forces.

Property arrangements are often relatively stable, but they are never static. Forms of property proliferate. Engineering and technoscience develop new objects, interests, and agents, as for example with irrigation schemes that transform land into capitalisable assets, cellular devices that allow parts of the radio spectrum to be auctioned to phone companies, algorithms that extract privately controlled data from online searching, or the 'great economisation of the ocean' that transforms the Atlantic cod into an asset to which claims of ownership can be attached and then traded.[8] The very nature of property is continually up for renegotiation, requiring new forms of enframing and disentangling, and the management of new frontiers.[9]

But property relations have another kind of limit. While the airwaves or the oceans can be rearranged into new forms of property, and thus transformed into marketable assets and capitalised, there are other sites that have known property relations for decades or generations yet have kept the rules of the market at arm's length. In fact, most people in most of the world live lives in many ways protected against the rules of private property and the process of assetisation. They may live in self-built housing on plots they do not formally own, or farm agricultural land that is protected against outside purchase, or produce food intended largely for household consumption. Marriage or inheritance arrangements may operate to ensure that land, livestock, or other goods are not lost to the market through the turnover of generations.

Formal rules of property help determine what is legal and illegal and what is public and private. Such distinctions define the market but also create ways to survive outside it or in the pathways and spaces that these distinctions open up. People may produce and trade illicit goods, such as hemp or opium, in which legal property rights are typically unavailable. Their main resource might be the public space of an urban street, as in the wealthier neighbourhoods of car-choked third-world cities, where people establish a territory and earn a living as informal parking attendants, or the corridors of a government office, where the difficulty of obtaining legal permits and approvals enables informal expediters to thrive.[10]

These kinds of informal or nonmarket practice used to be thought of as something residual, representing pockets of resistance to economisation or transitional regions occupying some intermediate space. But as examples such as informal urban housing suggest, many of these arrangements are constantly expanding, possibly faster than formal property arrangements. Although they are thought to exist outside the market, and outside the laws of property, they are far from lawless. Even when relatively new, such as the squatter settlements in Calcutta described by Partha Chatterjee, they are characterised by complex moral claims, social rights, and political obligations. As Julia Elyachar shows, they can embody alternative modes of economic success and offer some of the most vibrant forms of economic life. As Lawrence Liang's account of 'porous legalities' suggests, they also can penetrate the pathways of the law, seeping into the more closely governed arrangements of property and the market.[11]

Despite the porosity of such arrangements, the idea persists that the market, indeed capitalism in general, has a boundary. The boundary is thought to separate the market from the large areas of material activity and resources that seem to exist beyond its limit. For countries outside the West, the idea of a boundary provides a common way not just to think about these places but to diagnose their problems and design appropriate remedies. Under neoliberal development programmes, such remedies became increasingly popular. Older ways of approaching non-Western development typically referred to different sectors of the economy, such as the traditional and the modern, the rural and the urban, or the agrarian and the industrial. These sectors had moving and sometimes overlapping boundaries. Development could be planned as a series of transitions, in which people and resources were to be moved from one sector to another, or in which the expansion of the market transformed different parts of the economy at different moments. Since the changes occurred over time, the difference between the non-market and the market, or the non-capitalist and the capitalist, represented as much a temporal transition as a territorial one.

This chapter investigates this conception of the market and its limits. It proposes a different way of understanding the boundary of the market and the status of the difference between the market and the non-market. The chapter argues that the distinction between market and non-market, capitalist and non-capitalist, or assetised and non-assetised should be considered not as a thin line but as a broad terrain, in fact a frontier region that covers the entire territory of what is called capitalism. The region is the scene of political battles, in which new moral claims, arguments about justice, and forms of entitlement are forged. When economics helps to devise technical mechanisms to move people and assets across a line from outside the market to inside, it plays a part in this politics. Such mechanisms do not move things across a fixed line, but they do rearrange the control and distribution of assets. The line is something created by these mechanisms, as part of the battle over redistribution and control. For economics, this involves what might be called a work of misrepresentation, for the outside must be constituted in terms of its relation to the market – that is, in terms of its deficiencies, as the non-market, as something defective or dead. Yet it is inadequate to call this a misrepresentation, for it implies the possibility of an adequate representation of the non-market from within a capitalist discourse. In practice, the non-market is a realm of ignorance that is required for the truths of economics to operate. It is more useful to consider what kind of world the (mis)representation helps to organise.

Technologies of representation claim a double role in these arrangements. One the one hand, economic analysis tries to help reorganise technopolitical life by representing a world outside the market, as part of the process of seeming to bring it inside, or assetise it. On the other, in describing what this non-market world lacks, economics tends to diagnose its defects as an absence of techniques of representation. Things are stuck outside the market because they are not properly represented – by property records, prices, incentives, or other systems of reference. What economics does, however, is not to represent what was previously unknown and unrepresented, but to reorganise the circulation and control of representations. The power of economics does not necessarily lie in getting people to adopt its (mis)representations; rather, in helping to constitute the apparent border between the market and the non-market, or the assetised and the non-assetised, economics contributes to the work of technopolitical mechanisms that reorganise how people live, the political claims they can make, and the assets they can control. Its particular role, I argue, is in formatting a form of exclusion-inclusion.

Under neoliberalism, the view of development as a transition from traditional to modern gave way to a more simple idea: the 'market' is surrounded by a boundary, outside which stands the non-capitalist, non-market world. The task of 'development' is to help extend the rules of the market and the process of assetisation into these other spaces. The popularity of such proposals can be gauged by the success of the work of Hernando de Soto, a Peruvian entrepreneur and economist, whose two books, *The Other Path* (1989) and *The Mystery of Capital* (2000), became the most widely cited studies of non-Western economic development in a generation. The Institute for Liberty and Democracy (ILD) in Peru, which carried out the research presented in these books, was at one point called the second most influential think tank in the world.[12] Both books described the forms of wealth and material activity that exist outside the capitalist economy, diagnosed the nature of the barrier that keeps them out, and proposed techniques of assetisation for bringing them in. The ILD worked in several countries in Asia, Africa, and Latin America devising procedures to enable what is trapped outside capitalism to be brought in. The technical design and implementation of these mechanisms became one of the most popular solutions to the problem of economic development in the Global South.

A Very Great Book

Presenting the findings from research in five countries, de Soto argued in *The Mystery of Capital* that the main reason why most countries outside the West had failed to emulate the West's economic development is that a large amount of their wealth lies outside the formal economy.[13] It is trapped in forms that cannot enter the market, and therefore cannot be assetised or capitalised to create further wealth. Described as 'dead capital', this wealth consists principally of land and housing. Most people in non-Western countries live in housing whose ownership is not formally registered with the state and thus lack proper title to their property. These people, said de Soto, 'are outside the global economy, are in fact outside the market economy, are certainly outside the capitalist economy'.[14] Because their land and houses are outside the economy, they are unable to use them as collateral to borrow funds. Their assets are locked in the material form of their homes and cannot be transformed into cash or credit. Credit is 'live' capital that can be accumulated, invested in business ventures, and turned into further income.

Live capital, according to de Soto's simple schema, is created by devising techniques of representation. Representations of material assets transform their value into abstract forms, which can live an 'invisible, parallel life' alongside their physical existence. The West has invented procedures to create these invisible forms. Individuals in the West can unlock the assets accumulated in physical property, transforming material wealth into abstract capital. Using a house as collateral for loans, wrote de Soto, is an important source of credit for launching small businesses, providing a large reserve of funds to stimulate economic growth. The most important difference between successful capitalist economies and the rest of the world lies not in the wealth they possess, he argued, but in how that wealth is held. The rest of the world holds its assets in 'defective forms'. The absence of property title and the mechanism of credit it enables are the principal reason for the failure of capitalist development outside the West.[15]

The solution to global poverty, it follows, is to construct in every country a simple apparatus of representation that will transform dead capital into live assets. The machinery consists of neighbourhood-based programmes to enable people to register ownership of their property and simplified rules for using the property as collateral for obtaining credit. De Soto organised wide support for these proposals. In 1994, he persuaded the government of Peru to launch a property titling programme under the ILD's direction that by 2002 gave formal title to 1.2 million urban households.[16] Numerous other countries subsequently recruited the ILD as a consultant, to measure the extent of their countries' untitled assets and devise mechanisms for moving them into the market.

De Soto's researchers found their most impressive results in Egypt. They estimated that as much as 92 per cent of the country's housing was held in defective form, representing $248 billion in underused assets. In one of their most widely repeated statistics, they said that this dead capital represented more than fifty-five times the amount of all the direct foreign investment ever recorded in modern Egypt, including constructing the Suez Canal in the nineteenth century and building the Aswan High Dam in the twentieth.[17]

Some of the reasons for the appeal of these programmes are easy to see. Compared with other diagnoses of the problems of development, de Soto offered a much more positive account of the resources and potential of the majority of the people of the Global South. His earlier book, *The Other Path*, a study of the informal economic sector in Peru, argued that unregulated economic activity was more dynamic and more efficient than the over-regulated formal economy. The size of the informal sector was not an indication of backwardness but a rational response to excessive bureaucratic regulation of formal economic activities.[18] *The Mystery of Capital* showed that the poor possessed not just entrepreneurial skills but also assets. The failure fully to develop these assets was not the fault of those who owned them and therefore could not be blamed on the backward or traditional culture of the poor. It was the fault of the formal sector. Governments in the Global South had failed to introduce the legal arrangements and financial mechanisms that were the unnoticed secret, de Soto argued, of the success of capitalist development in the West. The development of modern property rights during the previous two centuries was so successful that the West now takes this apparatus for granted and fails to notice its absence or incompleteness in most countries outside the West.

These arguments were useful alternatives to popular ideas about the poor as people incapacitated by their own traditions or culture. They presented the poor as competent economic agents who needed to acquire only the proper technical equipment to be brought into the market. But the success of de Soto's arguments was due to something more. They provided a way to bring the poor into the arguments and programmes of neoliberalism.

De Soto's proposals were developed and circulated within the network of political agencies and financial resources of the Euro-American neoliberal movement. From its beginnings in the late 1940s as a small association of conservative economists and political theorists supported by corporate benefactors, with the Austrian-born British economist Friedrich Hayek (whom we met in chapter 4) as their most influential thinker and organiser, the neoliberal movement grew over the following decades into a transatlantic network of think tanks, academic economists, and policy makers, funded by private foundations and corporations, that came to play an influential role in the shaping of government policy in the West.[19]

In 1981, Antony Fisher, a leading organiser and benefactor of the movement, established the Atlas Foundation for Economic Research, based near Washington, DC, to coordinate the growing network of think tanks and launch 'a concerted effort', in the words of a letter of support from Hayek, 'to create similar institutes all over the world'.[20] Hayek wrote the letter after returning from a trip to Peru, where he spoke at a conference funded by German neoliberals and organised by Hernando de Soto.[21] At Hayek's urging, the Atlas Foundation helped de Soto set up the ILD. Atlas provided financial support and trained de Soto and his collaborators in the economic ideas of neoliberalism and the techniques it had developed for political organisation and advocacy.[22]

The support from Hayek and the neoliberal movement enabled the ILD to carry out its research into property rights and the poor, published in *The Other Path* and *The Mystery of Capital*, and to introduce its first property titling programmes in Peru. The programmes failed to have any effect on household poverty. Other benefits that were supposed to compensate for this failure were equally illusory.[23] Yet the failures did not prove an obstacle to de Soto's success. Neoliberal writers and think tanks gave his books favourable reviews and recognised the author and his organisation with well-publicised awards. *The Other Path* carried endorsements from two former US presidents and received a prize from the Atlas Foundation. *The Mystery of Capital* received endorsements from Margaret Thatcher, Jeane Kirkpatrick, Thomas Friedman, Francis Fukuyama, and William F. Buckley Jr.[24] De Soto received the Milton Friedman Prize for Advancing Liberty from the CATO Institute, the Compass Award for Strategic Direction from *Forbes Magazine*, the Adam Smith Award, the Goldwater Award, and several other prizes.[25] At the same time, the populist tone of his arguments on behalf of the poor persuaded others to see him as a progressive. The ILD reported on its website that even the left-of-centre British journal *The New Statesman* had written that de Soto's work 'has put him in the pantheon of great progressive intellectuals of our age'. Although this was not what the journal wrote, de Soto managed to claim supporters well beyond the limits of the neoliberal movement.[26]

The ILD's work also benefited from the neoliberal networks connecting it to academic economics. De Soto was a practitioner, not a scholar, producing research in a think tank rather than a university. Academic economists praised his work not as a contribution to academic theory but as a practical demonstration of neoclassical economic truths about property rights. Two Nobel laureates in economics, Ronald Coase and Milton Friedman, endorsed *The Mystery of Capital*. It is 'a very great book', wrote Coase: 'Powerful and completely convincing'.[27] This positive reception enabled its arguments to flow back into academic economics. De Soto's work became the basis for further university research and provided the subject matter for teaching materials in economics, endorsed by economists such as Friedman, again, and Douglass North.[28]

Despite the continuing absence of reliable evidence that his ideas worked, de Soto came to be described as one of the world's leading experts on development. The World Bank funded the ILD titling programme in Peru, the International Monetary Fund (IMF) promoted its ideas, the United States Agency for International Development provided the think tank with continuous funding for its overseas operations, and the International Labor Organization appointed de Soto to its World Commission on the Social Dimension of Globalization.[29] In September 2005, the United Nations Development Programme (UNDP) supported the establishing of the High Level Commission on the Legal Empowerment of the Poor, cochaired by de Soto and Madeleine Albright, the former US Secretary of State, with the goal of generating further international support for de Soto's political programme.[30]

The success of *The Mystery of Capital* and its author's recognition, manufactured by de Soto's public relations efforts with the help of the

Euro-American neoliberal movement, enabled the ILD to transform its own activities into a form of property. In marketing his property rights programme to governments around the world, de Soto's method was to bypass government ministries and local development agencies and seek authorisation and support directly from a country's head of state. This enabled the ILD to win exclusive government consulting contracts, typically awarded without the normal request for competitive bids from rival consulting firms.

In Tanzania, for example, de Soto met the country's president when both of them were appointed to the World Commission on Globalization. He used this connection to obtain a no-bid government contract, even though other development agencies had been working on the reform of Tanzanian property rights for many years. The ILD justified its exclusive contract on the grounds that its method of ending global poverty, explained in *The Mystery of Capital*, was unique and could not be combined with existing programmes. Signed in November 2004, the contract included a requirement that the Tanzanian government 'respect ILD's intellectual property rights over the methodology and techniques it provides'.[31] Explaining why other development agencies had not been allowed to bid for the contract, the executive director of the ILD, Manuel Mayorga, said that

> in this particular area, we invented the wheel. Understanding the gap between the formal legal and extralegal sectors, analysing how these two parallel sectors operate, evaluating their problems . . . quantifying their economic effects, and how they might be integrated under one rule of law creating a modern, productive economy is the area that the ILD has pioneered.[32]

Thus de Soto's claim to ownership of knowledge about markets – his assetisation of that knowledge – became a means to the accumulation of capital. As always with assetisation, the accumulation depended on protecting his own activities from competition in the market.

In Egypt, the ILD gained the backing of Gamal Mubarak, the son of the country's president and the wealthiest and most powerful of a younger faction among large Egyptian entrepreneurs seeking to strengthen the powers of private property as a means of undoing an earlier generation of social reforms. This group used de Soto's arguments to help push new economic measures through parliament, including a mortgage law, a property titling programme, and new rules for licensing small businesses. The ILD helped draft the laws, along with a public relations campaign to win political support.[33] As in Peru, the outcome of the legislation was disappointing, but the publicity campaign was an immediate success. *Forbes Magazine* published a story in February 2004 predicting that the country's new property laws would 'dramatically transform its economy into a wealth-creating, wealth-distributing dynamo that will lead millions of Egyptians into a vibrant, increasingly democratic middle class'. The country was poised to become 'an economic miracle rivaling Ireland or Hong Kong', the magazine's editor wrote – adding that in doing so 'Egypt will deal a devastating blow to global terrorism'.[34]

Since Egypt provided the most dramatic instance of the defects that de Soto diagnoses, and overcoming these defects was said to promise an economic miracle, it is worth looking at the evidence from Egypt. What can it tell us about the force of de Soto's arguments? How are these ideas, despite the lack of evidence to support them, actually put to work?

Outside the Market

The first thing to note about Egypt is that the remedy the ILD proposed had been tried before. One hundred and fifty years earlier, beginning in the 1850s, a series of laws introduced a modern system of private property, implemented with its own courts, property registers, and mechanisms of enforcement. In most cases, these procedures recognised existing claims to the land under Ottoman and local law, but they also made possible mortgages for the acquisition and transfer of land, and as a source of credit for those who wished to use their property as collateral.[35] As we saw in chapters 1 and 2, there followed a 'Klondike on the Nile'.[36] Local landowners and European creditors extended loans for new irrigation schemes, land reclamation, and steam-powered pumping equipment in the countryside and for housing and modern infrastructure in the cities. By the turn of the twentieth century, the Egyptian stock market, whose largest shareholdings were in mortgage companies and property development, was one of the most active in the world.

The result was a disaster. In rural Egypt, small farmers faced rapidly rising prices. Tax payments increased sharply, to cover mortgage payments on the estates of the ruling family. A cattle plague in 1861 and 1863–4 killed most of the country's draft animals, causing further difficulties. The plague returned, as we saw in chapter 2, in the 1870s and

1880s. To obtain loans to survive such crises, farmers now had to mortgage their own land. Creditors were able to use the new powers of foreclosure to seize the assets of those unable to keep up debt payments. Farmers described the courts that enforced foreclosure decisions as 'a machine for transferring the land' from small farmers to the wealthy. Creditors could take possession of not only the fields but also draught animals and ploughs and could seize or demolish debtors' houses. The level of debt made it increasingly difficult for small farmers to erect waterwheels, a contemporary report tells us, on which the production of staple food depended.[37]

The disaster suffered by individual owners was repeated on a larger scale, using the same new arrangements of property and secured loans. As we saw in earlier chapters, the machinery of assetisation, debt, and foreclosure provided the leverage for colonial occupation. In 1874 a global economic depression began, brought on by the crisis of speculative European banking, including loans to the Ottoman viceroy in Cairo, for which he had pledged his large cotton and sugar cane estates as collateral. As the depression caused the price of cotton and sugar to fall, and made further credit unavailable, the government was unable to postpone the financial collapse. In 1876, the British and French banking houses established a Debt Commission in Cairo, which took control of the country's finances and used the new courts to take possession of the viceroy's estates. When he resisted the takeover, the British and French governments had him deposed in favour of his son. The latter began to lose power to a popular constitutionalist movement, led by junior army officers and disaffected notables. In 1882 the British invaded the country, established a military occupation to eradicate the populist movement, and reasserted European control over finances and mortgaged property, including the extensive viceregal estates.[38]

The British attempted to consolidate the new arrangements of private property. They carried out a land survey, mapping the ownership of every acre of agricultural land in the country and recording the information in property registers. Completed in 1907, the survey and property register were more comprehensive than anything known in that period in Britain.[39] At the same time the colonial authorities were forced to introduce measures to limit the application of the new property rules. In 1912, in an attempt to slow down the rate at which villagers were losing their land and their homes to creditors, they introduced the Five Feddan Law.[40] Modelled on measures the British had introduced to deal with similar problems in Punjab in northwest India, the law prevented creditors seizing from small farmers their last five acres (feddans) of land, their essential farm tools, two draft animals, and their house. Nevertheless, by the 1920s, it was estimated that more than one-third of the agricultural population in the Nile Delta had become landless.[41] The government introduced rent control laws during the Second World War, to combat steep price rises and profiteering. Similar measures protected tenants in urban housing. Following the 1952 coup d'état that ended the power of the largest landowners, Egypt adopted a land reform that gave more extensive security to tenant farmers and similar protections to urban property dwellers.

The Egyptian experience raises a number of questions about the ILD's more contemporary proposals. First, it is not the case that Europeans took their success in creating private property at home so much for granted that they overlooked its introduction in other parts of the world. If anything, it was the other way round. Much of modern European thinking about the nature and importance of private property was worked out in the context of colonial rule, whether by the British in Ireland, India, Egypt, and Australia or the French in North Africa and other territories.[42] What was different in the colonies was the local ability to slow down or prevent the process of assetisation.

Second, de Soto overlooked the fact that even the success at home was a qualified one. It was achieved at great cost. Those who consolidated formal property rights in eighteenth- and nineteenth-century Britain, for example, prospered. But a large majority of the population did not. The enclosure of lands held in common by rural communities and the creation of large private estates forced hundreds of thousands of households into poverty and drove them from the countryside. Across Europe, millions were forced into emigration, moving to the cities and, in even greater numbers, abroad. They moved to North and South America, Australia, northern and southern Africa, and other corners of the world. In contrast to Egypt and India, where the local population had the power and the political organisation to resist the complete loss of their land to European settlement and land acquisition, in settler colonies like North America and Australia the native population was pushed aside and gradually eliminated, to enable the expansion and formalisation of European property ownership.

Curiously, de Soto's discussion of this global history of private property examined only one case in detail – and most of the detail is missing. He describes the incorporation of informal owners into the formal property regime of the United States as the country expanded its colonisation westward, in particular the acceptance of the property claims established by squatters' and miners' associations in California. This case exemplifies for de Soto the successful process of turning the 'dead' capital of informal property into the live capital of a legal order recognised and regulated by the state. He disregards the fact that the problem of 'dead' capital was itself partly created by formalisation. Many settlers had been forced to seize land as squatters because land speculators had taken control of the better land just beyond the most recent legal land settlements. He ignores the fact that the claims of European squatters and miners in California, following the seizure of the territory from Mexico, were established by denying any equivalent rights to most of the native American and Mexican populations they pushed aside. And, by using as evidence the example of settler colonialism, he overlooks the larger problems of dispossession and eviction back in Europe to which settler colonialism was a response. The case of nineteenth-century California illustrates a success that contemporary reformers in developing countries might emulate only if one strips it of every layer of its history.

In contrast to the achievements of settler colonialism of North America, and the disaster it entailed for native communities, European colonialism in countries like Egypt and India was less successful. It encountered populations whose land tenure arrangements had been established over several centuries, whose claims to the land were more difficult to eradicate through assetisation and foreclosure, and whose physical presence could not be eliminated. Even in Palestine, where Zionist settlers had hoped to use assetisation, land purchase, foreclosure, and eviction to take over the country by stealth, the native population largely resisted, causing the settler movement, as we noted in the last chapter, to turn to dispossession by military force. In other parts of the world, although the attempt to establish European property regimes helped trigger a series of 'late Victorian holocausts' in several countries, in which thousands and, in some cases, millions of people lost their lives, most populations were able to resist complete dispossession and in the twentieth century managed to limit and even roll back the implementation of the powers of private property.[43]

While *The Mystery of Capital* ignores the history of dispossession that accompanied the development of formal property rights, de Soto is aware that the issues at stake in the present hinge on questions of security. He says that the problem with informal property arrangements is that they leave people too secure, with their rights against dispossession too strongly protected. Outside the West, he says, 'the law and official agencies are trapped by early colonial and Roman law, which tilt towards protecting ownership'.[44] One of the central elements of the reforms that he advocates, as Europeans demanded in Egypt in the nineteenth century, is the removal of this protection. Land and housing can only be assetised, serving as collateral for the extension of credit, if rules and powers are arranged to enable creditors to seize the property of debtors who default.

Although it acknowledges the need for these new powers, *The Mystery of Capital* fails to explore the kinds of consequence to which they can lead. Leaving aside the colonial history outlined above, even today these consequences are not hard to observe. Take the example of farming in the United States. Most of the white squatters whose land claims were incorporated into the formal property system of the United States during its nineteenth-century expansion lost their property in the course of twentieth century. Small and later medium farm owners were gradually eliminated. Between the 1930s and the 1980s, two-thirds of American farms went out of business and lost their property. This process of dispossession is entirely overlooked in the promotion of de Soto's ideas. 'In rich countries', notes an article in *The Economist* discussing the ILD's property titling proposals for poor countries, 'if a farmer wants to invest in better seeds or bigger tractors, he can probably borrow the necessary cash using his land as security'.[45] This is not usually a question of what the farmer wants, more an issue of necessity. By the 1980s, more than three-quarters of the American farm supply industry was controlled by four firms, and similar monopolies governed trade in most farm products. Squeezed by monopolies on both sides, farmers were obliged to use their property as collateral to borrow money to stay in business. The mechanism of foreclosure gradually forced them off the land.[46] The land was taken over by less efficient but more powerful corporate operations, which used their political resources to gain ever-larger government subsidies. The countries of the Global South could never afford to allow the mechanisms of private property, collateral, and credit to produce a disaster on this scale.

This outline of events suggests a different way of understanding the question of why so many people in countries such as Egypt hold land or housing that has not been incorporated into a formal property system, and the larger question of whether we should think of these assets as existing 'outside' the market economy, or outside capitalism. These arrangements have not arisen because people in such countries are ignorant of private property or because the West overlooked the need to export the system abroad. Unlike the victims of property formalisation and assetisation in Europe and North America, rural and urban populations were able to delay, divert, or limit the extent of its implementation. Their position 'outside' these property mechanisms was the outcome of a long, often violent, but ultimately relatively successful objection to undergoing the dispossession inflicted on the rural populations of Europe and North America or the complete marginalisation or elimination of native populations in parts of the world where settler colonialism was carried through.

It is curious to describe the outcome of these events as the existence of a world outside the market. For the ILD, this outside is a place shaped by ignorance and lack of access. It is a place to which the benefits of formal arrangements were never extended. It would be more instructive to think of the outcome as a frontier. People have seen formal property arrangements advance, recede, and advance again. They have sometimes evaded them and sometimes been overtaken by them. They have worked against them from the inside and sometimes turned them to local advantage. The frontier has been a battleground. It is not a thin line marking the barrier between market and non-market, or formal and informal, or assetised and non-assetised. It is a terrain of warfare spread across the entire space of the market, the entire length of what is called the history of capitalism. If it is an outside, then it is an outside found everywhere, a scene of battle that seems to define every point at which the formal or the capitalist can be identified. It is therefore a zone of 'inclusive exclusion', since what is declared to be outside the market already plays a role within it, through the declaration of exclusion and the continuous battles over its inclusion.[47]

Homely Assets

The proposals for transforming the Global South through property titling and assetisation envision three distinct steps for including what is declared to be excluded: turning property into collateral, collateral into credit, and credit into increased income.[48] What is the evidence that property titling unlocks credit and that the newly available credit has this set of consequences? What is the mechanism that turns assets 'outside' the market into financial prosperity within?

The Mystery of Capital has little to say on this question. The main argument presented in the book is a passing reference to the idea that, in the United States, many people launch small businesses by borrowing funds using their homes as collateral. How significant is this source of credit? De Soto cites no evidence for the claim, and the data available on small business credit in the United States does not offer much support. Among very small businesses, 40 per cent borrow no funds at all and the most common source of loans for those that do is a personal credit card.[49]

One reason why assetising and mortgaging one's home may play a less significant role in financing business investments than the ILD believes is that, in the West, most people do not have homes to mortgage. While home ownership rates range from 40 per cent in Germany and Sweden to 54 per cent in France and as high as 68 per cent in the United States and Britain, these figures include homes that are mortgaged and not yet paid for, including those whose occupants owe more than the value of the house.[50] Historically, rates of home ownership were even lower. In Britain, for example, before 1914, as much as 90 per cent of housing was privately rented.[51] For a majority of households in the West, using one's home as collateral for a business loan is not an option available.

In the Global South, rates of home ownership are often much higher, above 80 per cent in India and Mexico, for example.[52] Because of the minimal role of housing credit in these countries, most of these homes are owned outright. *The Mystery of Capital* argues that low-income homeowners in the South possess assets that, if they lived in the North, they would transform into live capital. In fact, they possess assets that, if they lived in the North, in most cases they would not own.

De Soto describes the informal property systems of the Global South as a defective form of capital. The figures on comparative home-ownership rates suggest that they might, instead, be seen as a significant achievement. They enable millions of households to occupy their own dwellings, free of mortgage debt and the threat of foreclosure, even though their incomes are a fraction of those of households in the West, where far fewer can afford to own property.

The advantages of informal housing have been recognised since at least the 1940s, when people like the Egyptian architect Hassan Fathy promoted vernacular housing as an alternative to the plans of postwar governments to pay commercial contractors to build inhospitable and relatively expensive blocks of concrete apartments.[53] By the 1970s, it was widely accepted among development practitioners (but not by governments with ties to large contractors) that informal housing had significant advantages, especially if measures were taken to overcome some of the drawbacks of informality. These drawbacks might include inadequate services, poor site layout, and insecure tenure – but not an inability to use the property as collateral.

Evidence drawn from observations of a village in southern Egypt in 2004 can illustrate how its advantages still operated in recent years.[54] First, informal housing is often self-built. In the village, houses range from simple huts built of palm stalks plastered with mud, occupied by a few of the poorest households, to substantial houses made of mud brick (adobe), to four- and five-storey structures built with reinforced concrete frames and fired brick, sometimes accommodating a separate household of the same extended family on each floor. Many households build their own dwellings, often with the help of male relatives. Others hire a builder, who may be a relative or neighbour and usually employs help from the household. Expenses are also reduced by using building materials available locally, sometimes at no cost. Bricks are made from the earth, concrete from local sand, and ceiling joists and laths from the trunks and branches of date palms (usually the male trees, which do not produce fruit).[55] Windows, doors, and furniture are made by local carpenters, while cheaper furniture is assembled from palm stalks by the men who trim the trees. Recycled oil cans are hammered together to make doors for the simplest houses. To install electrical and plumbing systems some households employ electricians and plumbers, but most are able to install these systems themselves or with the help of a relative. This mixture of self-building, the services of relatives, and local crafts and trades keeps housing affordable while also supporting a significant local construction industry.

Other advantages of informal, largely self-built housing include the option to build incrementally, adding extra rooms, floors, or fixtures as needs develop and income arrives, or even to rebuild the entire dwelling in more durable materials later on; the use of locally appropriate methods and building supplies; and the ability to design the layout of the house to suit the occupants' needs and to alter it as those needs change – converting a front room into a small store, for example, or a rear yard into a workshop.

The success of self-built rural housing depends on the fact that it seldom requires the purchase of land, which would be beyond the reach of most families. Space for new housing is found by pulling down older, less substantial structures or those in need of renewal, by adding to existing houses, often by reducing the land previously used for domestic animals, or by filling the gaps between dwellings. Those who own no land on which to build look for small plots of vacant land. Some find patches of higher, uncultivable ground, where the government recognises squatter housing and charges a small rent. Others find the unused margin of their own or a relative's plot of agricultural land, typically along the edges of canals and roads or at the border where the cultivation meets the desert. The completion of the Aswan High Dam in 1971 made it possible to build on agricultural land, much of which was previously protected from building by the annual flooding of the fields for irrigation. A law passed in 1983 to ban construction on agricultural land, and reinforced by a military order of 1996, reduced the loss of farmland but failed to eliminate the practice.[56] Following the revolutionary uprisings of 2011, there was a surge in unregulated building on agricultural land. Further legislation in 2018 and harsher penalties introduced in 2022 were unable to prevent the loss of farmland.[57] Most new construction takes place within existing village boundaries or on unused land. However, the shortage of unused plots as well as the inability of the government to facilitate self-build projects along the desert margin and on military property and other underutilised land exacerbates the problem of access to suitable land.[58]

The housing built in this way may be legal, semi-legal (infringing building regulations, for example), or illegal (such as housing built on agricultural land). All owners would prefer their homes to be legal, but not because they plan to use them as collateral. Legal housing does not carry the expense of the frequent summonses and fines imposed on unlawful construction and may be easier to connect to the water and electricity supply. The latter has been less of a problem since the 1990s, when a series of government decrees ruled that houses built in violation of laws banning construction on agricultural land or of building regulations were nevertheless eligible for connection to utilities.[59]

Legality, however, also has drawbacks. The main problem is that it makes land unaffordable. In 2004, agricultural land in the village sold for about E£1,000 per qirat.[60] Occasionally, the Ministry of Agriculture would remove a plot of land from the cultivated area (the *zimam*) of the village; for example, because it was surrounded by buildings and no longer received enough sunlight for cultivation. This made it legal to build on the plot. Legalisation increased the value of land by a factor of at least ten, to more than E£10,000 per qirat. In more valuable locations, such as agricultural land on the edge of a major town, legalisation could increase the value by a factor of thirty or forty.

Urban housing differs in several obvious ways from rural, yet many of the same principles apply. In Egypt, the rebuilding of the city of Ismailiya following the withdrawal of the Israeli army from the Suez Canal zone after the 1967–74 occupation included a successful project to demonstrate the advantages of 'site and service' programmes to facilitate self-building while avoiding the problems of inadequate services, poor site layout, and insecure tenure.[61] The self-funding project made available development sites laid out with building tracts and services, along with small loans and supplies of low-cost building materials. The recipients of the plots were given long periods to repay the unimproved value of the land. Provisions requiring immediate construction, owner occupancy, and delayed acquisition of title discouraged speculators and assetisation. The project recognised that offering formal title to the land can hamper the provision of low-cost housing, as it encourages property speculators, by assetising housing, to bid up its price. Conversely, the failure to discourage assetisation, through requirements such as owner occupancy, makes housing less affordable and less available. In 2016, it was estimated that 6.4 million housing units in urban Egypt were unused.[62]

Boundaries of Property

The ILD acknowledges many of the advantages of informal housing. Yet its proposals insist that formalisation through property titling is the only route to economic development. The insistence is based on an argument not about the relative benefits of the informal but about access to collateral and credit. Only a formal property system, it is claimed, can release the dead capital held in informal housing. It is time to turn more specifically to this argument – an argument about development through capitalisation.

Once again, the case of rural Egypt provides evidence with which to begin. Although the proposals the ILD drew up for Egypt were focused on the formalisation of urban housing, over the preceding decade some of the reforms they advocated had already been carried out in the countryside. The reforms (actually, the undoing of earlier reforms that had set limits on the agrarian property market) included measures allowing farmers to use their land as collateral for loans from the Agricultural

Development Bank, and a gradual lifting of rent controls on agricultural land, culminating in the abolition of all security of tenure in 2002. After that date, landowners were free to renegotiate rents or even evict their tenants – households that in most cases had been farming the land at nominal rents since before the land reform of 1952. Taken together, the two reforms seemed to promise exactly what *The Mystery of Capital* described, enabling owners to free up the capital locked in their land, either by pledging it as collateral for loans or renting it at increased levels of income.

There was insufficient evidence to draw any general conclusions about the impact of the reforms, in part because circumstances vary greatly from one village to another. In some parts of the country, for instance, entire hamlets farmed land rented from a single owner. In these cases, the abolition of security of tenure in 2002 transformed the whole community from effective possessors of the land into landless farmers, who either lost access to the land or were forced to rent at much higher rates. In the village about which I have been writing, the rental plots were much more widely dispersed. This tended to favour the tenants. One group of three households, all brothers, together farmed four acres of land, three acres of which they owned and one of which they rented. The owners of the rented acre lived in a neighbouring hamlet of the same village but were the second-generation heirs of the original owner and had no idea of the exact location of their land. It lay somewhere in the middle of the other three acres, but, since individual plots in Egypt have no hedgerows or other permanent boundary markers, there was no way to establish its position. The original rental contract specified its position in relation to other plots of land, but these were also unmarked. Even if the owners had managed to locate their plot, they would not have been able to bring irrigation water to it. They had established no rights to use the ditches that carried water across neighbours' fields, and no separate ditch supplied their acre. The owner demanded an increase in annual rent from E£300 (about US$50) to E£1,200. The brothers refused and eventually agreed to pay E£600. If the owners had not been from a nearby hamlet, the brothers would have paid even less.

This case, one of several similar cases that came to light, illustrates some useful points. First, to the extent that the new law increased the rights of property owners, its effect was not to turn dead capital into live. The owners increased their rent, but the increase came from the tenant farmers, whose income declined by E£300 a year. This marked a shift in wealth from the productive labour of farmers to the unproductive labour of a rentier. Second, this transfer of income was limited by the weakness of property records. The inability of the owners of the rented acre to establish fully their claim to the land enabled the tenant farmers to pay a lower rent. The local knowledge of those who farmed the land was more powerful than the incomplete 'abstract' knowledge of property entitlement that de Soto sees as the secret to capitalist development. The incompleteness of such knowledge was a means of limiting the diversion of income away from those who worked and knew the land, into the hands of relative strangers. Formal property systems undermine this local knowledge by linking ownership into what are called more abstract, certainly more distant, forms, which can be accumulated, managed, and made sources of rent by outsiders. Preventing the transfer of this kind of knowledge into the hands of outsiders or property registers was a resource for protecting local income.

This case was not an isolated one. There were numerous similar cases in the village. Together, they are corroborated by all the evidence we have about the accumulation of property and the creation of inequality. The creation of formal legal title and property registration becomes a machinery for assetisation and hence for transferring property from small owners and concentrating it into larger and larger hands. We have already noted the operation of this machinery in Egypt between the mid-nineteenth and mid-twentieth century, and the measures the government was forced to adopt to limit its effectiveness. Evidence from a land titling programme in rural Paraguay suggests that gains go only to relatively wealthy producers.[63] Further evidence comes from the impact on the village of the other part of the new agrarian laws – the new powers of the Agricultural Development Bank to take land as collateral for loans, and to seize the land if the borrower defaults. Within a few years there were several cases of farmers losing their land by this mechanism, and in every case the defaulters were small owners. The only people in a position to buy land that became available in this way were the small handful of large owners, no more than a dozen households out of several thousand. Other studies show other adverse consequences of using land as collateral. When households use their land as collateral they have to add to their calculations relatively inefficient insurance measures that help them deal with poor harvests or other unexpected shocks, by keeping to low-risk, low-return crops – sugar cane in this region of Egypt.[64] Given these drawbacks, it is more efficient to build up and draw upon forms of credit whose loss is less catastrophic to the household, such as producing and storing grain and raising animals.

A further problem with using property as collateral for credit is that the poor are seldom able to recover from asset losses. They are often forced to sell property at low prices (during a recession or after a bad harvest), when there is little effective demand, and have to buy back in normal times, when prices are higher.[65] Historically, such distress sales play an important role in the concentration of property ownership, and are connected with the loss of local ways of managing risk.[66]

What about the alleged benefits of being able to use property as collateral for loans? The evidence available shows that there is little or no positive impact. If those with informal property seek title, it is not to risk it in taking out loans. The titling programme the ILD itself devised and managed in Peru, one of the largest, demonstrated this clearly. Four separate studies of the programme found that it had no discernible effect on the supply of business credit.[67] As one study concluded, 'loan acceptance rates of both standard commercial banks and informal lenders are unaffected by residential ownership status'.[68] A large property titling programme in Thailand was also found to have no effect on the likelihood of receiving bank loans.[69] Other studies of rural titling programmes have found that 'the title might make it easier for large producers to access credit but would not make small landowners creditworthy, a situation that would deepen preexisting inequalities'.[70] Titling programmes can have adverse consequences in other ways as well. Women may lose claims they have under informal property systems, in situations where men deal with the formal system. Titling on demand 'has often had disastrous consequences for the poor because individuals with good political connections can often bypass the land rights of indigenous people, women, or other vulnerable groups'.[71] Research, publications, and conferences organised by development economists at the World Bank brought together the evidence from dozens of case studies indicating that the ILD's proposals would not work.[72]

We are no longer dealing here with a simple image of dead capital lying outside the market, beyond the boundary of the formal economy. We have a different picture, in which this boundary turns into a terrain of negotiations, the claiming of rights, relations of power, attempts at encroachment and exclusion. Rather than a problem of transferring assets from outside the boundary to inside, there are rearrangements of power, inequality, and poverty at stake.

Having to Lose

It is time to consider, then, not the mystery of capital, but the mystery of *The Mystery of Capital*. Its arguments appear to exist in defective form, ignoring historical experience and unsupported by any of the available contemporary evidence. What is the mysterious process that transforms such defective analysis into live political projects?

There appears to be no evidence to support the ILD's argument for property titling as the key to unlocking the problem of capitalist development outside the West. There exists a large amount of evidence to indicate that this will do nothing to improve the situation, and a strong case can be made that for the majority of the population it will make things worse for their futures. These studies provide the kind of detailed evaluations of the success and failure of particular projects that is entirely missing from the ILD's sweeping proposals for global transformation. Given this evidence, where do such arguments acquire their power? How do they come to circulate so widely?

The marketing of de Soto's project through the political networks of neoliberalism, described earlier, provides part of the answer, as does de Soto's claim to proprietary techniques for solving the mystery of capitalism. But these alone do not explain the rapid adoption of the particular programme de Soto advocates, which is one of several kinds of neoliberal political reform. The reports of the World Bank, USAID, and other agencies that have funded de Soto's programme suggest that what distinguishes the programme from other neoliberal reforms is that it appears to be extraordinarily cost-effective.[73] It seems to offer something for almost nothing. The problem of the Global South, we are told, is that people are poor. The ILD offers them, not riches, but a means of realising wealth in the future that no one knew they had. How is this effect achieved?

The possibility of getting something for nothing rests on the notion that the market has an 'outside'. The proposal that money can be created out of what presently counts as nothing would make no sense without arrangements whereby things can be said to exist outside the economy. Money is to be created out of nothing by the action of seeming to move resources from the outside to the inside. The production of a place outside the market creates the possibility of assets whose value is both existent and non-existent. The assets exist as material wealth, but not as capital. The act of bringing this defective wealth inside the economy transforms it from something inert into something active, from death to life.

The power of this account of a boundary between an outside and an inside does not arise from the accuracy of its description of economic relations. It comes from the tools and arguments that are made available for establishing this inclusive exclusion and performing the transfer of assets. The ILD helps organise the data on unrecorded assets, identify obstacles to their movement, and specify legal and financial mechanisms needed for this transfer. Its work helps pull together particular alliances of local politicians, international financial institutions, property developers, and even spokespersons for the poor, to carry out the transfer. At the same time, it attempts to silence other kinds of claims and other forms of politics.

Let us consider more closely how the action of moving assets from 'outside' the market to the inside brings capital into being. When owners of irregular housing or land acquire formal title, the value of the property tends to increase. In the village, as noted above, making land legal for building increases its value by a factor of ten. Where informal housing is already built, the increase in value that comes with legalisation is less, but still significant. In Brazil, a property titling programme led to a doubling in the value of land.[74] Another study found an increase of 25 per cent, and other estimates fell between these two figures.[75] These increases are to be expected. As a World Bank study explained in relation to the titling of agricultural land, 'the value of the ability to use unmortgaged land as collateral would be capitalised into land prices'.[76]

In principle, all owners of irregular property benefit from this increase in value. But very few can sell their properties, since that would leave them homeless. In any case, the need to purchase another property would eliminate any gain. Only those holding property not for their own needs but for speculation would benefit. Likewise, for reasons already discussed, only wealthy owners could take the risk of using their dwelling as collateral for a loan and turn its increased value into credit for investments. Over time, titling leads to the concentration of property in the hands of those able to purchase it at the higher values it now commands, and creates speculators, who also benefit from the opportunities for income from the rent that such property now offers.

The increase in property value comes from two sources, neither of which represents 'dead' capital brought to life. In the short term, it comes from speculative investment. Such investment simply draws credit creation away from more productive ventures, exacerbating broader problems caused by the lack of investment in activities that create employment. But the bulk of any increase in property value is realised only in the longer term, when the next generation of individuals seeks housing. The rising cost of land makes future housing more expensive. It now carries the premium of paying the income of speculators and rentiers. So those saving in the present for a house they hope to build in the future must work harder and longer and save more funds.

The outcome is an intergenerational transfer of wealth. Large owners and speculators gain immediately from the increased value of property. Small owners of property see no benefit from increased values. The gains of large owners and speculators are paid for by a future generation of owners, who face the prospect of paying increasing amounts for housing. The 'gain' is a capture from the future.

How is it possible to reorganise this intergenerational transfer of wealth from the poor to the rich so that it appears as free money, as a simple act of turning dead capital into live? An important means of achieving this feat is to begin with the assumption that there is a 'mystery' to capitalism, a secret that is not visible until it is represented in de Soto's text. Even before the secret is revealed and argued over, the promise that there is such a mystery disarms the reader of the text, or the recipient of the ILD's policy proposals. Capitalism is said to have a hidden key or principle. The multiple forms of expropriation, claim, violence, organisation, and resistance whose diversity and motility we have been discussing are imagined to express an underlying principle, whose name is capitalism. Beneath the diversity and violence, we are told, there is some rule or law, whose hidden existence makes every historical case an expression of the same mysterious essential form. The mystery of de Soto's success begins with the notion that there is a mystery. This indeed is part of the alibi of capital.

In de Soto's case, the key to the mystery of capitalism is the figure we have met before, in the first chapter of this book: the entrepreneur. The more people there are with spare assets for entrepreneurship, the more capitalism there will be. Offering large numbers of people the opportunity to use their property as collateral for loans was expected to produce this capitalist transformation.

When successive studies of the ILD's property titling reforms in Peru all showed that they had not had the predicted results, and that there was no increase in the use of property as collateral, economists still found a positive outcome. A study of the impact of the reforms claimed to show that, while newly titled property owners failed to use their houses as collateral, they did realise an important benefit. They began to work harder. The data was said to show an astonishing 40 per cent increase in the number of hours worked outside the house. The author of the study argued that property titling must have freed householders from the need to stay home to defend their property, enabling them to seek more employment in the market. It was this research that was reported in the *New York Times* as evidence of the remarkable impact of the ILD's reforms and gave another leading economist hope for the future of economics.[77]

What made the paper so popular? It offers an image of householders suddenly able to put down the weapons with which they were forced to protect their rights in the lawless world beyond the market. With property titles secured, they head off into the profitable, peaceful world of the market, where hard work will now be rewarded. Unfortunately, however, the evidence for the paper's conclusions suggests that they are completely mistaken.[78]

An Invisible Parallel Life

The persuasiveness of de Soto's arguments can also be connected to crises in the formal property system in Egypt, Peru, and many other countries that had experienced neoliberal economic restructuring programmes in the 1990s. In Cairo, measures to stabilise the currency, reduce government spending, privatise state-owned enterprises, and stimulate private investment had been praised by the IMF for achieving a 'remarkable turnaround in Egypt's macroeconomic fortunes'.[79] In the same period, the government and the courts began to alter the laws protecting commercial and residential tenants, freeing property developers from the constraint of rent controls.[80] The credit creation stimulated by the reforms flowed primarily into real estate, as speculators threw up large luxury and middle-income developments, and into exclusive concessions to provide services, such as mobile phones or McDonald's restaurants, or supply imports of electronics, cars, and other luxury goods. In other words, capitalisation was re-routed wherever possible into sources of rent rather than into productive activity. The share of manufacturing in the economy declined, and no significant efforts were made to increase large-scale employment.[81]

The boom in property speculation and in luxury imports and services ended in a deep recession by the year 2000. The price of luxury apartments dropped by more than half, and property developers found themselves with tens of thousands of unsold apartments.

A solution to these difficulties seemed to lie in the ILD's proposal for a mortgage law.

In 2001, parliament passed the law, as mentioned earlier, following draft legislation and implementation proposals drawn up for the Ministry of the Economy by the ILD and its local partner, the Egyptian Center for Economic Studies.[82] It took a further two years to set up a body to regulate the new industry and to establish the first mortgage company, a subsidiary of the state-owned Housing and Development Bank. The Bank also established a company to survey and register informal property, co-owned with the government survey authority.[83] By 2005, only two mortgage companies had been established, and the number of transactions remained very limited.[84] The reasons for the delays and for the failure to establish additional mortgage companies are instructive. First, there was a problem in raising funds. The plan was to capitalise the industry by selling securities (mortgage bonds) on the Egyptian stock market. But neither the procedures nor the investors were available to do this. If the ILD's arguments were valid, this obstacle would not make sense. Property titling and a mortgage law were supposed to create funds, by unlocking the dead capital of irregular real estate. In reality, there was no dead capital to unlock. Instead, the new mortgage system attempted to draw other investors into real estate, to bail out the speculators. There was little capital available, so the reforms stalled.

The second and more important obstacle, and a reason for the lack of enthusiasm from investors, was said to be a fear that the government would not enforce new provisions giving mortgage companies the power of foreclosure.[85]

De Soto had acknowledged that the new property system was about making property owners less secure. People of the Global South remain 'trapped in the grubby basement of the precapitalist world' not because they have no property, he claimed, but 'because they have no property to lose'.[86] Without the power to evict borrowers who fell into default, the entire project of releasing wealth from dead capital would fail. But a long history of struggle against the powers of foreclosure in the colonial period had made it difficult to use any powers of eviction. One of the main achievements of the 1952 revolution in Egypt had been to consolidate the protection of tenants' rights. Numerous local struggles over the ensuing decades had reinforced a widespread political claim that the state should not be involved in making people homeless. Indeed, following the revolutionary uprisings of 2011, the new Egyptian

Constitution made explicit the right of citizens to 'adequate, safe and healthy housing'.[87] Nothing in the ILD's proposals explains how this long-standing political claim was to be overcome, or how the new forms of coercion on which capitalisation depends were to be put in place and made effective.

In place of an account of the relations of forces involved, de Soto offers a much simpler picture of the procedures required to turn unregistered assets into live capital. He describes the process as one of representation, a concept that plays a central role in his account. Representation is the operation that moves the assets of the poor into 'the market'.

If the poor hold their assets in defective form, the defect is that they are not visible – not visible, that is, to the market. In the West, de Soto explains, property ownership is represented in a document, which is 'the visible sign of a vast hidden process that connects all these assets to the rest of the economy'. Thanks to this process of representation, 'assets can lead an invisible, parallel life alongside their material existence'. Other countries do not have this representational process, and this explains why people have not been able to produce sufficient assets to make capitalism work. Only the West possesses the 'conversion process' needed to decode the capital that material assets invisibly harbour. 'This is the mystery of capital.'[88]

How does representation achieve this conversion? The answer is that, unlike physical assets, representations of assets are set free from material restraints and therefore can be easily moved around, combined, divided, and used to launch business ventures. By uncoupling the economic value of an asset from its physical form, a representation makes the asset fungible.[89]

We should note, first, that this description of the process of representation in the West is misleading. In many Western property systems, including those of the United States and Britain, there was no title document of the sort that de Soto advocates. Historically, the use of such a document was avoided, because it could be outdated or might be deliberately altered and was therefore an invitation to fraud. Representations can be dangerous things, as their use cannot always be controlled. What distinguishes different techniques of representation, as I argue below, is not their abstractness but the degree of their control. In the United States and Britain, property claims were established through the use of title deeds. A deed is not a representation of property ownership, but its performance. It is a legal instrument used to perform the transfer – not of an object (property), but of a right (ownership). To be legally valid, it had to include a term such as 'hereby', indicating that the deed is a performative instrument. And it had to be physically performed, by handing the instrument from the previous owner to the new owner in the presence of a witness. Property titles in Britain and the United States consisted of an accumulation of deeds, marking the successive acts of transfer of rights from one owner to the next.[90]

Even today, attempts to introduce property registers in the way de Soto advocates have been unsuccessful in the United States. Most states use a system of private title insurance, combined with public registration of transactions. These methods tend to be expensive, as they are open to price fixing and the capture of legislative regulation by private interests. But attempts to replace them with property recordation, in which the state records the ownership and transfer of all property, were unpopular because of the long delays these procedures introduced.[91]

It is also misleading to suggest that informal systems of property lack complex systems of representation. We can return briefly to the village discussed earlier. It is remarkable the extent to which villagers are constantly moving resources into and out of a variety of assets. The most common instance is the raising of domestic animals. Almost every household is involved in raising small animals (chickens, ducks, geese, rabbits, and pigeons), many have sheep and goats, and more than half own a cow or a water buffalo.[92] These investments are productive, providing food for the table and making their own offspring. But, when a need for cash arises, a sheep, some chickens, or a calf can be sold. When surplus income arrives, it can be converted into more secure assets, typically domestic animals (but also gold, stored in the form of jewellery). Households also organise rotating credit unions to raise funds for large purchases, such as a new stove or a water buffalo. Selling a large animal can help pay for a small business venture, such as acquiring a sewing machine to take in dressmaking work or setting up a kiosk to sell cigarettes or household items (sugar, soap, light bulbs) to neighbours.

Such kinds of saving and investment have several advantages over using land and housing as collateral for credit. First, they draw savings into productive activity rather than real estate investment. Second, they are typically controlled by women, who are more likely than men to direct income towards the basic needs of children and the household. Third, assets are fungible, but not too fungible. Resources held in the form of domestic animals can be converted to cash when urgent needs arise but are solid enough not to drip away on casual purchases.

What is the difference between these methods of accumulating and circulating assets and those that de Soto sees as the secret to the mystery of capital? The difference does not lie in the presence or absence of representations, or their degree of abstraction. De Soto may be correct that certain forms of wealth, or certain ways of representing assets, become more mobile, travel further distances, and are more easily combined and accumulated. However, this mobility does not derive from the degree of abstraction. It derives from establishing techniques of control, which make it possible to manage assets at a distance and over time and accumulate them in larger quantities.

In the village, most household wealth is held in assets controlled by direct supervision. The wealth is accumulated in domestic animals, the fields outside the house, an irrigation pump adjacent to a relative's house across the fields, or a plot of land in the next hamlet. All this is relatively easy to manage. Large landowners face much greater challenges of supervision. They hire guards to protect stores of grain, use client families to operate irrigation pumps in different areas of landholding, and rely on labour contractors to supply gangs of workers for harvesting. The much larger estates built up in the nineteenth and early twentieth century, dismantled after 1952, were possible only after innovations in methods of monitoring and disciplining large labour forces, based on purpose-built worker housing units in which workers could be locked at night and kept under continuous supervision.[93] The largest properties possessed 'a veritable brigade of employees whose sole occupation was to supervise the workers, continuously and in the closest and most rigorous fashion'.[94] The powers of supervision, including the new courts, bailiffs, prisons, and armed police forces, were all essential to the accumulation of 'representations' of wealth on a new scale.

The accumulation of wealth that de Soto has in mind depends, as I have noted, on the reintroduction of powers of foreclosure and eviction that were abrogated, in the face of popular protest and economic dislocation, in the course of the twentieth century. In other words, rather than a question of creating something more abstract, it is, if anything, a question of something more physical – more extensive powers of eviction and control. The delay in establishing mortgage banks, despite the appropriate legislation and regulatory body, the setting up of loan companies, and a property titling system, comes down to the failure to secure the most physical of powers.

De Soto's plans envisage a vast creation of wealth, by the transformation of so-called dead capital into live capital. In practice, the evidence suggests that this will produce not live capital out of dead, but a transfer of wealth from the less affluent to the more secure, and, in particular, serve to enrich the more prosperous among the present generation at the expense of the future poor.

This is to be accomplished not by moving what is outside the market to the inside, but by deploying this boundary mechanism in new ways. It is not to be accomplished by representing what is currently invisible, but by altering the circulation and management of representations. It is to occur through a reorganisation of forces among a variety of parties. And it requires the creation of mechanisms of compulsion.

The unrecorded and less closely governed forms of socioeconomic life that are described by neoliberal economic reform programmes as lying outside the economy are not in fact outside, but nor are they simply inside. They can be understood as a frontier or border, a status that is neither exterior nor interior to the market. They are partially outside, because these assets cannot be priced by the market, just as the labour of those who control them is not easily available for capitalist enterprise. Economics helps to manage this border, by producing and validating rules and procedures that demarcate certain forms of life as informal or non-market. Yet those forms of life are, in some ways, inside, because the arrangements taken by what is called 'capitalism' or 'the market', such as the ways of earning rent that are the principal means for an elite to reproduce its wealth, are the outcome of a long and continuing interaction with this so-called outside.

De Soto's programme, further promoted by bodies like the UNDP's High Level Commission on Legal Empowerment of the Poor, involves the movement of assets from the outside to the inside. This reordering offers a means to create wealth. But the wealth is not produced by the method de Soto's account implies. The process of property titling and the use of property as collateral bring into being opportunities for speculative gain from the future, for the concentrating of wealth, and for the accumulation of rents. The existing assets of the poor are not the material source of this new wealth. Rather, they are the objects of an inclusive exclusion. Through schemes such as de Soto's, they are turned into the apparatus through which this reorganisation and accumulation is carried out. The poor remain 'outside' this process, for the outcome of a process of property titling and the mortgaging of property is that land and housing become even less affordable to the poor. They are further excluded from opportunities for the accumulation of resources. Yet, at the same time, they are 'inside' the process. Their houses and their future lives must be transformed in order to carry out the capture of this future wealth. The starting point of this transformation is to render these ways of life defective, almost dead, by grasping them in terms of the economic rationality and forms of representation they are said to lack. Since their defectiveness is what makes the accumulation possible, it is an outside on which the so-called inside depends.

A conventional critique of de Soto's proposals would expose the neoliberal fallacies on which they rely for their plausibility and popularity. It would draw attention to the history of battles over property rights that de Soto's writings ignore and his inability to account for the moral claims, financial protections, practical flexibility, and affordability offered, whatever their defects, by alternative property arrangements. Having dismissed his writings as no more than the latest vulgate of the neoliberal canon, one could then reveal the real interests for which his ideas are such a flimsy screen.

The work of neoliberalism, however, does not lie behind a screen. If the fallacies of de Soto's arguments seem surprisingly easy to expose, that is because their effectiveness does not rest on the precision with which they capture the workings of particular property arrangements. The power of economics does not rest on the accuracy or inaccuracy of its representations. De Soto's writings form part of the equipment for neoliberal projects. They provide a means of mobilising certain facts of neoclassical economics in alliance with the planning of development agencies, the resources of property developers, and the political powers of local regimes. They form part of the novel apparatus by which speculators are to be rescued, the poor made to give up their houses, new forms of wealth created, and the impoverishment of the future extended.

7

Infrastructures Work on Time

Infrastructures promise to increase our control over space. Highways, ship canals, railways, power grids, and data networks embody the kinds of terraforming modifications through which, largely over the last century and a half, as we noted in chapter 1, humans have transformed the Earth.[1] By conquering space, in turn, infrastructures increase the command of time, speeding up the rate at which energy, data, goods, and people flow. This acceleration is perhaps the most common way in which we express the experience of capitalist modernity.[2]

However, infrastructures have a different role in relation to capitalisation. These large, Earth-shaping projects have another relationship to time, without which capitalisation, as we know it today, would collapse. Infrastructures are brought into being not to speed things up but to introduce a delay. The virtue of infrastructure is not the acceleration of time, from this perspective, but the ability to place the future further away. This approach changes our understanding of capital and the purposes for which modern humans modify their relations with the Earth. It also enables us to rethink the politics of climate change, the destruction of habitat, and the crisis of species extinction. To consider this, we need to start from a reversal of the way we think about how infrastructures work on time.[3]

The standard way of writing about infrastructure is to start from the question of space and treat time as a consequence. Infrastructures create channels and connections that link separate points, facilitating movement between them. China's Belt and Road Initiative, launched by the Chinese government in 2013 to finance the building of roads, rail lines, seaports, and power stations connecting locations across Asia, Africa, and Europe, exemplifies this association between infrastructure and the command of distance. As the largest of several 'mega-corridors' under construction around the world, its promoters emphasised the interlinking of continents and the interconnection of oceans.[4] Yet this focus on infrastructures as a mode of conquering space obscures what is often most important in building them.

From a common-sense perspective, increased control over time comes as a consequence of the conquest of space. Roads and railways, pipelines and container ports, transmission lines and communication networks all appear to offer the benefit of faster connections. Emblematic of the 'technical improvement' through which we understand the history of capitalism, they promise to improve the rate of flow. Food supplies, workers, electrical power, crude petroleum, news reports, data storing, surveillance, and armed force all depend on purpose-built networks to overcome distance. Each technical improvement and spatial extension of the network can reduce the time it takes to convey a given quantity of the material the line carries.

The claim of faster connections and improved rates of flow supports most economic arguments for the building of large infrastructures. A World Bank study of the Belt and Road Initiative estimated the benefits that the new infrastructure would produce by considering only the increasing speed at which products would move from one country to another.[5] The report echoed an old but newly popular argument among economists. Large investments in infrastructure reduce the time it takes to move goods, producing an improvement that is measured as an increase in human welfare, and contributing to the promise of a better tomorrow.

From this perspective, colonialism can be shown to have improved the life of the colonised. The railways that British entrepreneurs built across colonial India, for example, reduced the time and thus the cost of shipping agricultural products – allowing landowners to raise the price of the main market crops and increase land rents. In a recent study of this process, the rising prices were assumed to measure an increase in welfare.[6] This research, and a companion paper demonstrating the benefits that the building of railroads brought to the United States in the nineteenth century, won their author the John Bates Clark Medal of the American Economic Association, awarded to the person considered to be the best economist under forty. Contributing to the work of economisation on which the development of infrastructure depends, such research is as remarkable for the skill with which it measures the impact of railways on prices, as it is for its obliviousness towards other widely studied effects of their construction not easily measurable via prices, including famine, epidemic disease, impoverishment, ecological destruction, and genocide.[7] The economisation of time is presented as the key element in the benefits brought by such large infrastructure projects. Time and space are linked, so the economisation or compression of the first is naturally associated with the second.

The focus on faster connections allows other consequences of large infrastructures to be treated as side effects. It is this focus that enables the economists' models to measure welfare in terms of prices, and in the case of colonial railways to ignore questions of inequality and famine, or the spread of the cholera vibrio and other pathogens, or, for example, the extensive deforestation that their construction caused.[8] To reduce derailments and delays, wooden railway sleepers (crossties) had to be replaced on average every seven years. By 1860, an estimated 177 million ties were produced each year globally, which required the clearing of 885,000 acres of forest. The area cleared each year had tripled by 1880 and tripled again by 1900. Forests were also destroyed for fuel wood and to expand farming along the new routes served by rail. In India, 70,000 acres of forests were removed each year in the 1860s and 1870s, and double that rate by 1890s, leading to increased flooding and erosion.[9] As with the impact of irrigation works in Egypt, as we saw in chapter 2, such consequences were not just side effects. The new spaces and corridors that infrastructures open allow improved connections for multiple agents, human and nonhuman. They create routes for armies of men and armies of other creatures. The digging of the Suez Canal enabled hundreds of invasive species of marine life to begin to colonise the Mediterranean, just as canals and dams, as we saw, allowed water hyacinth to clog the waterways of Lower Egypt, as it did in the river basins of India, and the snails and nematodes carrying bilharzia and hookworm disease to infest the countryside.[10] Easier connections and faster travel times apply to all those who take advantage of a new route, not just to those whose movements are tracked by freight rates, market prices, and other numbers generated by the building of infrastructures.

Leaving aside what is missing from economists' hubristic accounts of the benefits of improved rates of flow, there is a different way to understand how infrastructures work on time. Infrastructures can produce a delay; and they do this not only in the way we commonly imagine, but as part of the work of capitalisation.

Any improvement in connections brings the possibility of increased delay. This can be an unplanned consequence, examples of which are well known. Opening or improving a route creates its own demand. Traffic that might have moved using other modes switches to the new route, while movements that enjoyed a local circulation take advantage of wider connections. The vehicles using the improved route may become larger and less manoeuvrable. For all these reasons, the delays that the new infrastructure had promised to reduce can be exacerbated.[11] On occasion, the delay created by improvement is dramatic, a phenomenon that might be termed the Ever Given effect. In 2015, the Egyptian government completed a partial enlarging of the Suez Canal, doubling one section of the waterway and widening and deepening other stretches. The maximum tonnage a vessel could carry grew from 210,000 to 240,000 deadweight tons (when the Canal opened in 1869, the maximum was 5,000 tons).[12] Container ships, which at the start of that decade carried up to 10,000 containers, were now built to carry more than 20,000. Larger ships are less manoeuvrable, especially in narrow channels, where the water displaced by the hull can suck the stern of the vessel towards the bank. In March 2021, this 'bank effect' may have contributed to the grounding of the giant container ship Ever Given, which blocked the canal for six days and stranded the global shipment of commodities worth billions of dollars.[13]

Very large infrastructure schemes are commonly associated with another form of unplanned consequence: the delay in completing their construction. In building the largest projects – dams, bridges, canals, tunnels, rail lines, deep-offshore oil fields – long overruns are the norm, typically blamed on technical complexity, changes in scope, unusual geological formations, or extreme climate conditions. Those responsible usually describe these elements as unexpected, but they occur so routinely that they reflect not an absence of knowledge or experience but the self-interested optimism of the promoters and planners.[14] Those who promote the building of infrastructures must also build a picture of a world that is simple enough for the infrastructure to be approved as planned. The unknown and the unexpected are not a frontier to be overcome, but another ignorance produced as a requirement of capitalisation.

Purpose and Price

In contrast to such unintended outcomes, however, a different kind of delay is often the very purpose of an infrastructure project. While the building of infrastructure may be justified by a need to move people or supplies, their principal use is often to provide vehicles for another form of movement: the flow of finance. The promoters of the great transcontinental railways of nineteenth-century America advertised the importance of the lines for shipping grain, cattle, and people, but their value was often as a means to construct financial networks. A main purpose of the railways, as Richard Wright puts it, was to move paper.[15]

Large infrastructure projects move investment paper, as we noted in chapter 1, in the form of corporate bonds, bank loans, and share certificates. They are built not to transport these financial instruments across space, but to provide the means to create them and carry them through time. In this case, creating the delay is not an unintended outcome of the building of infrastructure, but its goal.

Once again, we can distinguish two ways in which this happens. First, the building of infrastructure can be the means to carry out industrial sabotage, by first monopolising a route and then delaying its development or slowing the movement of supplies. For much of the twentieth century, for example, the oil infrastructure of the Middle East was built and run according to this principle. The major oil firms that acquired monopoly rights to the region's petroleum reserves sought to maximise not the flow of oil but the flow of profits. Higher profits came from ensuring that oil was never too readily available. In order to maintain the extraordinary levels of financial gain that could be generated by restricting the supply, oil firms delayed the drilling or completion of wells, blocked the construction or expansion of pipelines, and triggered disputes that allowed the shutting down of pumping stations and refineries.[16] Thorstein Veblen argued early in the twentieth century that these and other methods of sabotage, taking advantage of the ability of large firms to disrupt or disturb the complex interdependencies among industries, had become a principal means of making business profit under the modern 'industrial system'.[17] The term sabotage had first been used, around the same time, to describe the power of workers, in particular those who ran a country's railways, to introduce a minor technical disruption that might trigger the slowing or breakdown of an entire rail line or network. Workers were able to use this new power that came with modern infrastructures and energy systems to campaign for better conditions of work and increased wages. The most profitable use of sabotage, however, was in the hands of large business firms, whose threat or use of sabotage became in many industries a routine method of enrichment.[18] The trope of 'technical improvement' that operates as an alibi of capital distracts us from understanding sabotage.

Second, beyond the profits to be made from practices of sabotage and disruption, the financial goods that infrastructures bring into being enjoy a different relation to time from that of the other materials that infrastructures carry. The investment acquires its value not from the speed that brings things closer together, but from the delay that pushes things into the future. This quality arises from the fact that an infrastructure operates not only as a transportation device, but as a payment mechanism. It is an apparatus for the capture of future payments, in the form of freight charges, passenger fares, port fees, bridge and highway tolls, water rates, fuel costs, electricity bills, property rents, internet subscriptions, streaming payments, and more. The infrastructure manufactures a future flow, not simply of goods or services, but of revenues. The claim to those future revenues is a financial asset that can be made available in the present.

There are several ways to monetise the future claim, but, historically, the most common method has been to organise the infrastructure project as a joint-stock company. The shares in the company, sold to investors, represent claims on the future fees or revenues arising from the flows. While goods and services may often increase in value through faster delivery, the claim upon a future flow of revenue, as we will see, acquires its value through delay.

Any financial instrument can increase in value thanks to the passage of time. In fact, the strange ability to grow in value through the mere extension of time is part of what defines an arrangement as a financial asset. This quality gives what we call finance or capital a particular dependence on infrastructure. The infrastructure project is a device for stretching forwards the passing of time. Not all finance is created through infrastructure projects, but finance always requires an infrastructural process for the construction of postponed payments. Infrastructure is the paradigmatic apparatus for the creation of a delay.

To consider how exactly an infrastructure is built to create delays, we must first reorder the standard terms in which we think about large technical projects, and about capital in general. Infrastructure has shaped our conceptions of space and time, of materiality and money, and of human world-making, impeding us from thinking differently.

If we tend to think of infrastructure in terms of speed rather than postponement, in terms of the quickening of flows rather than their delay, perhaps the reason is that the building of a rail line, port, road system, pipeline, or utility network invites us to separate the avowed purpose from the cost. The proposals and plans for such projects typically present the material purpose first, and the monetary arrangement second. This distinction helps underwrite the optimism, even the hubris, that is a characteristic of infrastructure planning.[19] One part seems material, tangible, and visible, the other financial, intangible, and difficult to visualise. One is the end, the other merely the means.

The distinction between material and financial, real and monetary, visible and intangible, is partly an effect of infrastructure. In practice, each aspect of an infrastructure project is a mixture of both. The goods carried by a shipping line or the energy moved by a utility network are made up of various combinations of matter, force, management, value, and regulation, including quality controls, price mechanisms, and labour costs; likewise, the investment products that a project brings into being depend on the materials and mechanisms that produce such 'paper'.[20] Whether printed as stock certificates or generated in computer networks, financial instruments involve methods of transmission, registration, auction, and, above all, enforcement of future payment. And they cannot exist without the future good or service for which that payment will be made. But, although both the goods and the finance are hybrids of matter and value, of the tangible and the intangible, the building of infrastructures appears to separate them, conjuring a world that seems divided into material goods versus money, ends versus means, purpose versus price, the real versus the financial, the infrastructure versus its cost.

In effecting this seemingly divided world, infrastructural projects also give it a hierarchy. The promoters of new projects usually justify them to the public by material needs, not financial opportunities. The movements and supplies that will flow faster thanks to new infrastructure are presented as a requirement to be met, a necessity for life, a condition of well-being, an improvement, or, as economists call it, an increase in welfare, or 'growth'. The finance – although secretly the only element that actually exists in order to grow, the one factor whose very possibility lies in its own expansion – appears as something secondary. Claiming this alibi, capital pretends to arrive after the fact, to enable needs to be met. In contributing to the effect of a world divided into the real and the financial, infrastructure projects reinforce the apparent primacy of the real or the material, of physical needs and human wants, and place finance as the servant, the subservient element existing only to help meet such needs. They generate a metaphysical effect, the real versus the financial, that is reproduced in theories of finance and economics, and in social science more generally.

Delay of Capital

Let us reverse this order of priority and, at least provisionally, put the question of finance first – although we will later rephrase the question. Why does capital appear to want delay? The answer is that postponement brings capital into being and allows it to multiply. In fact, the delay – the technical ability to push something into the future yet continue to control it – can be understood as the very nature of capital, and of the liquid form of capital known as money.[21]

By convention, as we noted in chapter 3, orthodox economists usually describe money as just a means to facilitate exchange, store value, or record accounts. However, we now know, or rather have rediscovered the knowledge, that while money can serve these purposes, for the most part it comes into being as credit. Credit is an accounting mechanism – that is, both an accounting, and a mechanism or technical apparatus. Through the working of the mechanism, promised future revenues are made accessible in the present. If the creditor can use that promise-to-pay to settle other debts, then the credit account operates as money.[22]

Although the understanding of money as something created as a credit mechanism, rather than simply a means of storing value or facilitating exchange, has recently begun to receive new attention, much of the writing on the topic is focused narrowly on the study of banking and other financial institutions. While there have been important studies of, for example, the legal apparatus that supports the creation and operation of money, there is little exploration of the wider technopolitical apparatuses of credit and capital creation presented here.[23] Money is still approached with an emphasis on trust, social relations, and legal regulation – not as a technopolitical relation. Even the seminal texts proposing that capital is to be understood not through value theory but as a power relation often rely on rather general notions of domination.[24]

They do not usually consider in material and technical terms how a specific apparatus of credit production is constructed.

If an accounting mechanism does not refer merely to practices carried out on paper, or electronically, but to a technopolitical, material apparatus, then the more reliable the apparatus for promising and capturing future revenue, both technically and politically, and the further the mechanism can be extended into the future, the greater the opportunity for the creation of credit money.

This is how to understand the relationship between infrastructure and finance: infrastructures are not simply material investments that happen to require the borrowing of large amounts of money. They are the arrangements through which large amounts of credit money are created – arrangements that, like all money, come into being through an extended material apparatus.

From this perspective, money can itself be approached as an infrastructure. However, we must be careful with this analogy. In *The Wealth of Nations*, Adam Smith compared metallic money to a road or railway, allowing the circulation of goods. Money, he wrote, is like 'a highway, which, while it circulates and carries to market all the grass and corn of the country, produces itself not a single pile of either'. He used this metaphor to explain the role of banks in issuing paper money: 'The judicious operations of banking, by providing, if I may be allowed so violent a metaphor, a sort of waggon-way through the air, enable the country to convert, as it were, a great part of its highways into good pastures and corn-fields.'[25] Waggon-ways were wooden railways carrying horse-drawn wagons, a new infrastructure built extensively in Scotland in Smith's time, mainly to carry coal from collieries to harbours. Smith employed this metaphor from contemporary infrastructure schemes to grapple with the problem of the proliferation of bank credit in the 1760s and 1770s, and, in particular, a banking crisis in 1772 when the collapse of Ayr Bank caused a 'commercial earthquake' across Scotland and threatened Smith's own benefactor, the Duke of Buccleuch, one of the largest landowners in Britain, with bankruptcy. The calamity caused Smith to delay the publication of *The Wealth of Nations* by five years as he rethought the problem of bank-created credit and the need for its regulation by government.[26] Portraying credit money as 'a sort of waggon-way through the air' has remained an inadequate image ever since. Finance is not built out of air, as some ethereal counterpart to real goods. And it is not merely a means to facilitate the circulation of those real materials. It is itself a material infrastructure, and one that exists not merely to facilitate real movement, but as the movement to be facilitated by the proliferation of infrastructures.

If the construction of an apparatus to capture revenue that lies in the future creates the possibility of money, it also gives money its peculiar ability of appearing to 'grow'. In other words, the infrastructure of delay gives rise to the possibility not just of money, in the sense of a claim on a future asset, but of capital: of what appears as the self-expanding power of money. This appearance comes about through processes that take advantage of two features that define the making of the future: it is subject to delay, and its course is not certain.

Those who engineer and acquire control of a set of claims on future payments, such as those promised by a durable infrastructural project, but also something as simple as the promised income from a future sale, may decide to sell a part of those claims or promises in the present, perhaps to settle earlier debts or to support the building of new payment machines; or simply to realise the windfall to which such control may give rise. To repeat a straightforward instance from chapter 3, the owner of a credit note – a promise from someone else to pay a certain sum on a fixed day in the future, in exchange for the delivery of a certain good – might sell that promise-of-payment to a third party. Due to the postponement of the payment into the future, and the element of uncertainty that attaches to any arrangement for governing prospective events, such claims to future revenue can usually be purchased at a reduced price. This makes the future income available at a discount.

Let us recount out briefly how this process works. The claim to a payment of $100 that will be available one year from now might be purchased in the present at a discount, say of 10 per cent, or for the price of $90. If the date of the promised payment can be pushed further into the future, the discount can be compounded each year that the revenue is postponed, multiplying its effect. Thus the claim to a revenue of $100 in seven years' time might be purchased in the present, at the same annual discount rate, for about $50.[27] Provided the apparatus of postponement is politically and technically reliable and the residual uncertainty can be managed, this arrangement allows an investor to purchase the claim to an income of $100 at the price of $50. In other words, at a discount rate of 10 per cent, the sum of $50 can appear to double in size in seven years, through no effort on the part of the investor, or even of the funds themselves. (The funds do not represent the cost of building the apparatus of postponement, even if some of the income may be used to defray those costs; rather, they represent the discounted purchase price of its future revenues.) In this way, the mechanism of delay allows the creation of credit money and endows it with its curious capacity to grow.

Most investors, and the wider public to which they belong, come to think of this transformation of $50 into $100 over a period of seven years as the power of money to 'earn interest'. They see it, as it were, through the other end of the telescope, staring at the future from the present. From the perspective of a person in the future, looking back, an asset was shrunken in size, because an income of $100 was reduced in value to $50. But, seen in reverse, from that present moment looking forwards, the sum of $50 has enlarged to become double the size. The concept of interest makes the increase appear as a natural property of money. This appearance diverts attention from the infrastructural work that must be done – and the powers that must be organised – to build and govern a reliable apparatus of postponement, revenue capture, and credit creation.

Historically, the practice of calculating the difference between the value of a future revenue and its price in the present was known, as we have seen, as 'discounting', a term still commonly used in the world of finance. What today appears as the 'growth' of an investment can more usefully be understood, as the term discounting suggests, as the 'reduction' of the future. Payments can be extracted from the future at a fraction of their eventual value. But this depends on building the mechanism that makes that future available. The ability of a certain quantity of money to apparently double its size does not rest on any power of physical expansion of money. It is a consequence of a technopolitical apparatus that creates an obstruction, interruption, and shrinkage. What we call capital, or money with the power to grow, is this kind of arrangement for building and governing the reliable postponement and discounting of payments.

In asking earlier why capital *appears* to want a delay, the term 'appearance' was important. But it is better to rephrase the terms of the question. Capital may appear here as the agent, seeking the forms of reliable postponement that allow it to expand. However, since capital is the *outcome* of the processes of delay, it does not make sense to consider it simply as the agent of those processes. Capital is not something stored up from the past, seeking expansion into the future. It is acquired from the future, through the building of payment devices and enforcement mechanisms that make prospective income reliably available at a discount in the present.[28]

The revenue acquired from the future is paid largely by the users of the infrastructure. In the case of an infrastructure project organised as a joint-stock company, those who purchase shares in the company establish a perpetual claim on a portion of the company's profits. Going forwards, the claim of the shareholder becomes a burden to be carried by those who use and run the infrastructure. It is a cost added to charges for shipping goods, or the fares paid by passengers, or the rates charged for water and electricity, or the fees for subscribing to a service, and it is an expense imposed on the wages of those who manage and operate the process. Those fares, fees, rates, and wages carry not only the cost of operating and maintaining the infrastructure, but the additional claim sold in advance to the shareholders.

The Durable

The creation of credit money through claims to delayed payments is not in itself a recent practice. However, as chapter 3 explored, in earlier centuries the infrastructures of delay were not built of iron and steel, concrete and masonry, copper and fibre-optical cabling, or the other materials typically used today to build durable apparatuses for the capture of future payments. Delay was less durable. Its most common apparatus before the imperial age was the crop cycle, in which the promise of next season's harvest constituted the future for which credit could be created. Another kind of delay, as we noted, could be built from extended commercial relations. Large trading networks, such as those organised across the medieval Indo-Islamic world and beyond, allowed the building of credit arrangements such as the promissory note or bill of exchange. These techniques of creating credit and discounting it were later extended to Europe.[29] As a promise to pay for future goods, due for delivery within a certain number of weeks or months, the note or bill could function as money. It became the most widely used system of credit and capital creation across the medieval and early modern world.

Larger opportunities to manufacture credit arose with the European diversion of Middle Eastern–Asian trade from overland and coastal routes to circuitous seaborne passages. The building of reliable, ocean-going ships and military trading posts and provisioning points enabled Europeans to gradually dominate this trade. To recall a point made in chapter 1, before the age of railways, the armed sailing ship and its fortified route became the most complex and expensive apparatus for generating and discounting future revenue. 'A vessel is an extremely composite machine, or rather it is a combination of most known machines', wrote the French shipbuilder Pierre Forfait in 1788.[30] As a historian of naval architecture notes, armed sailing ships 'combined the heavy wooden construction of the hull and masts with a dizzying array of standing rigging to support the masts, hundreds of lines and blocks to control the yardarms and sails, capstans for hauling up the anchors, tillers and wheels to turn the rudder, bilge pumps, and such'.[31] This combination of strength and controllability formed an apparatus that was reliable enough, manoeuvrable enough, and sufficiently armed to operate at a great distance, remote from those who sponsored the trading venture, creating a profitable 'delay' in revenue, via the new time lag of colonial trade. The Dutch and English merchants who organised the shipping created the first joint-stock companies, through which shares in prospective revenues could be bought and sold at a discount. The same methods of credit creation were extended to Europe's North Atlantic and Caribbean colonies, and to the trade in enslaved populations and the sugar, tobacco, and other goods they produced.

To support and expand this colonial trade, and their own revenues, European states built naval forces and engaged in protracted wars. Colonial warfare gave rise to a larger system of delayed payment, the credit advanced by merchants to fund the deployment of military force. This provision of credit formed the new and more reliable payment delays known as national debt, which led, in turn, to the creation of national currencies. These currencies would mistakenly come to be thought of as the very essence of money. As governments developed the more extensive modes of taxation required to finance these payments, and the accompanying powers of seizure of goods and imprisonment for debt, the longevity and increasing stability and scale of state power gave rise to capital creation on a new scale.[32]

Prior to the development of modern, colonial-scale trade and warfare, profiting through financial discounting of the future was constrained by the limited duration and extent of the processes that created a reliably delayed income. The delays that could be created were generally measured in the months laid out by crop cycles and trade routes, not in years or decades, so were restricted in the possibility of so-called 'growth'. There was little value to a claim to revenue that lay far in the future. They were also constrained by practices of reciprocity, obligation, ethical improvement, and generosity – characteristic no doubt of all moral traditions in various combinations, but especially marked, perhaps, in some of the centres of great commerce, such as those of the Islamic world – that placed limits on the extraction of unearned income and the enforcement of debt. Landed estates might guarantee long-term income, but since the rights to that future revenue typically depended (with the exception of slave plantations) on claims to ancient possession, or on guarantees of protection, they could not easily be sold to strangers.[33] There was no readily available mechanism for trading shares in such futures in the present. Apart from the modern state itself, and the colonising corporations and plantations through which it coevolved, there was no apparatus for generating revenue that was reliable enough that it might be profitably delayed over several years, and liquid enough to trade at a discount in the present.

As we explored in chapter 3, and can summarise here, this dependence of capital creation on colonial expansion and warfare, and on enslaved labour and other forms of violence, brought risks of disruption. The response to this disruption created what became a misunderstanding of the role of delay and postponement in the creation of capital, a misunderstanding that still prevails widely today. Over roughly four decades from about 1775 to 1815, the entire system of credit creation through long-distance colonial trade was disrupted and almost brought to a halt by political transformations in South Asia, North America, the Ottoman Empire, and the Caribbean, and then by war across Europe and the Near East. Dislocated patterns of credit and capital creation were reorganised by means of a new detour: through the water-driven cotton mills of Lancashire. The detour took advantage of, and rapidly developed, a new credit system, one with a different relation to temporality. Rather than the long-standing practice of merchants loaning goods to those who made them into textiles, the credit was extended by workers, who loaned their labour to the merchants, now building mills.[34] This detour was hard work, not only for the tens of thousands forced into factory labour, but for the merchants deprived of the older and easier method of creating wealth through the monopoly of infrastructures. Within a few decades, new opportunities for building monopolistic infrastructure emerged. But, in the meantime, Marx had read the reports of English factory inspectors and derived from them a theory of capital based on this new time machine, the factory.

From the later nineteenth century, much more durable apparatuses for capturing the future could be built. Most of these took the form of infrastructure projects. They were typified by the building of long-distance railways. Such large, long-term construction projects offered new opportunities for the creation of credit.

Two features of the new infrastructure projects contributed to making possible the much larger postponement of time. First, as we noted in chapter 1, compared with other means of capturing revenue, even today, infrastructures are unusually durable. Railways, dams, ports, pipelines, highways, power lines, communication networks, and data centres are built of concrete, steel, copper, thermoplastics, and other resilient materials. Technical resilience and durability are constituent features of such infrastructure arrangements. By design they must be more resilient than the transient goods, vehicles, liquids, data, and persons whose flow they channel and enable. While the materials carried by infrastructures are often intended to be consumed within months, days, minutes, or less, creating in themselves little in the way of long-term revenue streams, the life of infrastructure is typically measured in years or decades. The future revenue that can be purchased at a discount is, in part, the product of this necessary difference between the decomposability of supplies or streams intended to be consumed and the non-decomposing quality of the apparatus through which they flow.

Second, more than many other means of capturing revenues, infrastructures are typically built with another kind of durability, a form of political and legal guarantee. As a shorthand, they can be described as monopolies. But that term simplifies the different elements of the guarantee. As an apparatus for the ordering and control of flows, an infrastructure tends to be fixed in place, built out, a form of real estate (in some uses, infrastructure becomes the label for almost any material structure larger than a building). Like all real property, in contrast to movable goods, the structure is monopolising, occupying space exclusively, blocking other users from that space, or charging them for its use. The charge can be as minor as the household rates paid to a utility company for the supply of domestic water, or as large as the $500,000 to $800,000 transit fee that a container ship or oil tanker was paying, at the time of the Ever Given disaster, to navigate through the Suez Canal.[35] Even digital infrastructures monopolise space, not only figuratively and legally, through transforming digital platforms into propertied assets, but by occupying key geographical locations or privileged access to electricity supplies, transoceanic cables, and other requirements.[36]

Moreover, since infrastructures are large, materially extended arrangements, building them usually involves powers of government and forces of coercion. Dams, ports, railroads, highways, power stations, and communication lines typically require a concession or authorisation from an authority, awarding and enforcing exclusive control of a land route, river channel, harbour, transmission corridor, or communication node. This is not simply a question of governments granting monopoly rights. The very powers of government, in the era of the modern state, both local and national, emerged through the development of legal systems, legislative committees, municipal and port authorities, and other bodies that claimed powers to build, license, police, and manage infrastructure.[37] Government and infrastructure coevolved as interdependent forms of spatially defined power, collaborating in the technopolitical work of credit creation.

The coevolution of government and infrastructure allowed certain infrastructure sites to operate as zones in which ordinary government power is modified or suspended. The fortified enclaves created by colonial trading companies were an early example of this, as were the 'factories' or trading stations populated by the various 'nations' of European merchants in the Levant.[38] Lauren Benton shows that imperial jurisdiction and sovereignty expanded unevenly through 'corridors', such as river and road networks, and 'enclaves', such as settlements and outposts.[39] Today, as Deborah Cowen notes, these 'logistics spaces' continue to exemplify an intensification of organised violence, where the 'deadly life' of supply chains is concentrated.[40] The most common example today is the curiously named 'free zone'. This might be a port area or similar enclave, typically close to an international border and connected to transportation routes, in which an authority independent of domestic law authorises and facilitates the movement of goods, capital, policing, and labour. The zone is typically exempt from national taxes and excise duties, suspends the application of national labour laws and environmental legislation, and operates its own security and surveillance systems.[41] In Egypt, the Suez Canal Export Zone provides a typical example. In some parts of the Gulf, including Bahrain and Dubai, much of the wealth and power of the state has been organised around the creation of such zones.[42]

There is a third feature of infrastructures that enhances their durability: they become the object of technical, economic, and business reporting, which measures, tabulates, stress-tests, monitors, and indexes. Poor's *Manual of the Railroads*, discussed in chapter 4, was a notable example of such practices. This reporting, measuring, and monitoring of infrastructure requires and encourages the development of what Keller Easterling terms 'extrastatecraft': standards, repeatable formulas, and protocols which form the 'operating system' for the governing of space, especially spaces seemingly outside the direct government of the state, such as free zones and other fortified and exceptional sites that, as we noted, large infrastructures enable and encourage.[43]

The Liquid

Durability, however, both technical and political, is only one aspect of the expanding power to profit from the future. Extending the length of time over which revenue is controlled cannot become the source of large prospective gain by itself. After all, even the ancients, over time, could build large, enduring material structures and political orders. The organisation of time as a means of creating capital on a large scale requires the combination of durability with its obverse: liquidity.

Future payments can become a source of expanding profit, of differential gain, only when the claim to them becomes a right – a legal claim to property – that can be quickly put in motion. The owner of the claim on future revenue must be able to readily transfer it to another owner. This transferability allows the creation of credit in the first place, setting up the possibility of purchase at a discount and thus the profit from discounting.

So, the development of durable apparatuses for capturing future payments is accompanied at every stage by the development of a counterpart process: a mechanism for converting claims on the future into something immediately exchangeable. The forms of financial 'paper' to which durable infrastructures give rise – bills, bonds, shares, derivatives, and other kinds of monetary instrument – are the lightweight, 'liquid' corollary of the hardened, resilient apparatuses of capture.

These lightweight legal devices, whether in paper or electronic form, carry out an act of translation, converting all the different modes of infrastructural capture of future revenue into instances of the same thing: credit money. Making claims on the future translatable, and thus instantly transferable from one owner to another, and discountable, is not simply a matter of marking them with a number, or a price. A stock certificate or bill of exchange may have a face value, but its effective value can only be established by building a pricing mechanism, or, in other words, the financial apparatus known as a 'market'. The building of infrastructures was accompanied, and made possible, by the building of money markets. The militarised shipping routes and warfare of the colonial age, the transportation, energy, hydraulic, and communication infrastructures of the last century and a half, and the digital pathways and platforms of the most recent decades each facilitated the forming of new institutions for creating, exchanging, and discounting credit money.

This creation of liquidity requires something more. The lightweight quality of the monetary devices allows them to be bundled together in large quantities. It is, in part, this bundling that enables bankers to make any one claim on the future available in advance, at a 'discount'. The credit money issued by banks in the present is secured by the sheer quantity and variety of future payments with which the bank guarantees its credit.[44] Simultaneously, as Geoffrey Ingham points out, capitalist money is distinctive in the mechanism that converts these private debt-credit contracts into 'money' – a universal promise, recognised and implemented by the state (via its acceptance for the payment of taxes). This happens through another production of 'difference', between private banks and a government monetary authority or central bank, and between the government and its creditors (bondholders) and debtors (taxpayers). The central bank purchases the private debts of banks – promises to pay – with state money. As a result, private credit money is monetised, and made acceptable anywhere. This first happened when the Bank of England discounted the bills of country banks. The state does not 'control' this but attempts to regulate it.[45]

The great carbon-fuelled infrastructure projects of the later nineteenth century helped create modern commercial banking, as organisers of the liquidity and credit mechanisms facilitated by the durability of the new infrastructures. They also spurred, as we have seen, the development of another institution for the production of credit via the discounting of future earnings: the shareholder-owned corporation. Over the following 150 years, the corporation became perhaps the most effective arrangement for organising the accumulation and concentration of wealth. It was the building of railways, especially the transcontinental and other long-distance colonial lines of the later nineteenth century, that gave rise to this new way of organising capital. Shares of railway companies were the most common investment traded on the expanding stock exchanges of London, New York, and other financial centres. Previously, shares represented simply the part ownership of the current material assets of a business. In the new age of large infrastructure, shares were transformed into the ownership claim on the revenues that infrastructures placed reliably in the future.

Those with access to credit could purchase a share of the future revenue – from the developers of a railway line, for example – at a discount. The developers would realise an immediate profit, in fact often a great windfall, since the future revenues might be worth many times the cost of building the infrastructure. The share-buying investor would acquire the future revenue, discounted to reflect the fact that the revenue would not be available for five, ten, or fifty years.

Large infrastructure projects were not the only apparatus for the postponement and capture of the future. The process gradually took many other forms, including other large shareholder-owned industries, speculative real estate and the development of mortgage financing, forms of consumer credit, and eventually the building and control of electronic platforms, as assetisation spread into more and more areas of life.[46] But, in earlier decades, starting in the later nineteenth century, railways, the nascent oil industry, and other large infrastructure firms pioneered these forms of constructing and capturing the future and were the largest examples; they continue to exemplify many of the ways in which infrastructure, in its changing forms, allows us to understand the modes of living at the expense of the future.

The legal system helps to give claims to future revenue both their durability and their liquidity, but also does more. In adjudicating among rival claims to a stream of income, for example among those who own shares, those who are owed debts, and those who are owed wages or retirement pensions, it establishes priorities. In most jurisdictions, the law came to treat shares in a business venture as a form of property, giving their holders not only rights over the physical assets of a company but an ownership claim to revenue in the future.[47]

While an infrastructure promises durability, in practice, it is seldom likely to enjoy the stability that the promise might imply. Scholars used to argue that one of the qualities of infrastructure was its invisibility. It typically operates in the background, or below ground, keeping most users of its services unaware of its functioning. As with much history of technology, this approach is written from the standpoint of successful projects, rather than the standpoint of those designing a technical system, when the outcome is uncertain.[48] Infrastructure is better seen not as invisible, but as an organisation of processes of testing, reporting, and accounting that attempts to manage the forms of visibility. Every technical apparatus requires maintenance and repair, and is subject to the possibility of malfunction, decay, and failure.[49]

From the possibility of failure, it follows that infrastructures are a means not only to create capital, but to destroy it. But this does not mean that the destruction of capital simply balances its creation, cancelling out the power of money creation. It is important to understand the forms of wastage, redundancy, sabotage, obsolescence, and decay, all of which are made visible by infrastructures in a way that is less obvious with modes of thinking that focus on capitalism as simply the 'consumption' of 'goods'; and to understand the elements of risk and loss that accompany any organisation of acquisition from the future. The more important issue is how the technopolitical organisation of apparatuses of capture distributes the costs, risks, and rewards.

The Climate of Infrastructure

Infrastructures work on time, but not only in the ways we commonly assume. While they may increase the speed at which goods are transported, people travel, or energy flows, this acceleration of time may not be their most important attribute. Their technical durability and political strength give them another purpose. They introduce an interruption, a gap between the future and the present out of which the present extracts its wealth from the future.

Understanding capital in this way, as a technopolitical process of extraction from the future, exemplified by the work of infrastructure, opens up several questions for the politics of the climate crisis. These will be taken up in detail in the final chapter. But let us lay out the issues here.

First, how should we think about the question of growth? Discussions of economic growth in relation to the climate crisis typically argue either that growth must be measured differently, to take account of ecological costs, or that it must be abandoned as a goal, replaced by a politics of degrowth. How do these arguments change if we consider growth differently? Rather than approaching it as an (inadequate) measure of economic expansion, what happens once we grasp growth as an alibi that has been constructed, not just to exclude certain costs from political consideration, but to categorically misrepresent the nature of capital and our mode of living at the expense of the future?

Second, how can we reframe questions of materialism, extractivism, and the politics of nature? As chapters 2 and 3 explored, much of our political thinking about material life remains caught in the framework of productivism, in which the extraction and transformation of materials represents not just a pervasive mode of despoliation but the key to the process by which value is created and humans are governed and exploited. Can the alternative suggested here, in which the realisation of surplus is the product of methods of detour and delay, and the technopolitical apparatuses through which these are organised and controlled, allow us to see material production as one among those detours, much of it driven not by human need but by opportunities for sabotage, waste, breakdown, deterioration, and delay.

Third, the inevitability of decay, breakdown, and failure raises a further question concerning the way in which infrastructures work on time, a question of particular importance for thinking about capital and the climate crisis. Many attempts to address the climate emergency revolve around the repurposing of infrastructures or the building of new apparatuses on a gigantic scale. Infrastructures, as we said at the beginning of this book, exemplify what seems the distinctively human capacity to transform the Earth. Their longevity and size appear to defy nature, remaking the world according to human purposes. For more than a century, the scale of modern infrastructure has helped redefine the human experience of nature – the ability to tame rivers, illuminate cities, cut through landscapes, conquer distances, and command energies. Yet, despite creating this effect of humans versus nature, and, indeed, helping to conjure the very concept of nature, the building of infrastructure was never a human accomplishment alone.

It is useful to remember that what we think of as human world-making requires a lot of nonhuman elements, from raw materials and construction equipment to energy sources and calculating devices. These elements are not just passive inputs, but contribute their own properties, powers, and propensities. They form their own alloys and compounds, create their own connections and layers of disconnection, and contribute tendencies to stabilise and endure or become volatile and cause combustion. Even what we think of as the human parts of any complex and enduring apparatus are composed of widely distributed powers to calculate, record, recall, signal, and command. Every such competence is an amalgam of technical, mental, and physical components, in which the exact divide between what is human and nonhuman, between what is intentional and the product of other propensities, is never more than an uncertain effect.

But does there remain a distinctively human element in the durability and scale of infrastructural projects? Perhaps that is a view that the climate crisis places in perspective. Earth systems turn out to be more complex, interconnected, and durable than any apparently human intervention. And what we represent as large-scale, mega projects are simply localised elements and arrangements that enjoy particularly strong and durable interconnections.[50]

Finally, can we reconsider the question of intergenerational justice? The question is usually posed as an exceptional issue brought about by the climate emergency. What sacrifices have to be made urgently in the present to allow the possibility of viable forms of collective life in the future? However, the approach presented here to understanding our relation to the future frames the question differently. Our mode of living is a threat to those who come later, not just as a by-product of our own careless exploitation of the resources of the planet, but as a form of life built upon the taking of livelihoods from the future.

8

A Better Tomorrow, Tomorrow? Climate Crisis and the Alibi of Growth

The climate catastrophe demands that we develop new conceptions – of freedom and the political, of the human and nonhuman worlds, of justice between generations and peoples – to provide better tools for coping with the unfolding crisis. But the catastrophe also asks us to account for the political failure to have acted in time. Why have existing conceptions of collective life appeared so impervious to critique? One concept stands out for its resistance to repeated calls for its recasting or abandonment: the idea of the economy. In countless critical texts on climate politics, we are urged to escape from the economy, with its narrow definition of well-being, to replace it with a more ecological conception of our material needs and limits, and to break with its logic of limitless growth. But the idea of the economy, as a way of grasping and measuring our material life, and the public preoccupation with its continued growth, have been largely immune to such criticisms. Why is that so?[1]

The answer is twofold. First, we have lacked an understanding of what the economy does. The economy is usually understood as a natural feature of all modern societies, designating the sphere of material production and exchange. It needs to be better managed, we are often told, and measured in better ways, to more accurately capture forms of social injustice and ecological harm. The economy should be seen instead as an apparatus or technical device, one that operates as a mode of subjection – or what I shall call, with a nod to Foucault's reading of Jeremy Bentham's Panopticon, the 'EconoCon'. As a device, the economy is formatted in ways that help produce the forms of political subjection and collective subjectivity that enable a minority in the present to live at the expense of most people in the future – including, as this book has argued, their subjection today to the repayment of previous extractions from the future, a debt the majority must continually and increasingly bear.

Second, we have lacked an understanding of growth. For most people, economic growth appears as the natural movement of capitalist society, driven by an ineluctable process of technological improvement or a continuous accumulation of capital from the past seeking further opportunities for expansion. It is almost never connected to capitalisation, or the capitalist mode of extraction from the future. Instead, we should approach the question of growth not as a historical force – one that, if we are to avoid even more calamitous consequences than those that global warming already has in store, we must somehow limit or reverse – but as a mode of indemnification. Growth is a device for securing the repayment of debt. This approach does not make growth any less damaging to collective well-being. But it does allow us to grasp it not as the inevitable movement of historical development or capitalist expansion, but as a more recent arrangement that arises from – rather than simply causing – our mode of extraction from the future.

A third argument emerges from these two. In explaining the phenomenon of accelerating economic growth, we have misunderstood the role of fossil fuels. Their role in causing the climate emergency is not in question. But what is the relation between the device of the economy and our preoccupation with its growth, on the one hand, and the increasing use of carbon-based energy, on the other? Here, again, our ordinary idea of the economy as the sphere of material life common to all modern societies, and of its growth as an expansion natural to capitalism, stands in the way of understanding the climate crisis and our dependence on fossil fuels.

Let us begin with this question of carbon-based energy and its relation to growth, then turn to the economy and how to understand its role in managing modes of extraction from the future. We will return at the end to a different understanding of fossil fuels, the economy, and growth as an alibi of capital.

The Economy and Its Growth

The era of fossil fuels has been defined by the idea of growth. But what is growth and how is it related to carbon energy? How did we come to believe that burning fossil fuels created a collective world whose defining character has been to grow? If we want a world without growth, or with a different form of growth, less destructive of Earth systems and species of life, we must ask how growth came to be associated with the extraction and burning of fossil fuels.

The answer seems at first obvious. Over the last two centuries, as earlier chapters of this book explored, the exploitation of coal and later oil and natural gas made available a concentrated and rapidly increasing source of energy that fuelled manufacturing, industrial agriculture, mechanised warfare, the expansion of cities, and the territorial enlargement of trade and empire. This growth could be measured as something material. It appeared in the form of the expanding quantities of goods, numbers of buildings, miles of transportation, size of populations, and areas of territory under the control of capitalist production.

But the obvious answer is not a useful one. It misunderstands both the history of energy and the nature of economic growth. It suffers from a technical determinism, as if history could not have been otherwise – as if it were driven by the very unavoidability of human progress and technical development. We cannot build a politics to address the climate crisis on these terms.

Let us begin, instead, where the climate crisis itself can be said to begin, not 200 years ago but in the mid-twentieth century. The crisis has a particular relationship to the great acceleration of the last six or seven decades, and thus to our accelerated consumption of fossil fuels, and especially to the more rapid acceleration of just the last three decades, since the 1990s. More than 80 per cent of all fossil fuels burned since the start of the industrial age have been burned since 1960, and more than half of them since 1990.[2] But that relation – between the growth of 'the economy' and the expanding production of fossil fuels and atmospheric carbon – needs a new explanation.

We should locate the problem of growth not in the burning of fossil fuels or the history of technical apparatuses that demanded them, such as the steam engine, the electrical power station, or the automobile, but in the invention of a different technical apparatus, the economy. As we saw in chapter 5, economists began to use the term 'the economy' only in the 1930s (older meanings of the word 'economy', used without the definite article, as chapter 4 explained, were very different). It came into widespread use only in the 1950s. Statisticians devised a set of protocols and assumptions for estimating the size and expansion of this new object, summarised as the measurement of the gross domestic product, or GDP. The economy was not just a new word for what already existed. It designated a new object of politics and calculation, understood for the first time as an entity in itself, a stable domain composed of identifiable and measurable relations. So rapidly was this device assembled that, by the mid-1950s, economists, political scientists, and historians started to imagine that 'the economy' had always existed, even though they had learned of its existence only ten or twenty years before.

The emergence of the economy, it should be noted, was largely a Euro-American story. When Europeans and Americans started managing their political life around the production and measurement of their economy, not every country agreed at first to follow. These disagreements could be far-reaching. As chapter 5 noted, the 1956 Suez Crisis, triggered by a dispute over the financing of the Aswan High Dam, and leading to the invasion of Egypt by Israel, Britain, and France in an attempt to remove its government from power, was, in part, a disagreement over the new forms of economic calculation that the imperial powers, through the World Bank, demanded that the Egyptian government adopt. Gradually, the World Bank and other international institutions required the countries of the Global South to adopt the methods that produced what seemed comparable figures for GDP and other measures of economic life. But the suitability and comparability of these measurements was always open to question. And the mode of governing populations through the measurement and management of the economy and its growth did not take the same form or significance in different places.[3]

The GDP is not a measure of material wealth. Rather, it measures the circulation of money through accounts. The accounts are of three kinds: those of business firms, government, and households. Households are imagined as miniature businesses, each with an annual level of income, expenditure, and profit – a surplus that is seen misleadingly as the source of 'investment' or 'capital' through which the economy expands, on the assumption that capital is something saved from the past, not extracted from the future. The accounts are presented as a statistical abstraction, and thus as a representation, and not of material wealth, but of income. The change from older attempts to calculate national wealth to the calculation of gross income appeared to introduce a separation – between the measurement of growth and physical expansion. With the invention of the economy, a country could now 'grow' without expanding its territory, enlarging its population, or even, perhaps, producing more material goods. One might expect this decoupling to raise questions about the idea of growth. But, on the contrary, the notion that capitalist modernity was defined not by cycles of expansion and contraction, of impoverishment and improvement, but by the logic of endless, uninterrupted, and exponential growth became self-evident. Economic growth without end became the very measure of political well-being and progress.

From the start, the statisticians, planners, and budget officials who began to apply these new metrics were aware of their limitations. The idea of infinite exponential growth is absurd, even just mathematically. Drawn as the curve of a graph, the ascending line that represents the growth of the economy rapidly approaches the vertical. As mentioned in chapter 5, its absurdity was partially disguised by adopting the convention of plotting the vertical axis, showing the increase in GDP, on a logarithmic scale. The log plot converted an ascending curve into a straight and much flatter line, making its path appear almost reasonable. Flaws in the way GDP is calculated were also noted. The list of these is astonishingly long, and increasingly familiar. Over several decades, ecological economists such as Herman Daly, feminist scholars such as Kathy Gibson and Julie Graham, heterodox economists developing alternative measures of wealth such as Jean Gadrey and Florence Jany-Catrice, and many others, have explored these issues.[4] More recently, an important series of works by Giorgos Kallis, Jason Hickel, and several others have proposed the idea of degrowth, or modes of living well with less, by 'prioritizing wellbeing, equity and sustainability', as a means of breaking with the use of growth as the goal of collective life.[5]

As these works point out, GDP fails as a measure of well-being: it excludes most unpaid labour, in particular the labour of householding, childcare, and human reproduction – a gendered and racialised exclusion; it includes as contributions to prosperity harmful expenses, such as addictive or unhealthy consumption or the cost of arms spending, police violence, and incarceration – so is racialised a second time. It does not value leisure, except when this is paid for. It does not value the reliability of goods or the avoidance of waste, or account for the wear and tear that reduces the value of productive assets. It treats as positive the expenditure on medical treatment and psychological care; on the legal services and accounting procedures required to maintain grossly unequal access to property and income; and on repairing damage from the effects of global heating – and it does this in ways that do not adequately value improvements that might have arisen from better provision for care of the person, the community, or the planet. It attributes no value to the exhaustion of finite resources, beyond the increased cost of extraction as reserves are run down. And it includes no valuation of the loss of biodiversity, of species extinction, or other costs to life on Earth beyond those priced by human claims to the control of processes or materials as financial assets.[6]

Orthodox economists, already forced to respond more than half a century ago to criticisms from environmentalists, defended the usefulness of the measure. If GDP failed to measure environmental and other costs, this was due to 'a defect of the pricing system', wrote William Nordhaus and James Tobin in 1972. The remedy would be to produce better prices. As for the danger of 'global ecological catastrophes', they added, such as the risk 'that we will melt the polar icecaps and flood all the world's seaports', there seemed to be 'great uncertainty about the causes and likelihood of such occurrences'. As a result, they proposed, there was 'very little that economics alone can say'.[7] The certainty with which orthodox economists could speak about the economy was dependent on producing this necessary uncertainty and ignorance.

These long-familiar shortcomings in the measurement of the economy and its growth make our normalisation of the concept even more remarkable. Such flaws can be distinguished from a second kind of criticism of economic growth, usually drawing more closely on Marxian traditions of thought: growth as the increasing extraction from nature.

For the latter, the problem of growth lies not in practices of measurement or mismeasurement, of how we value or disvalue nature and other noneconomic goods, but in the underlying nature of capitalism. What distinguishes capitalism, in many versions of this view – the work of Nancy Fraser or Jason Moore would be good examples – is its extractivism.[8] Capitalism came into being and expands as a specific mode of organising the relationship, not only among humans, but between humans and the natural world. If capitalism is understood as the expansion of value and appropriation of surplus through the command of labour to produce commodities, this depends on the endless expansion of the processing of materials into goods. Capitalism addresses its own crises through further spatial expansion and increased exploitation of materials from nature, which provide a constantly expanding reserve of what it treats as 'cheap' or free gifts – a theme in political thought that Alyssa Battistoni has explored.[9] Since capital relentlessly seeks its own self-expansion, capitalism depends upon material growth, and growth is always at the expense of natural resources. No amount of redefinition of GDP or better measurement of costs will free capitalism from this destructive and unsustainable relationship to nature. Growth is not a problem of improper ways of representing and valuing resources. It is the exploitative dynamic at the base of the capitalist mode of production.

The Effect of the Economy

One could summarise the two approaches to growth as a question of whether 'the economy' as measured by GDP is a *mis*representation, or a *mere* representation. On one side are those – for example, the influential Kate Raworth – who propose that we should replace the metric of growth, as measured by GDP, with alternative calculations of well-being, and become, not necessarily anti-growth, but at least agnostic about it.[10] On the other side, the theorists of capital – see the postcolonial historian Andrew Sartori for a recent example – who suggest that the twentieth-century conception of 'the economy' is relatively unimportant, because it is *only* a representation. It can be contrasted, he argues, with a different kind of ideational process, that of 'abstraction' and specifically the so-called real abstractions – labour, property, value – that emerged mainly in the nineteenth century and are fundamental to the working of capital.

Whichever side one takes, both positions in this argument suffer from a weakness – and the same one. While it refers to something real, they argue, the economy is a representation. Or rather, it is a representation, a mere convention, *because* it refers to something else, to a material reality – to a process of material production.

Despite their power, neither of these arguments is the one followed here. Neither examines the reason why 'the economy' was invented so recently, the work it does, or its form of political subjection and control. The economy is neither a way of depicting a material reality, as most mainstream scholarship assumes, nor its (mis)representation, a mere product of (mismeasured) statistics. It is, rather, as chapter 4 examined, a specific way of producing the *effect* of the material world versus its representation: namely in the form of the real versus the financial, or the material world and the system of accounts, prices, and financial exchanges that merely reference it and manage it. This effect, this mode of ordering things and appearances, this apparatus or diagram, is precisely the mechanism through which the capitalist, or rather the capitalising, mode of capture has come to operate.

Note the use of two terms here, 'effect' and 'capitalising': the term 'effect' is more useful than a grander term like abstraction, partly because it can direct our attention to many of the minor, seemingly insignificant technopolitical practices – for example, the compilation of financial statistics in government offices, or, in a previous era, the record keeping by cash registers of large retail companies; or, as chapter 3 recounted, the gradual process by which banking houses were separated from 'industry', architecturally, legislatively, and professionally, after starting as two interconnected aspects of merchant life. The term 'effect' carries the meaning of both mere appearance and effectiveness. It draws our attention to the always unstable difference between an abstraction and what it abstracts from – something the term abstraction, once it ceases to be used as a verbal noun indicating a process, usually fails to do.

This argument is analogous to one made previously about the state, as an effect.[11] But it is different, because the effect is produced differently. We are governed differently by the effect of the economy, compared with the effect of the state. While the economy is yet another mode of abstracting, like state, or society, whose process of assembling we fail to see in the shortcut of referencing it, there is an effect of abstracting, of constructing differences, that is particular to the production of the economy. In the specifics of that mode, we have to understand and disassemble the question of growth.

Second, the term capitalising. Our aim, as always, is to replace terms that refer to large objects or structures, like capitalism, with words that describe repetitive processes, such as capitalising. Referencing large, seemingly intractable structures reinforces the powers of abstraction, whereas focusing on mundane, reiterative procedures can draw attention to points of vulnerability or potential disruption. As seen throughout this book, the term 'capitalising' refers to the mode of capture of surplus, understood not as predation upon nature, nor even upon human labour per se, although it may make use of both resources, and of others: it is predation upon the future. To capitalise an asset refers to the politico-legal mode of establishing control of flows of revenue stretching into the future – rents, purchases, profits, fees, interest payments, and more – and then extracting them at a discounted price in the present, thereby impoverishing the future, on which the burden of full repayment is placed.

How does the predation work? The relation between present and future – essential to grasping the 'economy' as an apparatus of domination – offers a double mode of capture, to which we will add a third, below. First, those who establish a viable hold on future assets, through rights of property or other power relations, enjoy a steady stream of rents, sales, dividends, or debt repayments. However, as we explored in chapter 3 and again in chapter 7, the possessor of the claim to future revenue does not have to await the future. If the claim is reliably established and supported with legal and market devices, especially devices that make the claim easily and instantly transferable from person to person, and thus 'liquid', the owner can transfer it to others and enjoy an immediate windfall. Each successive transfer of the claim, as when corporate shares are traded, or when derivatives are formed, creates an additional claimant on the future. Second, and more important for making sense of the economy and its apparent growth, those who then purchase the claim to future revenue can typically acquire it at a discount. To repeat the example from previous chapters, an income of $100 available in a year's time – from dividends, sales, or rents – might be purchased today for, say, $90. The discounted price of a payment acquired from the future reflects the delay or postponement engineered in establishing it, and the resulting premium generated through the mechanism that makes it available in advance. The purchase of a future revenue allows the $90 to appear to 'grow', over the course of a year, by more than 10 per cent, into the sum of $100.[12]

We inhabit a world that is defined, not by capitalism as a structure, but by capitalising as a process. The engineering of postponement, dependent on forms of violence, predation, infrastructure, law, and liquidity, gives rise to the very possibility of money capital, awarding it its curious power to grow. Thus, money is not merely a means to facilitate exchange or to store wealth, but is produced, thanks to the order of postponement, mainly as a claim on a future good or revenue. Money, in other words, is almost always credit money: it always refers to and claims something further. Issued largely by private commercial banks (backed by government guarantees), it organises not merely exchange or storage, but the encumbering of the future. Thanks to the effectiveness of postponement, the process of discounting that it enables, and the restricting of money creation to private, profit-making banks and financial firms, as we saw in chapter 3, the production of money comes with the power to impose a charge for 'interest'. (We commonly refer to this as the money 'earning' interest, naturalising its ability to increase.) Almost all money is credit, and all credit, due to the power of charging interest, at a compound rate, has to be repaid, at a cost that increases exponentially. At 3 per cent, for example, the cost of money doubles in twenty-four years; at 6 per cent, in twelve years. But this is just another way of saying that those who manufacture or have access to the manufacturing of credit money can purchase the future at a discount – an income of $100 available twelve years from now can be purchased today at the discounted price of $50.[13]

The interest rate and the discount rate are two ways of describing the same phenomenon, seen from opposite ends of the same telescope (to repeat the optical metaphor from the previous chapter). The interest rate looks forwards from the present and calculates the value of a present sum of money at a future point. The discount rate starts from that future point and works backwards, calculating what that future sum would be worth today if one allowed for the power of money to 'grow' in value over time. In a process of circular reasoning, one concept is often explained in terms of the other. Thus, discounting the value of future income is ascribed to the 'natural' process by which money earns interest. But none of this, in fact, is natural. Both rates arise from relations of power, in all the ways this book has explored. The power to reliably claim a future asset, dependent on the arrangement of postponement and the ability to exclude or reduce the claims of others, combined with the mechanism that 'pools' such claims through the banking system to make them liquid, is the source of the ability to sell the claim in the present.[14] The discount at which the future claim is sold is then ascribed, not to these mechanisms of capture, postponement, and pooling, but to the power of money to earn interest.

The obscuring of this process of capturing the future is aided by our naïve, everyday conception of money or capital, a conception reproduced in standard economic theory. The standard view thinks of capital not as a form of control of future payments but, as we saw in chapter 3, as funds saved up from the past. These savings are stored in banks, it is imagined, which then lend them to entrepreneurs, typically to invest in the 'machinery' or 'technology' required to make more money. In some imaginations, and in many economists' models, the machinery itself comes to represent 'capital'. The payment of interest can then be described as the reward paid to those whose apparent 'savings' have been used to grow more wealth – rather than the rent charged by the financial institutions that, thanks to their state-granted privileges and protections, monopolize the creation of credit.

There is, in fact, a third mode of capture that arises from the command of the future, besides the gain for those who organise the claim and the profit from acquiring it at a discount. Claims to future payments are always open to fluctuations in value, as the future becomes more certain or more subject to risk. Of course, the risk may reflect the fact that a capitalising claim might collapse, destroying the entire capital that a power over the future created. More frequently, as a claim becomes more liquid, thanks to the building out of banking, money markets, and calculative devices, profits can be made by purchasing, not the claim itself, but the prospect of its increase or decrease in value. Thus, the liquidity itself, derived from the underlying asset, can become an asset to be captured. Such 'derivatives', as pointed out in chapter 1, are in themselves nothing new. The deployment from the 1970s of a pricing formula similar to the 'heat' equation familiar to physicists, and worked out with pen and paper, the Black–Scholes–Merton model, made it possible to 'price' the risk attached to any financial instrument separately from the underlying asset. The value of these derivatives was not necessarily calculated with great accuracy, but with sufficient persuasiveness to allow marketing such risks as assets on an even larger scale.[15] As their manufacture increases, the opportunities multiply to profit from 'arbitrage', meaning the difference in price of the 'same' thing at different moments in time or different locations. Manufacturing the very gradients in liquidity, as Robert Meister explains, becomes the source of this difference, and thus of the opportunity to capture surplus.[16] As with all such capture, the majority of the gain will be imposed as a tribute or tax, as we saw in earlier chapters, to be paid by those who produce, use, or maintain the future asset whose revenue has been captured.

This proliferation of financial assets is not properly measured in calculations of 'the economy'. But this does not reduce the usefulness of the production of the economy. On the contrary, even as the economy becomes less appropriate as a measure of most people's well-being, for this and all the other reasons we have mentioned, it grows in importance as a stabilising force that secures the predictable future on which capitalising and discounting increasingly rely. Indeed, as Meister shows, the risk that the 'economy' might 'explode' becomes, in episodes such as the 2007–8 global financial crisis, precisely the threat that can be wielded by those who profit from the buying and selling of risk. A better understanding of this threat to blow up the economy, and the uncounted value that the stabilising effect of 'the economy' represents for the financial industry, Meister suggests, could provide the basis for a more democratic contestation of the power of the capture of the future.

The focus on capitalising as a process allows one to avoid the pitfall that faces many studies of the proliferation of modes of capture of the future – those that grasp them through the idea of financialisation. The lives of increasing numbers of people are governed by the growing burden of debt, whether incurred predominantly at the national level in many countries of the South, and imposed on individual households through programmes of government austerity, or mainly at the personal level, in most countries of the North. Among the latter, home mortgages, student loans, car leases, credit cards, private healthcare charges, and the costs of childcare and elder care – all these set up so much of life as a continuous and expanding repayment schedule.[17] However, the notion of financialisation usually depends on a distinction between finance and the 'real' economy. We will see shortly how this distinction between what is real and what is merely financial is one of the most powerful effects produced by the invention of the economy. The effect operates not so much by distinguishing the two, but by conjuring a very one-sided understanding of finance, divorced from all the kinds of work of postponement and product manufacture just described, and an equally one-sided conception of the real.

Financialisation expresses what we can already see to be only one aspect of a wider process. The dominant apparatuses for the capture of surplus operate in real, material ways, by arranging human and material dependencies as 'assets' from which a continuous investment income is derived.[18] The capture works not only through overt forms of indebtedness, but through the command of working life, the supply of electrical power, water, and other infrastructure, social media, entertainment, travel and leisure, healthcare, and more, all increasingly under the principle of investor- or shareholder-owned firms; and by the transformation of housing in many countries into unaffordable assets, for which the mortgage payments consume an ever-larger share of earnings. Under these arrangements, every process of capturing future payments, from mortgages to rents, utility charges to entertainment subscriptions, higher education fees to medical costs, to every kind of consumer purchase, becomes an asset, capitalised and converted to an income stream, available to investors at a discount in the present, to be repaid at full price from the future – a profit imposed as a charge on later users.[19]

Life in the present, moreover, is increasingly burdened for the majority with the repayment of earlier extractions. Meanwhile, households with some access to assets, such as those able to pay off the cost of purchasing a home or to build ownership of shares in stock markets through retirement funds, come to depend on the inflation of the value of such assets – whether through rising house prices or booming share values. With levels of income from employment seldom compensating for the increased charges imposed by capitalised assets, expanding the value of a household's assets becomes a principal means of increasing wealth, thereby raising the cost of housing and other services for the next generation. Even the modestly well-off are thus encouraged to protect their own standard of life through the intensification of methods of living at the expense of those who come later.[20]

Better Tomorrows

In 2011, the US Supreme Court removed the restrictions on independent spending on elections by business corporations. They could now pool their unlimited funds through political action committees, or PACs. The new super PACs, as they became known, began to coordinate corporate payments and run political campaigns on their behalf. The US television comedian Stephen Colbert parodied the Supreme Court decision by setting up a political action committee of his own, named Americans for a Better Tomorrow, Tomorrow. 'It's the way [the] founding fathers would have wanted it,' he said, celebrating the event on his television show, 'if they had founded *corporations*, instead of just a country.' The parody connected two themes: the problem of political life when extraordinary influence is exercised by what Veblen, even a century earlier, had already identified as its 'master institution', the business corporation; and a political rhetoric of postponement. The rhetoric promises well-being and improvement through the pledge of a better future, but its effect is one of deferral. This deferral is what I have referred to as the alibi of growth.[21]

How do you organise and control a social order built upon the deprivation of the future, in which the present is itself impoverished by previous predations, and the only means of building even moderate wealth is to participate in the same process of deprivation? How do you keep promising a better tomorrow, tomorrow? The answer is that you establish, as it were, an effect called the economy and recast the relationship to the future – even as the processes of predation and impoverishment expand – as the inevitable movement of its 'growth'. This understanding of the capitalising of assets, and of the economy as a politico-technical effect, will allow us to better grasp how and why the 'economy' was invented. To understand the role of the economy as an apparatus of government, we must first briefly recap the longer history of capitalisation, the excesses to which it gave rise, and the collapse in the mid-twentieth century of earlier methods of managing that excess. The creation of the economy as an apparatus of government, as we will see, provided a response to that collapse.

First, the recap. Methods of capitalising future revenue are not in themselves recent. Today's so-called financialised capitalism, where these techniques are perhaps more widespread, is not somehow fundamentally distinct from industrial capitalism, or the commercial capitalism that preceded it. Even a millennium ago, merchants in Fatimid and Mamluk Egypt, as we discussed in chapters 2 and 3, collaborated with the ruling household to manage tax payments by farmers on similar principles. What has changed is the scale on which the process happens, and the apparatuses through which future revenues are captured, assetised, and capitalised.

On the periphery of that Islamic world, in early modern Europe, the capitalising of future revenues was engineered most extensively in long-distance trade, a colonial and slave-based order that operated not simply to supply goods, but to generate credit money, through commerce and slavery, mainly in the form of bills of exchange. The shipping and sale of goods and enslaved persons was the apparatus for this capture of surplus from credit. A crisis in that system, as chapter 3 explored, led to the development of a different credit apparatus, the industrial factory.

The emergence of industrial capitalism, in the late eighteenth and early nineteenth century, should be understood in a new way. Manufacturing did not represent a fundamental break with the world of commerce and credit. More than that, the forcing of wage labour into coal mines and factories enables us to see in surprisingly clear form the problem of credit creation, for which 'the economy' was a later solution.

The invention of the economy was a twentieth-century answer to a challenge that had a longer history – how to secure and stabilise the production of credit money.

Even factory production and coal mining were a method of capitalisation – a form of predation on the future through the elaboration of credit money. The diverting of production into mechanised and later coal-powered factories presented a singular problem: how to pay large numbers of workers a weekly or monthly wage – in a context in which physical money (coin or specie) was little used in everyday transactions, almost every payment being handled through book credit, promissory notes, and bills of exchange. As we saw, the payment of wages turned out to be not a problem but an opportunity – perhaps even becoming a main advantage of concentrating workforces into factories. A common solution for paying workers in the absence of coinage was the system of 'truck', as it was called. The merchant-manufacturer paid wages with a credit entry on his books, or with chits or tokens. These could be exchanged for goods at local shops, often owned by the same merchant. Merchants could profit from the difference between the cost of making goods and the price at which the employee had to purchase them, and from other aspects of such arrangements: a monopoly on the supply of consumption items; the charging of interest; the indebting of workers by providing them with necessities in advance that were then charged on the books at full price; and supplying them with housing whose rent was also deducted from wages. The very delay in time between the work that was being paid on credit and the purchase or rental charge with which the credit money was returned to the merchant was a source of profit. This made the use of truck and book credit highly exploitative and led to the passing of laws in an attempt to restrain them.[22]

Legislation did not eliminate this machinery of extraction through credit. Wage labour today, one might say, is still a truck system. Its distinctive feature is that workers are still paid in credit money. The form of the wage payment changed from merchant credit or tokens to banknotes, then to paychecks, and today to electronic transfers between bank accounts. But wages are still brought into being as credit money. And they still depend, so to speak, on the 'company store' through which the credit money is paid back into the banking system, with interest.

However, the store is no longer just a single shop under the control of the same merchant. It is now part of a much larger credit apparatus, namely the array of apparatuses, discursive and non-discursive, that organise the capture of wages, not through the simple device of a shop or marketplace, but through the technologies of product design, trade shows, fashion, advertising, brand creation, consumer advice, and comparison.[23] These mechanisms were already appearing in the eighteenth century, for the well-to-do and the middling classes, in tandem with the development of manufacturing.[24] Today, transformed by the internet, with its interfaces, platforms, social media sites, and influencers, they constitute the vast apparatus through which wage earners repay their 'loaned' wage money. The prices workers pay for goods, services, and their marketing and advertising include the mark-up for the loan, the charge for the use of money upon which a capitalising monetary system operates. The interfaces and platforms are themselves now the main source of profit for the handful of corporations that recently surpassed the major global oil firms (other than Saudi Aramco) as the most highly capitalised companies in the world – Amazon, Apple, Alphabet, and Meta. With revenues arising from the near monopoly of online advertising, data, and e-commerce, these firms increasingly control access to the 'store' through which loan wages are recaptured. Indeed, the cloud-based megacorporations rely upon their users not only to pay for online goods and services, but to produce the data that these 'fiefdoms' own. Thus, users of the cloud contribute their labour a second time, in this case unpaid or even at the cost of a subscription charge. Their online interacting, browsing, purchasing, and reacting manufactures both themselves as the desiring subjects of consumption and the vast data on those selves that the megacorporations own and capitalise.[25]

The difference between the creation of credit by the merchant banker of the early industrial system and the arrangements operating today is that from the mid-nineteenth century governments introduced regulations to separate manufacturer and banker, workplace and bank. In fact, one could argue, the distinctive element of the form of economic life that began to emerge in the industrialised regions of the world from the second half of that century was no longer simply the separation of workers from the ownership of the means of production, nor the apparent separation of production from the market, nor the so-called economic from the political. It was the separation of the places of work and consumption from the source of money. Merchants who had previously combined the control of how goods were made and marketed with the supply of credit were separated into those who organised goods and those who created money. And it was not simply that the manufacturing and retailing of goods depended upon a supply of credit, but equally that the manufacturing of credit depended on a widening system of large-scale employment, combined in turn with advertising, taste-making, and consumption, with these increasingly monopolised today by the corporate-owned cloud.

Modern money is bank-produced credit money. Banks today create more than 95 per cent of the money in circulation, almost all of it as digital transfers between accounts within the banking system. The ability of banks to manufacture credit money depends upon a reasonable guarantee that the funds designated as credit will later be redesignated within the system that creates them as repayments for purchases. In fact, the 'paper' issued by a bank, in the form of notes, cheques, or electronic ledger entries, would not function as money if there were no requirement to return it through the payment for goods, services, subscriptions, and other rents.

However, banks cannot just manufacture money internally. To become a real asset, bank-made money must pass through some apparatus of capture, through which the debtor transfers an increment to the creditor. Thus, the money loaned out in the form of wages is produced and given effect in relation to the purchases and charges on which the wage must be spent. Only the guarantee of that future spending enables the bank's creation of assets to proceed.[26] Credit money can operate only as part of a technical and material arrangement where at a future point the money is spent, and 'returned' to the bank with interest. With the money manufactured to pay wages, the interest payment is not imposed on the worker as a separate fee, but is built into the prices of commodities, the charges for services, and the cost of housing. These prices, charges, and costs, measured in the calculation of GDP, include both the value of the product, and the surcharge imposed as corporate profit and interest on the credit money created as wages.[27]

The Economy Replaces Gold as the 'Invariant Standard'

This ability to capture surplus through a credit mechanism, by setting up a gap in time that separates present from future, labour from the purchase of its products, and thus loan from repayment, was not limited to the industrial system and wage labour. If anything, as chapter 3 proposed, modern industry was a detour. The manufacturing of goods – involving hard work, low margins, and the constant threat of competition – soon gave way to other apparatuses of capture, which were more long-term and less subject to threats of competition, especially those operating through the renewed powers of imperial expansion, including the building of railways, utilities, land development schemes, and housing mortgages, alongside older forms of capture, in particular the imposing of government taxes. All of these, as explored in previous chapters, operated as apparatuses to create credit, by establishing a particular stream of reliable future payments, in the form of rents, fares, freight charges, utility bills, taxes, and mortgage payments, and then selling claims upon those future payments in the present. The modern joint-stock company emerged, in tandem with imperial expansion, as a principal device for the manufacture and mobilisation of such claims. Subsequently, even industrial life, previously the preserve mainly of family-owned businesses or partnerships, was reorganised on the same principle, adopting the form of the joint-stock company in order to capitalise and profit from future earnings.

There remained a difference, however, between the production of credit to purchase these specific claims on the future and the production of credit to pay wages. The selling of claims to specific revenues, in forms such as company shares or government, corporate, and mortgage debt, was understood as the sale of 'financial' products. Such arrangements operated explicitly on the principle of capturing income from the future, in the shape of dividends, interest payments, or other modes of using money to earn money. The payment of wages seemed to operate differently. It appeared to belong to a 'real' world rather than a world of finance, in which the money required to pay wages was simply a cost of production. Despite this apparent difference, both financial products and wages depended on the creation of credit money, in which securing a reliable future expenditure and a return of the funds was the basis for the monetary system.

To make the modern system of money creation work, it was not enough to separate businesses from banks. Private banking required a means of enabling its promises to pay to function as public money. This required the powers of the modern state, to guarantee those promises. Such powers came into being, not by the state intervening to regulate and guarantee the private system of credit money, but through the coevolution of private commercial banks and public or quasi-public central banks and governments. Capitalist money and the power of regulation developed symbiotically; modern banking and modern government formed one another.

This codevelopment had two elements. England provides the paradigmatic case, although English merchant bankers drew upon methods of creating and governing finance previously developed by the Dutch, and before them by merchants and imperial city-states in the Mediterranean.

First, private money creation, built up through profitable colonial trading and slaving ventures, was given a public form. At the end of the seventeenth century, powerful merchants in London invited an ally of the Dutch merchant bankers, Prince William of Orange, to invade England. Dependent on them for his new throne and leading the country into its first major war in Europe in 400 years, William accepted an arrangement where the merchants formed a new joint-stock company, the Bank of England, to provide credit to fund the war-making. The loans for wars, fought increasingly to expand English overseas commercial monopolies, no longer required repayment within the year. They were given the new form of 'national' debt (rather than a private debt of the monarch, or a commercial debt of the colonial trading companies) and were guaranteed by hypothecated revenues from customs and excise, especially duties on alcohol and tobacco, paid disproportionately by the poor. Selling discounted claims to those future state revenues, the merchant bankers could now create private credit money with a public guarantee.[28] Well before the emergence of the 'nation-state', the 'nation' was conjured into being – produced as the artificial body upon which this obligation for war debts was imposed.[29]

Second, the bank was granted a monopoly on the discounting of private bills of exchange and the issuing of bank bills payable on demand.[30] When the growth of industrial employment brought into being small provincial banks, the notes they issued could be discounted only in London, with the Bank of England guaranteeing that all privately created credit money could be redeemed, in the last resort, in gold sovereigns – although in practice the Bank accepted local banknotes at a discount in exchange not for gold coins but for its own notes, which gradually took on the character of a national paper currency. The principle that paper notes were convertible at a fixed rate to gold, a precious metal with a market value, gave rise to the working fiction that all bank-made credit money derived its value from an underlying national commodity-money. The stability of the price of gold generated what appeared as a 'standard' of value, the gold standard, that supposedly governed the quantity of paper money that could be issued. In fact, however, the payment of workers' wages, under the new system of bank money, rested on this arrangement of private money creation backed by a quasi-public national bank. The credit was secured notionally against reserves of gold, but, in practice, against the state's guarantee of maintaining the profitability of the Bank of England, and the reliability of its notes, through the bank's claim on 'the nation's' future tax payments.[31] The private creation of credit money, now operating through the extensive system of wage labour and mass consumption, was secured by its connection to the public power of taxation.

Despite this guarantee, the widening power of private banks to create credit money had to be further managed, to limit the crises to which private credit creation was prone. This was especially the case when large financial speculations failed, or when prolonged warfare brought wider uncertainty about tax burdens and the future stability of consumption. In practice, only Great Britain operated on a gold standard, with the value of other major currencies tied to sterling through international trade, most of which was financed through London.[32] Eventually, in the political and economic crises of the interwar decades of the twentieth century, the fiction of a gold standard collapsed. The Bank of England and other national banks were forced to abandon the claim that reserves of gold represented the invariant standard on which the value of national money was based.

The invariant 'standard' was then reinvented in the postwar period by devising a new referent: 'the economy'. The generalised arrangement of truck, in which wages were paid in privately made credit money, could no longer depend on the fiction of reserves of gold as the standard establishing the value of money and the means of placing limits on the creation of credit. Instead, the monetary order would be stabilised by the production of a predictable future, in the form of the 'objective knowledge' of economic indicators.[33] A wartime device for planning the production of armaments, the system of national income accounts, was repurposed after the war as the means of managing the production of credit. The estimating of national income or GDP and its rate of growth, and of price inflation, employment levels, and interest rates, provided a set of numbers whose interpretation established, in place of gold, the creditworthiness that allowed the private creation of money to proceed. Through the metric of growth and other 'macroeconomic' indicators, central banks could now regulate and place limits on the creation of credit money – as well as on wage demands and price increases. Produced through these techniques of measurement as an object in itself, the economy had become the instrument on which extraction from the future would now depend.

The Financial and the Real

It may be useful to think of this arrangement not simply as the economy, but, by loose analogy with Jeremy Bentham's Panopticon, deployed by Foucault as a diagram of the operation of power, as the EconoCon. Like the Panopticon, it is a device that produces forms of knowledge, through ways of regulating the body and organising desire. To operate as an invariant standard, the economy could never just be a representation of a material reality, or just a statistical artefact or set of truths. It operated materially, as an apparatus of control, decomposing and recomposing persons, households, materials, desires, and futures, creating an effect of stability, structure, and objectivity.

To work in this way, the economy required a number of features. These requirements allow us to understand how the 'flaws' in the way the economy measures or represents well-being are not oversights, to be identified, revised, or replaced, but the key to its political effectiveness. They explain our inability, in the face of the climate crisis, to escape the dynamic of 'growth'.

First, the economy must organise the very relation between present and future. It achieves this, not by measuring dynamic processes, such as credit or capitalisation, that operate through the production of futures, delays, and detours, but by composing a series of static images, at regular intervals in time, then superimposing one upon the next to compare their difference. The economy is calculated as an annual snapshot, not as something that can measure relations constructed across time. This static form allows a separate total to be calculated for each year, which can then be compared with the previous year, and the following. As we saw in chapter 5, the sequence of snapshots is what is measurable as 'growth'. For the economy, present and future are not moments in the unfolding of an internal temporality, but coeval products of a political technology.[34]

Second, to make these snapshots comparable, a determining role is given to the elusive factor of 'technology'. When the national accounts for one year are compared with those for the next, the figures are adjusted for inflation. This requires deciding whether the increased cost of a good or service reflects a 'real' change or merely a 'monetary' change. Is the increased cost of a medical procedure, a mobile phone, accommodation, an entertainment subscription, or an Uber ride attributable to some technical improvement, representing a 'real' increase in the product, or is it merely price inflation, to be discounted so as to measure only the 'actual' growth of the economy? This determination must be made across every sector of economic life, and for every category of product, a challenge that from the very beginnings of national income accounting was recognised as 'insuperable', and progressively so, since minor variations in the nature and qualities of products increased through time.[35] No a priori rule can make the determination. Rather, a series of ad hoc judgements must be made about the contribution of a change in 'technology'.[36] Technical improvement appears to provide the force or momentum that accounts for growth. In practice, it provides an increasingly pervasive alibi for extraction from the future.

Third, the economy, measured by GDP, is composed of 'products'. In contrast to older terms like capitalism, understood as speculative, finance-driven, bound up with imperialism, and essentially unstable, the economy renders collective life as essentially stable, because it is 'productive'. It measures and represents material relations not in terms of speculative investment, unequal relations of property, unstable flows of capital, and currency movements, but as the circulation of products, meaning the commodities whose prices are aggregated to estimate GDP. It is a measure, specifically, of production of economic 'value' – based on counting how much people spend on new stuff, in the form of goods and services. This 'product' functions as the future against which credit money can be created. Without it, there could scarcely be any money.

Configured as a realm of 'products', the economy inscribes a distinction: between products, considered material, and money, considered simply as a means to measure and compare their values, and to purchase them. Even though money exists largely, not as a store of wealth, but as something effected through the products and services on which it is spent, and thus as something that can exist only materially, the economy presents the material and the monetary as separate realms, one rendered real and the other nonreal or fictitious. This distinction is taken for granted, rather than understood as an effect of the production of the economy – and continues to inform even very capable studies of 'financialisation'.

The distinction between product and price, between the material and the monetary, is itself something continually manufactured. It is an outcome of the measurement of the economy, not a prior reality to be measured. This mutual reinforcement can be appreciated from a historical angle, but in more analytical terms it can be seen both from the side of the material, and from the side of money. On the material side, on the one hand, as Michel Callon shows us, what we call products are actually more like processes than things.[37] A car, for example, is assembled out of ongoing operations of design, evaluation, performance, desire, attachment, repair, and obsolescence, not to mention the arrangements of roadways, fuels, vehicle loans, traffic laws, permits, and policing without which the 'product' could not exist as an item of value. In most countries, furthermore, between half and three-quarters of the products that are measured as the economy, unlike a car, do not even present themselves as material things. They fall into the category of so-called services, originally a residual grouping but now a catchall term for the proliferation of processes, attachments, obligations, payments, and performances that constitute the major part of 'the economy'. The economy functions, here, not to measure the material world and its growth, but to resolve this heterogeneity into what appears as real things and their price.

Money, on the other hand, is not the opposite of what is real. The financial is not something separate from the productive, for the simple reason that money, as a relationship of credit, always involves a means of materialisation. While almost all money is created in banks by opening a line of credit and corresponding debit, that individual act of accounting would have no effectiveness if there were not a future material process, some apparatus for capturing payment, through which the money were to be realised, and brought into being. Thus, a mortgage loan to purchase a house functions as money only through the presence of a house on which repayments of the loan and the interest charge can be collected and enforced, along with the processes of design, planning regulation, property rights, policing, zoning, racial segregation and redlining, and foreclosure through which the value of housing is produced and guaranteed.[38] Likewise, the credit money that a business firm requires to pay its workers can function as a means of payment only if there is some business product on which that credit money wage must later be spent. This hybrid form of finance, as credit payment in the present and as an apparatus built out into the future that compels repayment, is what the production of the economy, as a mode of governing the arrangement of present and future, secures.

Producing the difference between the real and the financial requires work. It is built into the very measurement of the economy. As we have just seen, for the measurement of every product and service, a decision must be made between the 'real' component of its cost and the merely 'monetary'. More than this, the economy as measured by GDP excludes the cost of the manufacture and purchase of financial products, such as stocks, bonds, and derivatives, some of which, in a recent recalibration of national accounts, were requalified not as costs but as final 'products'.[39] It also excludes income arising from the appreciation of the value of assets – that is, capital gains – because capital gains are not usually visible in an annual snapshot. This unmeasured process represents as much as 25 per cent of national income, the bulk of it going to the very rich, but increasingly important to maintaining even moderate forms of affluence.

Conversely, the national income accounts do include payments for unproductive activities, if they are widespread and can be considered a 'service'. For example, the payment of rent for housing is considered a payment for housing services, even if the home was built years or decades before, and the rent has no connection to the original cost of building the structure – of providing the service – but is merely a charge imposed by those who monopolise access to landed property. Curiously, owners who occupy their own homes are considered in the national accounts to be receiving an equivalent income. They are imputed to enjoy the 'service' provided by the opportunity to live in their own house. (So reproductive and other unpaid household labour of humans is not counted, but the house itself is recognised in the national accounts for the *un*productive 'labour' of housing its owners.)

Property rents and interest are not the only modes of extracting payment from the control of assets construed as part of the 'productive' economy. The most important feature of the measurement of the economy is that even the cost of material goods and useful services includes a significant financial charge. Estimates in the case of Germany, for example, suggest that 35 to 40 per cent of the price of everyday goods represents the cost of the interest payments that manufacturers and suppliers pay to banks and other creditors.[40] This vast and expanding extraction of interest income from the future, passed on as a cost to the final consumers of goods and services, has become an essential component of the 'growth' of the economy.

The economy is presented as a measure of material, productive life, of the economic 'product' that appears as the measure of our well-being. But its measurement not only fails to distinguish between products that enhance well-being and those that are harmful to life, or to account for environmental costs or unpaid labour, as others have pointed out; it also includes an entire system of so-called unproductive payments for rent, interest, advertising, and other charges, which are recorded, not as a mode of living at the expense of the future, but simply as the cost of the production of things.

The economy operates in relation to these modes of capture of the future by securing, in fact, a different relation to the future. It builds in and reinforces a certain place for practices of credit and capitalisation. But it does this by disavowal. Credit money is everywhere in the system of accounts, and at the same time nowhere. It is everywhere, because what is measured is not resources, or material products, or labour, or inequality, but prices. It is disavowed, because money is seen as nothing except an intermediary. In other words, the definition and measurement of 'the economy' not only establishes this object, but also inscribes and reinforces a fundamental but entirely problematic distinction between the 'real economy' and 'the financial system' – between 'real production' and mere finance.

The calculation of GDP and the measurement of growth can now be seen as the creation through the economy of the structural effect that was required to both stabilise and disguise the mode of living at the expense of the future. It does so by inscribing in the very object of politics (the economy) the apparent separation between the 'real' and the 'financial'. This is the key to the EconoCon, to understanding the economy as an apparatus of control. It is an apparatus that orders human action in partial ways – as producer, consumer, householder, business owner. But it operates at a more molecular level, generating relations between economic and noneconomic, between real and financial, and between past, present, and future.

This approach to understanding the economy explains not only its invention as a political device, but its defining feature, growth. Growth is something forced by the operation of capture from the future. The credit supplied as wage, or as mortgage, or as production loan, carries an interest charge. It must be paid back, not in the same amount, but with a premium – the cost of so-called interest.[41] This growth has the character of an alibi. For it actually occurs as impoverishment.

Carbon Growth

If the growth of the economy, as measured by GDP, is neither a misrepresentation, nor merely a representation, but a device for ordering the capture of the future, organising bodies and lives and materials in order to impoverish those who come later, and providing the moral justification, the 'oikodicy' (to repeat a term from chapter 4), for present impoverishments, how does this relate to the opening question of this chapter, about the relationship between fossil fuels, climate crisis, and the phenomenon of growth?

The answer does not lie in some underlying 'need' for energy. Rather, fossil fuels, especially oil in the second half of the twentieth century, were uniquely suited to this mode of capture. Just as the transition from water mills to the coal-driven steam engines of the 'first' industrial revolution is best understood, following Andreas Malm, not in terms of increased mechanical power or greater efficiency but as a more effective means of controlling labour and, as suggested here in chapter 3, as part of a detour allowing the increased manufacture of credit; and just as the great driver of the 'second' industrial revolution, the construction of railways, did not reflect a need for speed or communication, but new opportunities to move financial paper and conquer territory, opportunities that were soon superseded; so the same with the age of oil and automobiles of the last seven decades. The private car and commercial road transportation were vehicles built for driving the capture of the future.

The 'great acceleration' of the postwar decades was a period of an astonishing increase in dependence on fossil fuels, a span of time responsible for almost all the excess carbon in the atmosphere.[42] The acceleration is even more marked since 1990, the date of the first Assessment Report of the Intergovernmental Panel on Climate Change. That year marked the start of the three decades responsible by 2020 for more than half the excess CO_2 emissions since 1750.[43] But the critical term is the nature of this 'dependence'. In what way was 'growth' dependent on oil? We know there is no simple relation between energy consumption and GDP. The United States, for example, consumes carbon energy at more than double the rate of most other industrialised countries that enjoy a comparable or better standard of living.[44]

The dependence on oil reflected a general dependence of modern life not on high levels of energy use, but on the production of the economy, and thus on opportunities for capitalisation – for building out a future of revenue capture that was available in the present, for which the economy itself was an instrument of stabilisation. Oil mattered on this enormous scale, not as a requirement for life, but for these interconnected purposes. On the one hand, as outlined in chapter 5 and argued in greater detail elsewhere, the creation of the economy as an object of politics in the mid-twentieth century was assisted by a sudden abundance of oil. The postwar access to comparatively cheap and seemingly inexhaustible supplies of carbon energy, largely from the new oil regions of the Middle East, helped to underwrite the novel idea of 'the economy' as an entity capable of infinite growth.[45]

On the other hand, petroleum provided an exceptional means of commanding payments taken from the future. In the postwar decades, as through most of the twentieth century, oil firms were the most reliable source of extraordinary, long-term profits. A handful of companies monopolised the global production of oil. So long as the development of alternatives to hydrocarbons, such as nuclear power, renewables, or storage batteries, could be obstructed and delayed; so long as the production of oil, especially from the abundant fields of the Middle East, could be slowed and sabotaged just enough to keep future profits high; so long as ways of living could be developed to drive up the 'need' for oil; and so long as the prospective revenues of oil-producing states in the Middle East could then be assetised, largely through astronomical commitments to purchase arms, for which the alibi was to secure this same 'need' for oil; then the scale and above all the longevity of oil production guaranteed, more than any other industry, a stream of payments far into the future. The forms of suburban life, mortgage payment, vehicle finance, and consumer debt that expanded with dependence on owning private cars and using petroleum products, along with the arms manufacturing, war-making, and military debt that ostensibly 'secured' these ways of life, multiplied the manufacture of credit.[46] That stream of payments, from the revenues of oil companies and the financing and 'protecting' of oil-dependent ways of living, could be capitalised and consumed in the present.

Very little of this assetising of the future represented the 'cost' of energy production or house building or vehicle manufacture or a secure form of life.[47] It represented the discounted present value of an extraordinary future revenue stream. What the oil companies invested in – along with the property developers, car firms, and other enterprises manufacturing the 'need' for energy, and the weapons firms dependent on allegedly 'securing' the oil regions of the Middle East, through public relations campaigns, think tanks, consumer advertising, political lobbying, the financing of military sales, the demonising of Islam, the so-called war on terror, the defence of Zionism, and many other kinds of promotion, state subsidy, sabotage, war-making, and opinion forming – was a future of continued dependence on carbon fuels.[48] The economy was an object largely built to stabilise and manage, through the device of its growth, the production of this future. The great acceleration of recent decades, and the climate collapse it has brought upon us, were driven not by the general development of capitalism, nor by its need for energy, but by the specific forms of capitalisation that could be manufactured through a carbon-heavy form of life.

The Ending

In conclusion, let us come back to the larger issues with which this chapter, and this book, opened. We have built modes of living that operate through the capture of wealth from the future, in which a minority is able to live off extraordinary levels of unearned income in the present. The majority, meanwhile, are increasingly burdened with the repayment of previous impositions and with managing the forms of precarity and risk through which their lives are governed. As those impositions increase, even moderate affluence can be achieved only through further assetisation, via the acquisition of property and the building of shares in retirement funds, ensuring the further impoverishment of the future.

The book has examined forms of capitalisation across several centuries and over a variety of countries and continents. The aim has not been to write a history of the development of the capitalising process. Such a goal would not only be too ambitious – it would bring the risks and misreadings inherent in any approach to history as development. Many of the examples from other times and places offered in the book have brought into view the disjunctive nature of previous episodes and stressed the significance of detours and interruptions. These reflect something important to our understanding of capitalisation. If the future occurs as a set of processes to be captured and assetised in the present, it should not be seen as simply the unfolding of present developments, but as a set of projects, now and in previous times, always open to diversion and setback. This disjunctive character arises from the uncertainty to which all capitalising calculation is exposed, without which the process of discounting the future and profiting from its diminishment could not operate.

The diversions and detours reflect the variety of powers and agencies, human and more than human, that capitalising schemes must engage. Capitalisation depends upon control of the future, but that future is composed of forces, organic and inorganic, stabilising and destabilising, that have their own powers, purposes, and modes of interaction.[49] The uncertainties and disjunctures generated by this dependence on the future were experienced across many centuries, in the risks encountered in long-distance trade, for example, or in the unpredictability of the advanced purchase of agricultural production, or in the hazards of organising credit for extended warfare. As opportunities for capitalisation expanded, seemingly with increased control over natural processes and human communities, the possibility for setback and unanticipated outcomes expanded too. In such terraforming projects as plantation agriculture, urban expansion, mineral extraction, dam building, communication infrastructures, and river diversion, disruptive effects followed from the consequences for the flood pulse of a river, the biotic and abiotic processes of soil formation, the movement of new species and pathogens, and, above all, the catastrophe of placing excess carbon in the atmosphere and warming the planet.

In exploring the modes of diversion, disjuncture, impoverishment, and enrichment that operate through the capture of the future, the book has been equally concerned with another kind of diversion – the production of an alibi, literally the 'not there', the 'elsewhere', the misdirection that encourages us to look to some other place for explanations of our form of life. The alibi of capital refers, first of all, to just that: to capital, as alibi. Capital *is* the alibi, in the sense that in its original meaning, derived from chattel or goods, and in its everyday meaning, as fund or asset, the word misdirects us to think in terms of a thing, not a process, and in terms of accumulation from the past or the production of commodities in the present rather than capture from the future. The alibi connotes another misdirection, the tendency to refer to the West, to the history of the North Atlantic world, in accounting for the origins and development of capitalism. To counter this tendency, the book has returned repeatedly to examples and episodes from elsewhere, and especially from one particular place, the country of Egypt. These detours serve as reminders that none of the processes we have examined here are specific to the history of the West. They pre-date the so-called rise of the West and have always operated across multiple locations, in fact often depending on spaces of colonial occupation or imperial contest. They also allowed us to shift attention from large structures and processes, like capitalism, to specific sites of capture, control, and comprehension, allowing us to see the work of capitalisation up close. Again and again, as it happens, Egypt provided a site for disjunctive and distinctive moments, whether in Muhammad Ali's building of an empire on the Nile in the early nineteenth century; or the unprecedented scale of dam building and river diversion later in that century, all explored in the opening chapters; or the extraordinary struggles over development, dispossession, and recolonisation in the decade of 1948–58, examined in chapter 5; or the experiments in assetisation as an answer to global poverty at the turn of the twenty-first century, described in chapter 6. But Egypt, at the same time, offers a place to consider what disappears from these accounts of the capitalising process, in tracing in chapter 2 the silencing of the waterwheel. The very history of the 'development' of 'capitalism' eliminated, not only another mode of life, epitomised by the waterwheel, but the wider history of the forms of life that the waterwheel enabled, recovered here in fragments.

The alibis of capital refer, finally, to a set of key concepts whose powerful role in blinding us to the capture of the future we have examined throughout the book, including technology, the economy, and its growth. In place of 'technology', which operates as a principle of innovation accounting for the course of history and the capture of great unearned wealth, we have preferred the term technopolitics, as a reminder that so-called technology is always an arrangement of diverse materials, ideas, and agencies, both novel and reused, held together by relations of force and justification. The 'economy', we have shown, is a surprisingly recent invention. It is not a misrepresentation, disguising or miscounting the real costs of capitalism. It is the apparatus that saved the process of capitalisation, allowing its reconstruction not as speculative instability, but as a seemingly stable mode of governing through the future. By organising the relation between the present and future, between today and tomorrow, it continues to govern us through the promise, today, of a better tomorrow, tomorrow. 'Growth', finally, is not the ineluctable course of capitalist development, but a recent device for the managing of capitalisation. As a relation to the future, it coexists with a second, contradictory relation – that of climate collapse. Climate collapse, although unavoidable, is recent. It is not 'capitalism'. If the climate crisis demands of us new political concepts with which to understand it, it also requires a different understanding of the language bequeathed to us by the carbon age – the economy and its 'growth'. Approaching capitalism and its growth in this more circumscribed way offers us, perhaps, an alternative way to engage with the current crisis – and a reason for modest hope.

Notes

Uber Eats: How Capitalism Consumes the Future

1. Sridhar Natarajan, 'Goldman (GS) Missing Uber IPO, Gets Potential $600M Consolation', Bloomberg UK, 6 May 2019, bloomberg.com (accessed 16 July 2025); Hubert Horan, 'Can Uber Ever Deliver? Part 20', Naked Capitalism, 30 May 2019, nakedcapitalism.com (accessed 16 July 2025). This and some of the following paragraphs are based on my article, 'Uber Eats: How Capitalism Consumes the Future', in Bruno Latour and Peter Weibel, eds, *Critical Zones: The Science and Politics of Landing on Earth*, Cambridge, MA: MIT Press and ZKM Karlsruhe, 2020: 84–8.
2. See chapter 3.
3. Johan Rockström et al., 'Safe and Just Earth System Boundaries', *Nature*, 619, 2023: 102–11.
4. Thorstein Veblen, *The Theory of the Business Enterprise*, New York: Charles Scribner's Sons, 1904; Jonathan Nitzan and Shimshon Bichler, *Capital as Power: A Study of Order and Creorder*, London: Routledge, 2009; Jonathan Levy, 'Capital as Process and the History of Capitalism', *Business History Review*, 91: 3, 2017: 483–510; Tim Di Muzio, ed., *The Capitalist Mode of Power: Critical Engagements with the Power Theory of Value*, London: Routledge, 2013; Tim Di Muzio and Richard H. Robbins, *Debt as Power*, Manchester: Manchester University Press, 2016; Carolyn Hardin, *Capturing Finance: Arbitrage and Social Domination*, Durham, NC: Duke University Press, 2021; Carl Wennerlind, *Casualties of Credit: The English Financial Revolution, 1620–1720*, Cambridge, MA: Harvard University Press, 2011. This approach to capital is also inspired, in more general terms, by the argument of Julie Graham and Katherine Gibson for a more deconstructive approach to capitalism, in J. K. Gibson-Graham, *The End of Capitalism (As We Knew It): A Feminist Critique of Political Economy*, Minneapolis, MN: University of Minnesota Press, 2006; by Anna Tsing's illuminating study of

'the possibility of life in capitalist ruins' in *The Mushroom at the End of the World*, Princeton, NJ: Princeton University Press, 2015; and by Michel Callon's insistence, in *Markets in the Making*, Brooklyn, NY: Zone Books, 2021, that we not award to capitalism, or to capitalist markets, an abstract logic that must in fact be continually reinvented. For other recent approaches to capitalism, speculation, and futurity, see Ritu Birla, *Stages of Capital: Law, Culture and Market Governance in Late Colonial India*, Durham, NC: Duke University Press, 2009; Arjun Appadurai, *The Future as Cultural Fact: Essays on the Global Condition*, London: Verso, 2013; Jens Beckert, *Imagined Futures: Fictional Expectations and Capitalist Dynamics*, Cambridge, MA: Harvard University Press, 2016; and Sandra Kemp and Jenny Andersson, eds, *Futures*, Oxford: Oxford University Press, 2021.

5. The discussion of Uber's corporate valuation here is based on the figures provided by Aswath Domadaran, 'Uber's Coming Out Party', Musings on Markets (blog), 15 April 2019, aswathdamodaran.blogspot.com; and 'Lyft Off?', Musings on Markets (blog), 7 March 2019, aswathdamodaran.blogspot.com (both accessed 28 June 2019), where the dependence of Uber's profits on 'market dominance' is laid out.
6. On capitalisation, see Nitzan and Bichler, *Capital as Power*, and many of the studies collected at capitalaspower.com (accessed 16 July 2025); Andrew Leyshon and Nigel Thrift, 'The Capitalization of Almost Everything: The Future of Finance and Capitalism', *Theory, Culture & Society*, 24: 7–8, 2007: 97–115; Fabian Muniesa et al., *Capitalization: A Cultural Guide*, Paris: Presses des Mines, 2017; Eli Cook, *The Pricing of Progress: Economic Indicators and the Capitalization of American Life*, Cambridge, MA: Harvard University Press, 2017; Gunnar Heinsohn and Otto Steiger, *Ownership Economics*, New York: Routledge, 2012; and Suresh Naidu, 'A Political Economy Take on W/Y', in Heather Boushey, J. Bradford DeLong, and Marshall Steinbaum, eds, *After Piketty: The Agenda for Economics and Inequality*, Cambridge, MA: Harvard University Press, 2017: 99–124. See also the studies of the recent expansion of assetisation, including Lisa Adkins, Melinda Cooper, and Martijn Konings, *The Asset Economy*, London and New York: Polity Press, 2020; Kean Birch and Fabian Muniesa, eds, *Assetization: Turning Things into Assets in Technoscientific Capitalism*, Cambridge, MA: MIT Press, 2020; and Brett Christophers, *Rentier Capitalism: Who Owns the Economy, and Who Pays for It?*, London: Verso, 2020.
7. In 2023 Uber reported an operating profit for the first time, of $1.1 billion; a year later it declared a profit of $2.8 billion. Profitability was made possible by the dubious practice of estimating appreciation in the paper value of untradeable shares it held in companies that had driven Uber out of business in China, Russia, and Southeast Asia; by a 'valuation release' from deferred tax assets (tax credits and other assets it could not claim until there was a likelihood of positive income to offset); and by continuing to reduce the share of customer payments going to drivers. Hubert Horan, 'Can Uber Ever Deliver? Part Thirty-Five', Naked Capitalism, 25 February 2025, nakedcapitalism.com (accessed 16 July 2025).

8. On the term 'apparatus of capture', which is from Gilles Deleuze and Félix Guattari, see chapter 1.
9. As chapter 1 discusses, below, some speculations can fail, leading to the destruction of captured gains. However, the destruction of capital is minor in comparison to the capture of the future.
10. Adkins, Cooper, and Konings, *The Asset Economy*.
11. Thorstein Veblen, *Absentee Ownership and Business Enterprise in Recent Times*, London: George Allen & Unwin, 1923: 86.
12. See J. G. A. Pocock, 'Commerce, Credit and Sovereignty: The Nation-State as Historical Critique', in Béla Kapossy et al., eds, *Markets, Morals, Politics: Jealousy of Trade and the History of Political Thought*, Cambridge, MA: Harvard University Press, 2018: 265–84.
13. Patrice Derrington, *Built Up: An Historical Perspective on the Contemporary Principles and Practices of Real Estate Development*, New York: Routledge, 2021: 86–106.
14. Richard K. Green and Susan M. Wachter, 'The American Mortgage in Historical and International Context', *Journal of Economic Perspectives*, 19: 4, 2005: 93–114.
15. Thomas Piketty and Gabriel Zucman, 'Capital is Back: Wealth-Income Ratios in Rich Countries 1700–2010', *The Quarterly Journal of Economics*, 129: 3, 2014: 1255–310; in conventional analyses, the process of capitalisation is attributed to the increasing cost of 'land': Katharina Knoll, Moritz Schularick, and Thomas Steger, 'No Price Like Home: Global House Prices, 1870–2012', *American Economic Review*, 107: 2, 2017: 331–53.
16. The term is from Cas Gilbert, 'The Financial Importance of Rapid Building', *Engineering Record*, 41, 30 June 1900: 624. See Sharon Irish, 'A "Machine That Makes the Land Pay": The West Street Building in New York', *Technology and Culture*, 30: 2, 1989: 376–97; and Lucia Allais et al., eds, *Architecture and Land in and out of the Americas*, New York: Temple Hoyne Buell Center for the Study of American Architecture, Graduate School of Architecture, Planning and Preservation, Columbia University, 2023: 79–82, buellcenter.columbia.edu (accessed 16 July 2025).
17. Maurizio Lazzarato, *The Making of the Indebted Man*, Cambridge, MA: MIT Press, 2012; Costas Lapavitsas, *Profiting Without Producing: How Finance Exploits Us All*, London: Verso, 2013; Michael Hudson, 'The Transition from Industrial Capitalism to a Financialized Bubble', *World Review of Political Economy*, 1: 1, 2010: 81–111; On the longer history, see David Graeber, *Debt: The First 5,000 Years*, 2nd edition, New York: Melville House, 2011.
18. Leyshon and Thrift, 'Capitalization of Almost Everything'; Christophers, *Rentier Capitalism*; Doug Henwood, *Wall Street: How it Works and For Whom*, London: Verso, 1998; Randy Martin, *Financialization of Daily Life*, Philadelphia, PA: Temple University Press, 2002.
19. Hubert Horan, 'Will the Growth of Uber Increase Economic Welfare?', *Transportation Law Journal*, 44, 2017: 33–105.
20. Mike Isaac, *Super Pumped: The Battle for Uber*, New York: W.W. Norton & Co., 2019: 58, 84–7.

21. P. R. Morris, *A History of the World Semiconductor Industry*, London: Institution of Engineering and Technology in association with the Science Museum, 1990: 27–37.
22. Katrina M. Wyman, 'Taxi Regulation in the Age of Uber', *N.Y.U. Journal of Legislation and Public Policy*, 20: 1, 2017: 24–9.
23. Uber Technologies Inc., *Annual Report 2022*, US Securities and Exchange Commission, 2022, sec.gov (accessed 16 July 2025).
24. For an overview of this history of US corporate monopoly, see Matt Stoller, *Goliath: The 100-Year War Between Monopoly Power and Democracy*, New York: Simon & Schuster, 2020. For the widening use of predatory pricing, see Matthew T. Wansley and Samuel N. Weinstein, 'Venture Predation', *Journal of Corporation Law*, 48, 2023: 813.
25. Hubert Horan, 'Uber's Path of Destruction', *American Affairs*, 3: 2, 2019: 108–33; Hubert Horan, 'Can Uber Ever Deliver? Part Thirty-Three', Naked Capitalism, 9 August 2023, nakedcapitalism.com (accessed 16 July 2025); Horan, 'Can Uber Ever Deliver? Part Thirty-Five'.
26. David Edgerton, *The Shock of the Old*, Oxford: Oxford University Press, 2007.
27. Classic statements include, in economic history, G. N. von Tunzelmann, *Steam Power and British Industrialisation to 1860*, Oxford: Clarendon Press, 1978; and David Landes, *The Unbound Prometheus: Technological Change and Industrial Development in Western Europe from 1750 to the Present*, Cambridge: Cambridge University Press, 2003 [1969]; and in economics, Robert Solow, 'Technical Change and the Aggregate Production Function', *Review of Economics and Statistics*, 39, 1957: 312–20; and J. Bradford DeLong and Lawrence H. Summers, 'Equipment Investment and Economic Growth', *Quarterly Journal of Economics*, 106: 2, 1991: 445–502.
28. Michel Foucault, *Discipline and Punish: The Birth of the Prison*, New York: Pantheon, 1977; Joseph Vogl, *The Specter of Capital*, Stanford, CA: Stanford University Press, 2015.
29. Peter Cohen et al., 'Using Big Data To Estimate Consumer Surplus: The Case of Uber', NBER Working Paper No. 22627, September 2016, nber.org (accessed 16 July 2025).
30. Felicity Lawrence, 'Uber Paid Academics Six-Figure Sums for Research to Feed to the Media', *The Guardian*, 12 July 2022.
31. Sheila Jasanoff, 'Technologies of Humility: Citizen Participation in Governing Science', *Minerva*, 41: 3, 2003: 223–44, reprinted in Sheila Jasanoff, *Science and Public Reason*, Abingdon, UK: Routledge, 2012: 167–84.
32. Stephen J. Dubner, 'Why Uber Is an Economist's Dream', *Freakonomics* (podcast), 7 September 2016, freakonomics.com (accessed 24 May 2019).
33. Horan, 'Growth of Uber'.
34. Municipal regulation was far from perfect, but open to local protest, political organising, and reform in ways that a global company with near monopoly of the industry may not be. London, New York, and California subsequently attempted to subject Uber to labour law, with partial success.
35. Uber acknowledged that 'public transportation can be a superior substitute

to our Mobility offering and, in many cases, offers a faster and lower-cost travel option in many cities': Uber Technologies Inc., *Annual Report 2022*.

36. National Academies of Sciences, Engineering, and Medicine, *The Role of Transit, Shared Modes, and Public Policy in the New Mobility Landscape*, Washington, DC: The National Academies Press, 2021.
37. Schaller Consulting, *The New Automobility: Lyft, Uber and the Future of American Cities*, 25 July 2016, schallerconsult.com (accessed 16 July 2025).
38. John M. Barrios, Yael Hochberg, and Hanyi Yi, 'The Cost of Convenience: Ridehailing and Traffic Fatalities', *Journal of Operations Management*, 69: 5, 2023: 823–55.
39. Stefano Sgambati, 'Rethinking Banking: Debt Discounting and the Making of Modern Money as Liquidity', *New Political Economy*, 21: 3, 2016: 274–90.
40. Liliana Doganova, *Discounting the Future: The Ascendancy of a Political Technology*, Brooklyn, NY: Zone Books, 2024; Fabian Muniesa and Liliana Doganova, 'The Time that Money Requires: Use of the Future and Critique of the Present in Financial Valuation', *Finance and Society*, 6: 2, 2020: 95–113.
41. Michael Hudson and Dirk Bezemer, 'Incorporating the *Rentier* Sectors into a Financial Model', *World Economic Review*, 1, 2012: 1–12. Moreover, as we will see in chapter 8, even so-called costs of production include significant charges for the payment of interest, a rent imposed by the financial sector thanks to its monopolistic control of credit creation.
42. Greta Krippner, *Capitalizing on Crisis: The Political Origins of the Rise of Finance*, Cambridge, MA: Harvard University Press, 2011: 4. Her definition draws on Giovanni Arrighi, *The Long Twentieth Century: Money, Power, and the Origins of Our Times*, London: Verso, 1994. On the problematic distinction between real and financial, see Nitzan and Bichler, *Capital as Power*.
43. Karuna Mantena uses the term alibi to refer to another 'elsewhere', a process of displacement in reverse, where sources of justification for empire are moved from metropole to colony: *Alibis of Empire: Henry Maine and the Ends of Liberal Imperialism*, Princeton, NJ: Princeton University Press, 2010. I take up this aspect of the alibi of capital in chapter 4.

1 Groundbreaking

1. Sources for this and the following paragraph are given in chapter 2.
2. Lord Cromer, 'The Government of Subject Races', *The Edinburgh Review*, 1908: 22–3, reprinted in *Political and Literary Essays, 1908–13*, London: Macmillan, 1913.
3. David C. Coleman, Mac A. Callaham, Jr, and D. A. Crossley, Jr, *Fundamentals of Soil Ecology*, 3rd edition, London: Elsevier, 2018.
4. This and the following paragraph draw on Jan Zalasiewicz et al., 'Scale and Diversity of the Physical Technosphere: A Geological Perspective', *The Anthropocene Review*, 4: 1, 2016: 9–22.
5. Benjamin Fong Chao, 'Anthropogenic Impact on Global Geodynamics Due

to Reservoir Water Impoundment', *Geophysical Research Letters*, 22: 24, 1995: 3529–32.

6. Simon J. Price et al., 'Humans Are the Most Significant Global Geomorphological Driving Force of the 21st Century', *The Anthropocene Review*, 5: 3, 2018: 222–9.
7. Peter Haff, 'Technology as a Geological Phenomenon: Implications for Human Well-Being', in C. N. Waters et al., *A Stratigraphical Basis for the Anthropocene*, London: Geological Society, 2014: 301–9; see also Chris Otter, 'Socializing the Technosphere', *Technology & Culture*, 63: 4, 2022: 953–78.
8. George Monbiot, *Regenesis: Feeding the World Without Devouring the Planet*, New York: Penguin Books, 2022: 78.
9. Zalasiewicz et al., 'Scale and Diversity'.
10. Vaclav Smil, *Harvesting the Biosphere*, Cambridge, MA: MIT Press, 2012: 226.
11. Zalasiewicz et al., 'Scale and Diversity'; J. R. Ford et al., 'An Assessment of Lithostratigraphy for Anthropogenic Deposits', in C. N. Waters et al., *A Stratigraphical Basis for the Anthropocene*: 55–89.
12. Alexandra Arènes, Bruno Latour, and Jérôme Gaillardet, 'Giving Depth to the Surface: An Exercise in the Gaia-graphy of Critical Zones', *The Anthropocene Review*, 5: 2, 2018: 120–35.
13. Timothy M. Lenton et al., 'Selection for Gaia across Multiple Scales', *Trends in Ecology & Evolution*, 33: 8, 2018: 633–45.
14. Bruno Latour, *An Inquiry into Modes of Existence*, Cambridge, MA: Harvard University Press: 225. On a material approach to temporality, see Andreas Folkers, 'Fossil Modernity: The Materiality of Acceleration, Slow Violence, and Ecological Futures', *Time and Society*, 30: 2, 2021: 223–46.
15. Vaclav Smil, *Creating the Twentieth Century: Technical Innovations of 1867–1914 and Their Lasting Impact*, New York: Oxford University Press, 2005: 5–6.
16. Mark Williams et al., 'Underground Metro Systems: A Durable Geological Proxy of Rapid Urban Population Growth and Energy Consumption during the Anthropocene', in Craig Benjamin, Esther Quaedackers, and David Baker, eds, *The Routledge Companion to Big History*, London: Taylor and Francis, 2019: 434–55.
17. William Thomas Hogan, *Economic History of the Iron and Steel Industry in the United States*, vol. 5, Lexington, ID: Lexington Books, 1971; Thomas Kelly and Grecia R. Matos, 'Historical Statistics for Mineral and Material Commodities in the United States', *U.S. Geological Survey Data Series*, 140, 2014.
18. Stephen Gaukroger, *Civilization and the Culture of Science: Science and the Shaping of Modernity, 1795–1935*, New York: Oxford University Press, 2022: 331; Vaclav Smil, *Still the Iron Age: Iron and Steel in the Modern World*, Amsterdam: Elsevier, 2016: 56.
19. Alfred Chandler, *The Visible Hand: The Managerial Revolution in American Business*, Cambridge, MA: Belknap Press, 1977: 7.
20. Richard White, *Railroaded: The Transcontinentals and the Making of Modern*

America, New York: W.W. Norton & Co., 2007. On the failure of the Suez Canal to meet expectations for increased trade, see D. A. Farnie, *East and West of Suez: The Suez Canal in History, 1854–1956*, Oxford: Clarendon Press, 1969.

21. Joseph Schumpeter, *Business Cycles: A Theoretical, Historical and Statistical Analysis of the Capitalist Process*, 2 vols, Philadelphia, PA: Porcupine Press, 1939: 1: 260–8. Schumpeter describes (on p. 67) the process in the United States of 'promoters securing options of right of way, having the company chartered and endowed with land grants, selling the options to it and taking securities in payment, finally placing the bonds – the stock being commonly treated as a bonus – in order to provide the means for construction, and buying equipment on installments through equipment trust certificates'. This process, he remarks (on p. 261) 'inaugurated the colonization of a great part of the country'. On vendor shares, see Charles Kindleberger, *A Financial History of Western Europe*, Abingdon, UK: Routledge, 1984: 201.
22. Jonathan Nitzan and Shimshon Bichler, *Capital as Power: A Study of Order and Creorder*, London: Routledge, 2009; Jonathan Levy, 'Capital as Process and the History of Capitalism', *The Business History Review*, 91: 3, 2017; Liliana Doganova, *Discounting the Future: The Ascendancy of a Political Technology*, Brooklyn, NY: Zone Books, 2024.
23. Gilles Deleuze and Félix Guattari, *A Thousand Plateaus: Capitalism and Schizophrenia*, trans. from the French by Brian Massumi, Minneapolis, MN: University of Minnesota Press, 1987: 424–73. See also Maurizio Lazzarato, *Governing by Debt*, Cambridge, MA: MIT Press, 2015, and Carolyn Adkins, *Capturing Finance: Arbitrage and Social Domination*, Durham, NC: Duke University Press, 2021.
24. Max Weber, 'Stock and Commodity Exchanges ["Die Börse" (1894)]', *Theory and Society*, 29: 3, 2000: 315–16.
25. William Robert Scott, *The Constitution and Finance of English, Scottish and Irish Joint-Stock Companies to 1720*, vols 1–3, Cambridge: Cambridge University Press, 1910–12; Oliver C. Cox, *The Foundations of Capitalism*, New York: Philosophical Library, 1959.
26. John Law, 'On the Methods of Long-Distance Control: Vessels, Navigation, and the Portuguese Route to India', in John Law, ed., *Power, Action and Belief: A New Sociology of Knowledge? Sociological Review*, 32: 1 Suppl., 1984: 234–63.
27. Patricia Seed, *Ceremonies of Possession in Europe's Conquest of the New World, 1492–1640*, Cambridge: Cambridge University Press, 1995: 100–48; John Hobson, *The Eastern Origins of Western Civilisation*, Cambridge: Cambridge University Press, 2004: 134–57.
28. For a wide-ranging history, see Philip J. Stern, *Empire, Incorporated: The Corporations That Built British Colonialism*, Cambridge, MA: Harvard University Press, 2023.
29. Ron Harris, *Going the Distance: Eurasian Trade and the Rise of the Business Corporation, 1400–1700*, Princeton, NJ: Princeton University Press, 2020: 311–12.
30. For the London Stock Exchange in the early 1870s, government securities

accounted for more than half the securities quoted, railways almost one-third; commercial and industrial, only 1.4 per cent. In the fifty years before the First World War, 70 per cent of new share issues were firms in transportation, utilities, and communication. Caroline Fohlin, *Mobilizing Money: How the World's Richest Nations Financed Industrial Growth*, Cambridge: Cambridge University Press, 2011: 30. For France, see Michael S. Smith, 'Putting France in the Chandlerian Framework: France's 100 Largest Industrial Firms in 1913', *The Business History Review*, 72: 1, 1998: 52–3.

31. George Goschen, the leading British theorist of finance, whose banking firm, Fruhling and Goschen, became the largest supplier of credit to Egypt in that period, noted how 'the Joint-Stock companies system invaded every department of banking, commerce, and industrial enterprise'. (After the interest and fees on that credit drove the country into bankruptcy, Goschen then organised the system of foreign financial control of Egyptian revenues.) George Goschen, *Essays and Addresses on Economic Questions*, London: E. Arnold: 1. I am grateful to Alaa El-Shafei for this reference.
32. Paddy Ireland, 'Capitalism without the Capitalist: the Joint Stock Company Share and the Emergence of the Modern Doctrine of Separate Corporate Personality', *The Journal of Legal History*, 17: 1, 1996: 46. On the broader legal structure on which capitalisation depended, see Katharina Pistor, *The Code of Capital: How the Law Creates Wealth and Inequality*, Princeton, NJ: Princeton University Press, 2019.
33. J. Lawrence Broz and Richard S. Grossman, 'Paying for Privilege: The Political Economy of Bank of England Charters, 1694–1844', *Explorations in Economic History*, 41: 1, 2004.
34. Richard L. Grossman, Frank T. Adams, and Charles Levenstein, 'Taking Care of Business: Citizenship and the Charter of Incorporation', *New Solutions: A Journal of Environmental and Occupational Health Policy*, 3: 3, 1993.
35. Leonard W. Hein, 'The British Business Company: Its Origins and Its Control', *The University of Toronto Law Journal*, 15: 1, 1963: 140.
36. On economisation, see chapter 4. On scientific conceptions of probability and chance, see Ian Hacking, *The Taming of Chance*, Cambridge: Cambridge University Press, 1990; and Theodore M. Porter, *The Rise of Statistical Thinking, 1820–1900*, Princeton, NJ: Princeton University Press, 2020.
37. Jonathan Levy, *Freaks of Fortune: The Emerging World of Capitalism and Risk in America*, Cambridge, MA: Harvard University Press, 2012.
38. William Cronon, *Nature's Metropolis: Chicago and the Great West*, New York: Norton, 1992; Koray Çalişkan, *Market Threads: How Cotton Farmers and Traders Create a Global Commodity*, Princeton, NJ: Princeton University Press, 2010. On derivatives, see the further discussion in chapter 8, below.
39. See Timothy Mitchell, *Carbon Democracy: Political Power in the Age of Oil*, London: Verso, 2011: 43–65.
40. Eric Hobsbawm, *The Age of Empire: 1875–1914*, London: Weidenfeld and Nicolson, 1987. Cf. John Darwin, *The Empire Project: The Rise and Fall of the British World-System, 1830–1970*, Cambridge: Cambridge University Press, 2009: 64–111; and Charles Bright and Michael Geyer, 'Benchmarks of Globalization: The Global Condition, 1850–2010', in Douglas

Northrop, ed., *A Companion to World History*, Oxford: Blackwell, 2012: 285–300.

41. On the Suez Canal, see Ibrahim Elhoudaiby, 'Universalizing Egypt, 1854–1876: Suez Canal, Debt, Corvée, and the Rise of Modern Government', PhD thesis, Middle Eastern, South Asian, and African Studies, Columbia University, 2022; Valeska Huber, *Channelling Mobilities: Migration and Globalisation in the Suez Canal Region and Beyond*, Cambridge: Cambridge University Press, 2013. For the Baghdad Railway, see Mitchell, *Carbon Democracy*; for US railways, see White, *Railroaded*; for Argentinian railways, see Lewis, *British Railways in Argentina, 1857–1914: A Case Study of Foreign Investment*, London: Athlone, 1983. Lewis shows that railways in the early decades were unable to compete with bullock carts in price or convenience and were built for property speculation and for the ability to capitalise their modest but long-term, state-guaranteed returns. On Indian railways, see Dan Bogar and Latika Chaudhary, 'Railways in Colonial India: An Economic Achievement?', in Latika Chaudhary et al., eds, *A New Economic History of Colonial India*, London: Routledge, 2016.
42. On the Bengal Delta, see Iftekhar Iqbal, *The Bengal Delta: Ecology, State and Social Change, 1840–1943*, Basingstoke: Palgrave Macmillan, 2010. On the Orissa Delta, see Rohan D'Souza, *Drowned and Dammed: Colonial Capitalism and Flood Control in Eastern India*, New Delhi: Oxford University Press, 2006. On the Indus Basin, see David Gilmartin, *Blood and Water: The Indus River Basin in Modern History*, Berkeley, CA: University of California Press, 2020.
43. On railways, see chapter 7.
44. Christophe Bonneuil and Jean-Baptiste Fressoz, *The Shock of the Anthropocene: The Earth, History and Us*, London: Verso, 2016.
45. Marx had identified the 'capitalist mode of production' by the time he wrote *The Communist Manifesto* (1848), but the term capitalism was introduced later and at first used mainly in the more specific sense described here.
46. Alfred Marshall noted that the 'Industrial Revolution of the present generation . . . has far outdone the changes of a century ago, in both rapidity and breadth of movement': *Principles of Economics*, 5th edition, London: Macmillan and Co., 1907: vi; J. A. Hobson's *The Evolution of Modern Capitalism*, 2nd edition, London: The Walter Scott Publishing Co., 1906, described the 'radical change' brought by two recent shifts, the displacement of privately held firms by joint-stock enterprises and the new place occupied by 'a product of the joint-stock company', the financier (236–8); David Ames Wells argued that 'the economic changes that have occurred during the last quarter of a century . . . have unquestionably been more important and varied than during any former corresponding period of the world's history'. *Recent Economic Changes and Their Effect on the Production and Distribution of Wealth and the Well-Being of Society*, New York: D. Appleton and Company, 1890: v. On the influence of Wells, see Michael Perelman, 'Retrospectives: Schumpeter, David Wells, and Creative Destruction', *Journal of Economic Perspectives*, 9: 3, 1995: 189–97, and James Livingston, 'The Social Analysis of Economic History and Theory: Conjectures on Late Nineteenth-Century

American Development', *American Historical Review*, 92: 1, 1987: 75, which describes *Recent Economic Changes* as 'probably the most cited book of the period 1890–1910'.

47. In German, the appearance of the term capitalism is associated with Werner Sombart's *Der moderne Kapitalismus*, Leipzig: Duncker and Humblot, 1902. In French, the term *capitalisme* had appeared earlier, in the third decade of the nineteenth century, in a more specific sense – to refer to the new system of war finance emerging over recent decades, in which financiers or 'capitalistes' provided the state with credit for war-making. Michael Sonenscher, *Capitalism: The Story Behind the Word*, Princeton, NJ: Princeton University Press, 2022; Eve Chiapello, 'Accounting and the Birth of the Notion of Capitalism', *Critical Perspectives on Accounting*, 18: 3, 2007: 263–96.
48. Abdel-Maksud Hamza, *The Public Debt of Egypt, 1854–1876*, Cairo: Government Press, 1944: 181–6; Pat Thane, 'Financiers and the British State: The Case of Sir Ernest Cassel', *Business History*, 28: 1, 1986: 80–99; Kurt Grunwald 'On Cairo's Lombard Street', *Tradition: Zeitschrift für Firmengeschichte und Unternehmerbiographie*, 17: 1, 1972: 8–22; Jennifer Derr, *The Lived Nile: Environment, Disease, and Material Colonial Economy in Egypt*, Stanford, CA: Stanford University Press, 2019.
49. Pat Thane, 'Cassel, Sir Ernest Joseph (1852–1921)', *Oxford Dictionary of National Biography*, Oxford: Oxford University Press, 2004, and 'Financiers and the British State'; Paul H. Emden, *Money Powers of Europe in the Nineteenth and Twentieth Centuries*, London: Sampson Low, Marston, & Co., 1938: 331.
50. Matthias B. Lehmann, *The Baron: Maurice de Hirsch and the Jewish Nineteenth Century*, Stanford, CA: Stanford University Press, 2022: 223–46.
51. Naomi W. Cohen, *Jacob H. Schiff: A Study in American Jewish Leadership*, Hanover, NH: University Press of New England, 1999; Kurt Grunwald, '"Windsor-Cassel" – The Last Court Jew: Prolegomena to a Biography of Sir Ernest Cassel', in *The Leo Baeck Institute Year Book 14: 1*, 1969: 119–61.
52. Paul H. Emden, *Jews of Britain: A Series of Biographies*, London: Sampson Low, Marston, & Co., 1944: 335; Grunwald, '"Windsor-Cassel"': 129–32. The original Bessemer process could not remove phosphorous impurities from iron ore, but in the late 1870s Sidney Gilchrist Thomas developed a method of lining the converter with dolomite or limestone, solving the problem.
53. The Egyptian press drew parallels between the Kom Ombo plantation and Zionist settlements in Palestine. Samir Raafat, 'The Ephemeral Republic of Kom-Ombo', *Egyptian Mail*, 3 February 1996.
54. Emden, *Jews of Britain*: 337.
55. On the Mahdi and Ottoman-Egyptian colonisation, see Hengameh Ziai, 'Ploughing for the Hereafter: Debt, Time, and Mahdist Resistance in Northern Sudan, 1821–1935', PhD thesis, Middle Eastern, South Asian, and African Studies Department, Columbia University, 2021.
56. Mitchell, *Carbon Democracy*; Kurt Grunwald, '"Windsor-Cassel"': 147–8.
57. On Willcocks, see chapter 2, and Çanay Ozden, 'The Pontifex Minimus: William Willcocks and Engineering British Colonialism', *Annals of Science*, 7: 2, 2014: 183–205.
58. Cohen, *Jacob H. Schiff*: 171–4: 'Among the repeated arguments that Schiff

mustered against Jewish political nationalism, religion was the most important. Calling himself a "faith Jew" rather than a "race Jew", he adhered to the classical Reform position . . . that Judaism was only a religion and that Jews, the priest people, had been scattered among the nations by divine purpose to teach the unity of God and brotherhood of man. Ingathering into Palestine contradicted the very raison d'être of Jewish existence and, as Schiff thought, threatened the essence of Judaism' (p. 175). On the 'Mesopotamia Program' for Jewish colonisation in Iraq, see Moshe Perlmann, 'Paul Haupt and the Mesopotamian Project, 1892–1914', *Publications of the American Jewish Historical Society*, 47: 3, 1958: 154–75.

59. John Maynard Keynes, *The Collected Writings of John Maynard Keynes*, ed. Donald Moggridge, vol. 12, London: Macmillan, 1971–89: 3–8 (thanks to Stefan Eich for this reference). Cassel also created a trust to support education, including an endowed chair in economics at the London School of Economics.
60. On developments outside the main money markets of the West, see, for example, Ritu Birla, *Stages of Capital: Law, Culture, and Market Governance in Late Colonial India*, Durham, NC: Duke University Press, 2009.
61. On the influence of Say compared with the Scottish and English political economists, see Keith Tribe, *The Economy of the Word: Language, History, and Economics*, New York: Oxford University Press, 2015: 22.
62. On Smith, see Eugen von Böhm-Bawerk, *The Positive Theory of Capital*, New York: G.E. Stechert & Co, 1923: 26–30.
63. Jean-Baptiste Say, *A Treatise on Political Economy, or The Production, Distribution, and Consumption of Wealth*, London: Longman, Hurst, Rees, Orme, and Brown, 1821: 351, 356, 359. Capital, he added, should not be confused with money, the temporary form it takes 'but for a moment' as it is transferred from one productive purpose to another.
64. Leon Walras, *Elements of Pure Economics or the Theory of Social Wealth*, New York: A.M. Kelley, 1969: 84. Donald A. Walker, 'Early General Equilibrium Economics: Walras, Pareto, and Cassel', in Warren J. Samuels et al., eds, *A Companion to the History of Economic Thought*, Malden, MA: Blackwell Publishing, 2003.
65. Alex Preda, *Framing Finance: The Boundaries of Markets and Modern Capitalism*, Chicago, IL: University of Chicago Press, 2009: 113–43.
66. John Handel, 'The Material Politics of Finance: The Ticker Tape and the London Stock Exchange, 1860s–1890s', *Enterprise & Society*, 23: 3, 2022: 857–87.
67. William Stanley Jevons, *The Theory of Political Economy*, London: Macmillan and Co, 1931: 85.
68. Quinn Slobodian, 'How to See the World Economy: Statistics, Maps, and Schumpeter's Camera in the First Age of Globalization', *Journal of Global History*, 10: 2, 2015: 322–31.
69. Thorstein Veblen used the term 'neoclassical' as early as 1900: 'The Pre-Conceptions of Economic Science', *Quarterly Journal of Economics*, 14: 2, 1900.
70. Irving Fisher, *The Rate of Interest: Its Nature, Determination and Relation to*

Economic Phenomena, New York: Macmillan, 1907: 61. Fisher, a supporter of eugenics, further explained this impatience as the product, in part, of those with poor imaginations, and poor health and life expectancy.

71. Henry R. Seager, 'The Impatience Theory of Interest', *The American Economic Review*, 2: 4, 1912: 834–51.
72. Evert Schoorl, *Jean-Baptiste Say: Revolutionary, Entrepreneur, Economist*, New York: Routledge, 2012: 38–71. See also G. Koolman, 'Say's Conception of the Role of the Entrepreneur', *Economica*, New Series, 38: 151, 1971: 269–86.
73. Joseph Schumpeter, *The Theory of Economic Development: An Inquiry into Profits, Capital, Credit, Interest, and the Business Cycle*, Cambridge, MA: Harvard University Press, 1934. First published as *Theorie der wirtschaftlichen Entwicklung*, Leipzig: Duncker & Humblot, 1911.
74. Schumpeter introduced the term 'creative destruction' in *Capitalism, Socialism and Democracy*, New York: Harper & Brothers, 1947: 83.
75. For an influential discussion, see William Sewell, 'The Temporalities of Capitalism', *Socio-Economic Review*, 6: 3, 2008: 517–37.
76. David Landes, *Bankers and Pashas: International Finance and Economic Imperialism in Egypt*, Cambridge, MA: Harvard University Press, 1979.
77. Elizabeth Holt, *Fictitious Capital: Silk, Cotton, and the Rise of the Arabic Novel*, New York: Fordham University Press, 2017: 121, 133. The first quote is from the novel, Ya'qub Sarruf, *Fatat Misr*, 4th edition, Cairo: Maṭbaʿat al-Muqtaṭaf, 1922: 111; the second is from a French diplomat.
78. A. B. (Amédée Baillot) de Guerville, *New Egypt*, London: Heinemann, 1906: 138, reporting on a conversation with Princess Nazli Hanim.
79. 'La Krach des sucres: Le suicide de M. Cronier', *L'Humanité*, 29 August 1905: 1.
80. *Le Figaro*, 30 August 1905: 3.
81. See chapter 3 and the argument that industrialisation can be seen as a detour.
82. For details, see Samir Saul, *La France et l'Égypte de 1882 à 1914: Intérêts économiques et implications politiques*, Paris: Institut de la gestion publique et du développement économique, Comité pour l'histoire économique et financière de la France, 2013: 375–475.
83. 'The Bankruptcy of the Egyptian Sucreries', *The Economist*, 4 November 1905, 1754–5, describing the Report of Experts set up to investigate the collapse of the firm. The report stated that the local administrator, Mr A. J. Davey, wrote to Cronier on 15 August, having discovered that important matters were being kept from him. *The Economist* describes the fraud: 'For five years past the Egyptian sugar refineries have been apparently prospering in the midst of ruinous losses on their exploitations . . . Regular dividends were paid, excessive emoluments were given to the members of the board, and high salaries to an overgrown staff, while expensive plant of the most costly description was laid down in four factories, and all this out of a fictitious revenue. It was the simple ability and the prestige of the late M. Cronier that held these complex affairs in equilibrium, and with such success, that not only the outside world was convinced of the company's soundness and prosperity, but it seems that at least one of the members of the board in

Cairo, Mr. A.J. Davey, the English administrator, held the same delusion – up to August 15th last', when he wrote to Cronier.

84. 'La Krach des sucres', *L'Humanité*, 30 August 1905: 1.
85. Saul, *La France et l'Égypte*: 375–475. See also Derr, *The Lived Nile*: 76–92, and Jean Mazuel, *Le sucre en Égypte, étude de géographie historique et économique*, Cairo: Imprimé par E. & R. Schindler, 1937.
86. Roger Owen, *Cotton and the Egyptian Economy, 1820–1914: A Study in Trade and Development*, Oxford: Clarendon Press, 1969: 282–3.
87. Alexander Noyes, *Forty Years of American Finance*, 2nd edition, New York: G.P. Putnam's Sons, 1909: 325, referring to the chairman's address to the shareholders meeting of the [National] Bank of Egypt, 6 March 1908.
88. Roger Owen, *Lord Cromer: Victorian Imperialist, Edwardian Proconsul*, Oxford: Oxford University Press, 2004: 328–41.
89. Noyes, *Forty Years of American Finance*: 360. A recent study tends to confirm Noyes' thesis about the global origins and spread of the crisis: Mary Tone Rodgers and James Payne, 'Was the Panic of 1907 a Global Crisis? Testing the Noyes Hypothesis', Workshop on Monetary and Financial History, Federal Reserve Bank of Atlanta, 2015, frbatlanta.org (accessed 20 July 2025).
90. Schumpeter notes in the preface to the English translation of the *Theory of Economic Development* that 'Some of the ideas submitted in this book go back as far as 1907; all of them had been worked out by 1909, when the general framework of this analysis of the purely economic features of capitalistic society took the shape which has remained substantially unaltered ever since.' For about half that period he was in Cairo, from late 1907 to the end of October 1908. (The English translation of 1934 was based on the third German edition, a reprint of the second edition of 1926, for which Schumpeter omitted the seventh chapter and rewrote the second and the sixth ('Preface to the English Edition': ix).) On Schumpeter's sojourn in Cairo, see Casey Primel, 'Calculating Futures: Debt, Markets, and the Science of Prices in Colonial Egypt, 1882–1912', PhD thesis, Middle Eastern, South Asian, and African Studies Department, Columbia University, 2016.
91. Schumpeter, *The Theory of Economic Development*: 176.
92. Irving Horowitz, *Opening Doors: Life and Work of Joseph Schumpeter*, vol. 1: Europe, New York: Routledge, 2017: 67–8. 'On 2nd March 1908, he completed the writing of his first book, *Das Wesen und der Hauptinhalt der theoretischen Nationalökonomie*. This foreshadowed the theory of interest developed more fully in *Theory of Economic Development*'. He spent 1913–14 at Columbia University, as an exchange professor. Seymour Edwin Harris, ed., *Schumpeter: Social Scientist*, Cambridge, MA: Harvard University Press: 27.
93. William Lazonick, 'The Functions of the Stock Market and the Fallacies of Shareholder Value', in Ciaran Driver and Grahame Thompson, eds, *Corporate Governance in Contention*, Oxford: Oxford University Press, 2018, 117–51.
94. Schumpeter, *The Theory of Economic Development*: 151–2.
95. Thomas McCraw, *Prophet of Innovation: Joseph Schumpeter and Creative Destruction*, Cambridge, MA: Belknap Press, 2010: 260. McCraw does not name the princess, but it was most likely the Austro-Hungarian Marianna Török de Szendro, later known as Djavidan Hanum, author of the memoir

Harem Life, London: Noel Douglas, 1931. She became first the mistress then the second wife of Abbas Hilmi II, from around 1900 till their marriage in 1910. Samir Raafat notes: 'She lived in splendor in Mostorod Palace near Matarieh [outside Cairo]. The fantastic property came with a large garden and extensive agriculture domains, whose revenue was assigned to her. She was the good spirit for Europeans at the Khedivial court.' 'Queen for a Day', *Ahram Weekly*, 6 October 1994, republished on egy.com (accessed 18 July 2025). On the connections of the Ottoman-Egyptian with Vienna, see Adam Mestyan, 'A Muslim Dualism? Inter-Imperial History and Austria-Hungary in Ottoman Thought, 1867–1921', *Contemporary European History*, 30: 4, 2021: 478–96.

96. George Manville Fenn, *The Khedive's Country: the Nile Valley and Its Products*, London: Cassell and Company Ltd, 1904.
97. Schumpeter, *A Theory of Economic Development*.
98. See Rana Baker, 'Engineering Profit: Egyptian Railroads and the Unmaking of Prosperity 1847–1907', PhD thesis, Middle Eastern, South Asian, and African Studies Department, Columbia University, 2023. Towards the end of Cromer's time in Egypt one begins to see occasional references to development in a somewhat wider sense, to encompass land reclamation, irrigation schemes, and railway building, especially in Sudan. See, for example, Sir Auckland Colvin, *The Making of Modern Egypt*, London: T. Nelson & Sons, 1910. Schumpeter wanted the term 'development' to indicate change that was endogenous to the economic process, rather than exogenous, driven by the force of the entrepreneur. He would have preferred to use the term 'evolution', but disliked its Darwinian associations: *Theory of Economic Development*: 49–51.
99. Agricultural yields declined by 20 per cent between 1900 and 1920, and did not recover to their nineteenth-century levels until the 1930s. A. E. Crouchley, *The Economic Development of Modern Egypt*: 154–7; Bent Hansen, *The Political Economy of Poverty, Equity, and Growth: Egypt and Turkey*, Oxford: Oxford University Press, 1991: 104–5.

2 On Rivercide

1. William Willcocks, *Sixty Years in the East*, Edinburgh and London: William Blackwood & Sons, 1935: 212.
2. On the development of debt and banking in Egypt, see 'Abd al-'Aziz 'Izz al-'Arab, *European Control and Egypt's Traditional Elites: A Case Study in Elite Egyptian Nationalism*, Lewiston, NY: Edwin Mellen Press, 2002; and Ali Coşkun Tunçer, *Sovereign Debt and International Financial Control: The Middle East and the Balkans, 1870–1914*, New York: Palgrave Macmillan, 2015: 29–52.
3. On the nationalist uprising of 1881–2, known as the 'Urabi Revolution, see Alexander Schölch, *Egypt for the Egyptians! The Socio-Political Crisis In Egypt, 1878-1882*, London: Ithaca Press, 1981; and Juan Cole, *Colonialism and Revolution in the Middle East: Social and Cultural Origins of Egypt's*

Urabi Movement, Princeton, NJ: Princeton University Press, 1993; on the role of the Egyptian elite in supporting intervention, see Nadeem Mansour, 'A New Ruling Class and its Empires: The Case of Nineteenth-Century Egypt', PhD thesis, Middle Eastern, South Asian, and African Studies Department, Columbia University, 2025; on initial British motives, including the desire to defeat popular revolt out of fear of Irish nationalism, see John Darwin, *The Empire Project: The Rise and Fall of the British World-System, 1830–1970*, Cambridge: Cambridge University Press, 2009: 69–77; and John S. Galbraith and Afaf Lutfi al-Sayyid-Marsot, 'The British Occupation of Egypt: Another View', *International Journal of Middle East Studies*, 9: 4, 1978: 471–88; on the 1884 crisis, see D. A. Farnie, *East and West of Suez: The Suez Canal in History, 1854–1956*, Oxford: Clarendon Press, 1969: 325–33.

4. Jennifer Derr, *The Lived Nile: Environment, Disease, and Material Colonial Economy in Egypt*, Stanford, CA: Stanford University Press, 2019, offers the best recent study. For earlier accounts of British irrigation policy, see Robert Tignor, 'British Agricultural and Hydraulic Policy in Egypt, 1882–1892', *Agricultural History*, 37: 2, 1963: 63–74; and *Modernization and British Colonial Rule in Egypt, 1882–1914*, Princeton, NJ: Princeton University Press, 1966; Terje Tvedt, *The River Nile in the Age of the British: Political Ecology and the Quest for Economic Power*, New York: I.B. Tauris, 2004; John Waterbury, *Hydropolitics of the Nile Valley*, Syracuse, NY: Syracuse University Press, 1979; Alan Richards, *Egypt's Agricultural Development 1800–1980: Technical and Social Change*, New York: Routledge, 2019; and Claire Cookson-Hills, 'Engineering the Nile: Irrigation and the British Empire in Egypt, 1882–1914', PhD thesis, Queen's University at Kingston, 2013.
5. David Ames Wells, *Recent Economic Changes and Their Effect on the Production and Distribution of Wealth and the Well-being of Society*, New York: D. Appleton and Company, 1890: 458–9. On the influence of this text, see James Livingston, 'The Social Analysis of Economic History and Theory: Conjectures on Late Nineteenth-Century American Development', *American Historical Review*, 92: 1, 1987: 75.
6. The use of compulsory work gangs to build and maintain irrigation works (and previously, the Suez Canal) was widely unpopular. Its abolition was proclaimed as the first step in creating a more efficient, more enlightened, and less coercive political and economic order. See Nathan Brown, 'Who Abolished Corvee Labour in Egypt and Why?', *Past and Present*, 144, 1994: 116–37; for a different reading, see Samera Esmeir, *Juridical Humanity: A Colonial History*, Stanford, CA: Stanford University Press, 2012: 89–91.
7. Arthur Edwin Crouchley, *The Economic Development of Modern Egypt*, London: Longmans, Green, and Co., 1938: 3–4.
8. On the colonial ideology of flood control, see Rohan D'Souza, *Drowned and Dammed: Colonial Capitalism and Flood Control in Eastern India*, New Delhi: Oxford University Press, 2006: 1–2.
9. Figures from Roger Owen, *Cotton and the Egyptian Economy, 1820–1914: A Study in Trade and Development*, Oxford: Clarendon Press, 1969: 161, 184, 186.
10. Esmeir, *Juridical Humanity*; Alan Mikhail, *The Animal in Ottoman Egypt*, Oxford: Oxford University Press, 2014: 169–70. Found extensively in Egypt

before the nineteenth century, by the end of that century the hippo had been eliminated by loss of habitat and by sport hunting along the Nile, all the way south to Khartoum. See Nicolas Manlius, 'Biogéographie et Ecologie Historique de l'Hippopotame en Égypte', *Belgian Journal of Zoology*, 130: 1, 2000: 59–66.

11. Important recent studies that look at many of these issues include Zainab Abul-Magd, *Imagined Empires: A History of Revolt in Egypt*, Berkeley, CA: University of California Press, 2013; Derr, *Lived Nile*; and Aaron Jakes, *Egypt's Occupation: Colonial Economism and the Crises of Capitalism*, Stanford, CA: Stanford University Press, 2020.
12. Information from members of the Salim family and from Dr Wadie Boutros.
13. On storage, see Andreas Malm, *Fossil Capital: The Rise of Steam Power and the Roots of Global Warming*, London: Verso, 2016; and Andreas Folkers, 'Freezing Time, Preparing for the Future: The Stockpile as a Temporal Matter of Security', *Security Dialogue*, 50: 6, 2019: 493–511. On storage and the earliest states, see James C. Scott, *Against the Grain: A Deep History of the Earliest States*, New Haven, CT: Yale University Press, 2017: 128–37. For Martin Heidegger, the management of a reserve was a hallmark of modern technology, exemplified by the building of large dams and hydro-electric power stations. *The Question Concerning Technology and Other Essays*, New York: Garland Publishing, 1977.
14. The situation changed after the construction of the first Aswan Dam, when the British engineers became much more aware of groundwater, not as a resource but as a problem of waterlogging in regions where the water table was now much higher. They brought to Cairo a British geologist, H. T. Ferrar, who published two reports and then a book: Hartley Travers Ferrar, *The Movements of the Subsoil Water in Upper Egypt*, Cairo: National Printing Department, 1911. See William Willcocks and James Ireland Craig, *Egyptian Irrigation*, 3rd edition, London: E. & F.N. Spon, 1913.
15. Robert N. Proctor and Londa Schiebinger, eds, *Agnotology: The Making and Unmaking of Ignorance*, Stanford, CA: Stanford University Press, 2008.
16. On the British engineers in Egypt, see Cookson-Hills, 'Engineering the Nile'; on the duty of water, see also J. L. Jr Wescoat, 'Reconstructing the Duty of Water: A Study of Emergent Norms in Socio-Hydrology', *Hydrology and Earth System Sciences*, 17: 12, 2013: 4759–68; David Gilmartin, 'Imperial Rivers: Irrigation and British Visions of Empire', in Durba Ghosh and Dane Kennedy, eds, *Decentring Empire: Britain, India, and the Transcolonial World*, London: Orient Longman, 2006.
17. Willcocks and Craig, *Egyptian Irrigation*, 3rd edition: xx.
18. A partial exception is the important work of 'Abd al-Hamid Subhi Nasif, *Mashru'at al-rayy wa-atharuha fi al-mujtama' al-misri fi al-qarn al-tasi' 'ashar, 1805–1882*, Cairo: al-Hay'a al-Misriyya al-'Amma li-l-Kitab, 2018, which discusses at length the system of well irrigation outlined here, although mainly in connection with Muhammad Ali's attempt to expand the growing of cotton, which required summer irrigation water, not as part of a wider system of three-season irrigation. As explained below, with cotton

cultivation, due to the crop's long growing season, the fields could no longer be inundated in the winter.

19. Julien Barois, *Les irrigations en Égypte*, 2nd edition, Paris: C. Béranger, 1911.
20. Willcocks and Craig, *Egyptian Irrigation*, 3rd edition.
21. On the reduced upkeep of basins after the Black Death of the mid-fourteenth century and subsequent waves of plague, and the increased political power of Arab pastoralist communities using flood-recession land as pasture, see Stuart J. Borsch, 'Nile Floods and the Irrigation System in Fifteenth-Century Egypt', *Mamluk Studies Review*, 4, 2000: 131–46.
22. The sizes of basins and the sequence of their filing and emptying is given for every province of Egypt in 'Ali Mubarak, *al-Khitat al-tawfiqiyya al-jadida li-Misr al-Qahira wa-muduniha wa-biladiha al-qadima wa-l-shahira*, vol. 19, Bulaq: al-Matba'a al-Kubra al-Amiriyya, 1886–9.
23. Ghislaine Alleaume, 'Les systèmes hydrauliques de l'Égypte pré-moderne: Essai d'histoire du paysage', in Christian Décobert, ed., *Itinéraires d'Égypte: Mélanges offerts au père Maurice Martin s.j.*, Cairo: Institut Français d'Archéologie Orientale, 1992: 301–22.
24. On wheat versus barley consumption, see Pierre-Simon Girard, 'Memoire sur l'agriculture, l'industrie et le commerce de L'Égypte', *Description de l'Égypte: Etat Moderne*, vol. 2, Paris: Imprimerie Impériale: 557–9; John Bowring, *Report on Egypt and Candia*, London: reprinted by W. Clowes and sons for Her Majesty's Stationary Office, 1840: 66, and Helen Anne B. Rivlin, *The Agricultural Policy of Muhammad Ali*, Cambridge, MA: Harvard University Press, 1961: 168. On the use of barley to make beer, for which wheat was also used, see S. D. Goitein, *A Mediterranean Society*, Berkeley, CA: University of California Press, 1967–93, vol. 1, part 2: 118–19.
25. This refers to the Nile Valley south of Cairo. To the north, in the Nile Delta, the soil was usually turned with a plough before sowing, and then in the case of wheat watered twice, at sixty and ninety days after sowing, using water raised from wells using a waterwheel. Girard, 'Memoire sur l'agriculture': 515–17. Even today, where it is practised in other river basins, flood-recession agriculture remains the most labour-saving form of farming. Scott, *Against the Grain*: 66.
26. The winter crops were known as *al-bayādī* when grown in the flood basins, and *al-shitawī* when grown on the berms, using mechanical irrigation. The crops harvested in early summer were known as *al-gaidī* (القيظي) or *al-sayfī*, both terms referring to summer. The third season, in the months when the Nile started to flood, was known as *al-damīrī* or *al-nabārī*, or sometimes *al-nīlī*, terms referring to the swelling or flooding of the river. *Al-nīlī* crops grown on the berms and on islands in the river were irrigated by lifting water mechanically from the rising waters, while basin crops continued to use well water (see the next two paragraphs) where available. Girard, 'Memoire sur l'agriculture': 499–500; Barois, *Les irrigations*: 56–9. The three-crop season and use of wells is also mentioned in John Lewis Burckhardt, *Travels in Nubia*, London: John Murray, 1819: 22–3; and Edward William Lane, *Manners and Customs of the Modern Egyptians*, London: J. M. Dent, 1954: 334–7. There is a passing reference to waterwheels, used on both wells and

canals, in 'Abd al-Rahim 'Abd al-Rahman 'Abd al-Rahim, *al-Rif al-misri fi al-qarn al-thamin 'ashar*, Cairo: Jami'at 'Ayn Shams, Kulliyyat al-Adab, 1974: 179–80.

27. Most accounts of changes to irrigation before the 1860s focus on the digging of so-called summer canals, the deep channels that allowed Nile water to be pumped up into fields (using both animal and steam power) in the spring and early summer, before the river began to rise. The exception is Alleaume's important study, 'Les systèmes hydrauliques', which explains how engineers under Muhammad Ali and his successors also reorganised the diverse flood basins of pre-nineteenth-century Egypt into systematic chains of large basins connected by canals parallel to the river, arranged at different heights to allow the storing of excess water. The arrangement extended the use of lands outside the flood season, and, as we will see in the following paragraphs, worked in conjunction with the use of well irrigation.
28. Tignor, 'British Agricultural and Hydraulic Policy in Egypt': 63. Many historians, unlike Tignor, mention in passing that there was a second crop, and one or two mention the use of wells. But none explain how the system of well irrigation worked or its centrality to peasant livelihoods. Barois, *Les irrigations*: 28–9, 58–9, 99, briefly mentions wells dug in the fields. Rushdi Said, *The River Nile: Geology, Hydrology, and Utilization*, Oxford: Pergamon Press, 1993: 188–208, notes the use of wells but suggests they were placed mainly on the berms, not the field basins. Waterbury, *Hydropolitics of the Nile Valley*: 29, mentions the well system in passing. Stanford Shaw, *The Financial and Administrative Organization and Development of Ottoman Egypt, 1517–1798*, Princeton, NJ: Princeton University Press, 1962: 50–1, mentions that in some districts 'three or four different harvests were produced during every year', requiring 'artificial irrigation' in fields distant from the Nile, but does not mention wells or waterwheels. Alan Mikhail, *Nature and Empire in Ottoman Egypt*, Cambridge: Cambridge University Press, 2011: 10, mentions the second crop and refers to waterwheels and the animals that moved them several times in his richly detailed text, but does not explain the well system or the role that waterwheels played. Roger Owen makes no mention of wells and states that summer crops 'could be grown only near the river or one of the few canals deep enough to hold water throughout the summer, as they required constant irrigation': *Cotton and the Egyptian Economy*: 7–8. Kenneth Cuno, *The Pasha's Peasants: Land, Society, and Economy in Lower Egypt, 1740–1858*, Cambridge: Cambridge University Press, 1992: 18, mentions wells, noting more accurately that 'land . . . near the Nile, a canal or a well, was cultivated during the summer season', overlooking the fact that, as Willcocks reported, 'wells are dug everywhere', meaning that all basin land was potentially near a well. Willcocks and Craig, *Egyptian Irrigation*, 3rd edition: 32.
29. Karl W. Butzer, *Early Hydraulic Civilization in Egypt*, Chicago, IL: University of Chicago Press, 1976.
30. Willcocks and Craig, *Egyptian Irrigation*, 3rd edition: 85.
31. Ahmed Sefelnasr and Mohsen Sherif, 'Impacts of Seawater Rise on Seawater Intrusion in the Nile Delta Aquifer, Egypt', *Groundwater*, 52: 2, 2014:

264–76.

32. Smith archive.
33. Ahmad ibn 'Abd al-Wahhab Al-Nuwayri, *Nihayat al-arab fi funun al-adab*, vol 8, Cairo: Dar al-Kutub al-Misriyya, 1923: 253–4, cited in Hassanein Rabie, 'Some Technical Aspects of Agriculture in Medieval Egypt', in Abraham Udovitch, ed., *The Islamic Middle East*, Princeton, NJ: Darwin Press, 1981: 71–2; Rabie does not explain the source of water. The use of wells for summer plantations (and for orchards) is also documented in pre-Islamic Egypt, especially following the development of the waterwheel in the Hellenistic period. 'The water table was readily accessible throughout the Nile Valley, and a well or deep pond could be dug virtually anywhere'. C. J. Eyre, 'The Water Regime for Orchards and Plantations in Pharaonic Egypt', *The Journal of Egyptian Archaeology*, 80, 1994: 79.
34. The method of sinking the well is described in ʿAbd al-Laṭif al-Baghdadi, *Kitab al-ifada wa-l-iʿtibar fi al-umur al-mushahada wa-l-ḥawadith al-muʿayana bi-ʾard Misr*, written in 1204, ed. and trans. as *A Physician on the Nile: A Description of Egypt and Journal of the Famine Years*, New York: New York University Press, 2022: 69–70. Baghdadi describes the sinking of a series of masonry shafts to form an embankment or dyke (*zarbiyya*), but notes that this is the same method used to sink wells. There is also a brief mention of the method in Barois, *Les irrigations*: 266.
35. The details here are from Laïla Ménasse and Pierre Laferrière, *La sāqia: Technique et vocabulaire de la roue à eau égyptienne*, Cairo: Institut Français d'Archéologie Orientale du Caire, 1974.
36. Ibid.
37. Hekekyan Papers, British Library Add Mss 37,450, vol. 3 (1844–50), page 32 verso.
38. Charles Dudley Warner, *My Winter on the Nile, Among the Mummies and Moslems*, 18th edition, Boston, MA: Houghton, Mifflin and Co., 1895 [1st edition 1876]. See also Winifred Susan Blackman, *The Fellāhīn of Upper Egypt: Their Religious, Social, and Industrial Life Today with Special Reference to Survivals from Ancient Times*, London: G.G. Harrap, 1927: 171.
39. Aḥmad ibn 'Ali al-Maqrizi, *Kitab al-khitat al-Maqriziyya*, vol. 2, Cairo: Matba'at al-Nil, 1906–8: 35 (text also in Pellat, *Cinq Calandriers Égyptiens*, Cairo: Institut Français d'Archéologie Orientale du Caire, 1986: 115); Girard, 'Memoire sur l'agriculture': 500. See also the discussion of summer cultivation using well water raised by machines in the administrative manuals of al-Makhzumi and Ibn Mammati from the second half of the twelfth century: As'ad ibn al-Muhadhdhab Ibn Mammati, *Kitab qawanin al dawawin*, Cairo: Maktabat Madbuli, 1991; translated in Claude Cahen, 'Al-Makhzūmi et Ibn Mammātī sur L'Agriculture égyptienne médiévale', *Annales Islamologiques*, 11, 1972: 148–9.
40. Willcocks and Craig, *Egyptian Irrigation*, 3rd edition: 84.
41. Ibn Mammati, *Kitab qawanin al-dawawin*, in Cahen, 'Al-Makhzūmi et Ibn Mammātī': 148–9; the lower figure in Maqrizi may refer to the cultivation of cane, whereas Ibn Mammati refers to the cultivation of crops in general

using well irrigation.

42. By 15 Abib (9 July) 'water in the wells is scarce' – al-Maqrizi, *Kitab al-khitat al-Maqriziyya*: 38.
43. Al-Maqrizi, cited in Cuno, *The Pasha's Peasants*: 20.
44. Willcocks and Craig, *Egyptian Irrigation*, 3rd edition: 96. Moritz Schanz, 'Egypt and the Anglo-Egyptian Soudan', in Official Report, Ninth International Congress of Delegated Representatives of Master Cotton Spinners' and Manufacturers' Associations, held in Kurhaus, Scheveningen, 9, 10, and 11 June 1913: 203.
45. Willcocks and Craig, *Egyptian Irrigation*, 3rd edition: 85.
46. Ferrar, *The Movements of the Subsoil Water in Upper Egypt*, cited in Willcocks and Craig, *Egyptian Irrigation*, 3rd edition: 85, 90.
47. Girard, 'Memoire sur l'agriculture': 539.
48. Louis Linant de Bellefonds, *Mémoires sur les principaux travaux d'utilité publique exécutés en Égypte depuis la plus haute antiquité jusqu'à nos jours*, Paris: Arthus Bertrand, 1872–3: 420.
49. Alleaume, 'Les systèmes hydrauliques': 304–7. An English traveller, observing the Nile flood in 1850–1, wrote that the rise of the river 'is first felt in the lands at the greatest distance from the river; for these are the lowest. They gradually become moist, then soften into morasses, and in some places glisten for a short time as shallow lakes. There is no rush of water along the surface; but the inundation, as it were, oozes up from the ground. Deep holes, during summer with hard-baked cracked bottoms, change slowly into wells and ponds.' Bayle St. John, *Village Life in Egypt: With Sketches of the Saïd*, 2 vols, London: Chapman and Hall, 1852: 2: 104–5.
50. Willcocks and Craig, *Egyptian Irrigation*, 3rd edition: 388.
51. With the strong flood of 1889, for example, 'owing to the increased number of escapes there were only three breaches in the basins', according to the Scottish engineer recently appointed as Inspector General of Irrigation. 'The reduction of the water in the case of those basins which were filled too full was effected with great skill by the local engineers.' Justin Ross, 'Report of the Administration of the Department of Irrigation for the Year 1889', Cairo: Public Works Ministry, 1890: 4.
52. Girard, 'Memoire sur l'agriculture': 517; Bowring, *Report on Egypt*: 19; Barois, *Les irrigations*: 58–9, 397; Blackman, *The Fellāhīn of Upper Egypt*: 163–4. Wheat was cultivated more widely but was used mainly to pay the government tax and for making the bread of the well-to-do, with the surplus exported abroad. See Girard, 'Memoire sur l'agriculture': 516, 519. In the Delta, sorghum was often supplemented or replaced with maize (*dhurra shami*, or 'Syrian sorghum').
53. Karl W. Butzer, 'Landscapes and Environmental History of the Nile Valley: A Critical Review and Prospectus', in Ian Shaw and Elizabeth Bloxam, eds, *The Oxford Handbook of Egyptology*, Oxford: Oxford University Press, 2020.
54. Diego Ortiz and Maria G. Salas-Fernandez, 'Dissecting the Genetic Control of Natural Variation in Sorghum Photosynthetic Response to Drought Stress', *Journal of Experimental Botany*, 73: 10, 2022: 3251–67.
55. P. W. Orchard and R. S. Jessop, 'The Response of Sorghum and Sunflower to

Short-Term Waterlogging', *Plant Soil*, 81: 1, 1984: 119–32; Andre Piedallu, *Le Sorgho*, Paris: Société d'éditions géographiques, maritimes et coloniales, 1923: 140–2.

56. Willcocks and Craig, *Egyptian Irrigation*, 3rd edition: 334; Ross, 'Report of the Administration': 4.
57. Willcocks provides a sketch map to show how the basins 'are very skillfully placed' so that sorghum fields are protected from the flood and irrigated only at the end of the period of inundation, through the emptying of upstream basins. Unlike large landowners, 'the fellaheen prefer a crop of millets [sorghum] followed by a winter crop to a single winter crop ordinarily sown in the basins': *Egyptian Irrigation*, 1st edition, London: E. & F.N. Spon, 1889: 335–6.
58. Willcocks, *Egyptian Irrigation*, 1st edition: 247–9.
59. On this biotic exchange and the concept of the flood pulse, see Wolfgang Junk, Peter Bayley, and R. E. Sparks, 'The Flood Pulse Concept in River-Floodplain Systems', in D. P. Dodge, ed., *Proceedings of the International Large River Symposium*, Ontario: Canadian Special Publication of Fisheries and Aquatic Sciences 106, 1989: 110–27. See also Rohan D'Souza, 'Event, Process and Pulse: Resituating Floods in Environmental Histories of South Asia', *Environment and History*, 26: 1, 2020: 31–49; and Scott, *Against the Grain*: 53–4. Scott's longer, brilliant study of the Ayeyarwady (Irrawaddy) River of Burma, which appeared after this writing, illuminates how the flood pulse and other aspects of the complex life of a river enable us to write a less anthropocentric account of human and more-than-human history: *In Praise of Floods: The Untamed River and the Life It Brings*, New Haven, CT: Yale University Press, 2025. We have no study of this kind for the Nile, where the long history of underground storage adds another dimension to the complex processes that Scott illuminates.
60. On fish and flood cycles, see Gelego De Oliveira et al., 'Relationship of Freshwater Fish Recruitment with Distinct Reproductive Strategies and Flood Attributes: A Long-Term View in the Upper Paraná River Floodplain', *Frontiers in Environmental Science*, 8, 2020. On the importance of a great variety of bird life to both diet and medical remedies in medieval Egypt, see Heba Mahmoud Saad Abdelnaby, 'Treating with Birds: The Insights of Two Mamluk Sources about the Medical Benefits of Birds', *Mamluk Studies Review*, 25, 2022: 161–96. On the abundance of fish, see Vincent Stochove, Gilles Fermanel, and Robert Fauvel, *Voyage en Égypte*, Cairo: Institut Français d'Archéologie Orientale, 1975 [1631]: 36–7.
61. Hassanein Rabie, *The Financial System of Egypt A.H. 654–741/A.D. 1169–1341*, London: Oxford University Press, 1972: 88–9.
62. Muhammad Salim, personal communication, December 2023.
63. Bowring, *Report on Egypt*: 66. For the medieval period, see ʿAbd al-Laṭif al-Baghdadi, *Kitab al-ifada wa-l-iʿtibar fi l-umur al-mushahada wa-l-hawadith al-muʿayana bi-ʿard Misr*, English translation *A Physician on the Nile: A Description of Egypt and Journal of the Famine Years*, New York: New York University Press, 2022: 118–19; and Goitein, *A Mediterranean Society*, 2: 5. In contrast, large fisheries, such as those along the northern edge of the

Delta, were a monopoly of the Ottoman authorities: Stanford Shaw, *Ottoman Egypt in the Eighteenth Century: The Nizâmnâme-i Misir of Cezzâr Ahmed Pasha*, Cambridge, MA: Harvard University Press: 19, 47.

64. Mahmoud A. Zahran, 'Hydrophytes of the Nile in Egypt', in Henri J. Dumont, ed., *The Nile: Origin, Environments, Limnology, and Human Use*, Dordrecht: Springer, 2009; Yasser Mahmoud Ali and Ibrahim Salah El-Din Khedr, 'Estimation of Water Losses through Evapotranspiration of Aquatic Weeds in the Nile River', *Water Science*, 32: 2, 2018; Anna Tsing, 'What Is History? The Life and Times of Water Hyacinth', unpublished paper, 2023; Iftekhar Iqbal, *The Bengal Delta: Ecology, State and Social Change, 1840–1943*, Basingstoke: Palgrave Macmillan, 2010: 140–59. Water hyacinth is today the most widespread invasive alien plant in the world: Helen Roy et al., eds, *Thematic Assessment Report on Invasive Alien Species and their Control*, Bonn, Germany, IPBES, 2023.
65. Derr, *Lived Nile*: 103–13; on malaria, which caused a devastating epidemic in 1942, see Timothy Mitchell, 'Can the Mosquito Speak?' in *Rule of Experts*, Berkeley, CA: University of California Press, 2002: 19–53.
66. The area planted with maize grew from 14 per cent of the cropped area in 1879 to 24 per cent in 1913, while the proportion of land devoted to barley, beans, and vegetables all declined, even as the population of the country doubled. Crouchley, *Economic Development*: 161–4; George Manville Fenn, *The Khedive's Country: The Nile Valley and Its Products*, London: Cassell and Company, 1904: 72–3, 166–7. On fuel for pumps, see Barois, *Les irrigations*: 398.
67. On pellagra, see Derr, *Lived Nile*: 112–13; on the alarm at increasing dementia and its connection to pellagra, see Timothy Mitchell, 'As If the World Were Divided in Two: The Birth of Politics in Turn-of-the-Century Cairo', PhD thesis, Department of Politics, Princeton University, 1984: 35–6.
68. On pests, see also Esmeir, *Juridical Humanity* and Aaron Jakes, 'Boom, Bugs, Bust: Egypt's Ecology of Interest, 1882–1914', *Antipode*, 49: 4, 2017: 1049.
69. Jean Mazuel, *Sucre en Égypte: Étude de Géographie Historique et Économique*, Cairo: E. & R. Schindler, 1937: 99; J. H. Galloway, *The Sugar Cane Industry*, Cambridge: Cambridge University Press, 1989: 15.
70. Crouchley, *Economic Development*: 156.
71. Willcocks and Craig, *Egyptian Irrigation*, 3rd edition: 338. The disparagement of the engineers of the Khedevial period was a theme of British accounts. This somewhat rosy view of the old system can be tempered by Alan Mikhail's account of the hierarchies, both local and imperial, that governed rural Egypt before the nineteenth century. Nevertheless, the household-based system of well irrigation described here provides a corrective to the view of a collective order governed largely by the demands of large-scale irrigation works.
72. *Oxford English Dictionary*, s.v. 'cattle, n., Etymology'.
73. Blackman, *The Fellāhīn of Upper Egypt*: 154–5. In Upper Egypt and Fayyum, the wheels were also sometimes driven by camels. 'Abd al-Rahim, *al-Rif al-misri*: 188–9.
74. Bowring, *Report on Egypt*: 21. Report dated 1840, but written '8 or 9 years

before', and information is often from previous seasons. The attempt to develop a domestic silk industry was abandoned after 1835. Rivlin, *The Agricultural Policy*: 163–6.

75. See Tsugitaka Sato, *State and Rural Society in Medieval Islam: Sultans, Muqta's and Fallahun*, Leiden: Brill, 1997: 187, citing Ibn Mammati, *Qawanin al-Dawawin* and Abu al-Ḥasan 'Ali ibn 'Uthman al-Makhzumi, *al-Muntaqa min kitab al-minhaj fi 'ilm kharaj*, Cairo: Institut Français d'Archéologie Orientale, 1986.
76. Maqrizi, cited in Rabie, 'Some Technical Aspects of Agriculture in Medieval Egypt': 71–2.
77. Bowring states that it was the high cost of installing and maintaining waterwheels that persuaded Muhammad Ali to turn to the construction of the Nile Barrage. Bowring, *Report on Egypt*: 12–13; Rivlin, *The Agricultural Policy*: 248.
78. Ménasse and Laferrière, *La sāqia*, describes the allocation of tasks. On cooperation around water wells and the complexity of sharing arrangements, see Arjun Appadurai, 'Wells in Western India', *Expedition Magazine*, 26: 3, 1984: 3–14.
79. Nicholas Hopkins, 'Irrigation in Contemporary Egypt', in Alan K. Bowman and Eugene L. Rogan, eds, *Agriculture in Egypt: from Pharaonic to Modern Times*, Oxford: Oxford University Press, 1999: 372.
80. Anders J. Bjørkelo, *Prelude to Mahdiyya: Peasants and Traders in the Shendi Region, 1821–1885*, Cambridge: Cambridge University Press, 1989: 13, 36–7, 72, 86–8; El Haj Abdalla Bilal Omer, *The Danagla Traders of Northern Sudan: Rural Capitalism & Agricultural Development*, London: Ithaca Press, 1985: 26–36, 57–67; W. R. G. Bond, 'Some Curious Methods of Cultivation in Dongola Province', *Sudan Notes and Record*, 8, 1925: 96–103; William Willcocks, report to the Inspector General of Irrigation, 22 January 1890, in Public Works Department, *Nile Reservoirs: Appendices*, Cairo: National Printing Office, 1892, 18–26, reporting on the numerous masonry wells along the Nile Valley in Nubia, south of Aswan, and the local preference for well irrigation over proposals for digging canals.
81. Blackman, *The Fellāhīn of Upper Egypt*: 175. Cattle were an important source of energy and food in Ottoman Egypt, used for ploughing and transportation, and as a source of meat and dairy products. Cattle also provide a means of storing and building up wealth, in a form that is secure and yet movable enough to loan out or exchange. But they were also a critical component in the system of irrigation. Mikhail, *The Animal in Ottoman Egypt*.
82. Mikhail, *The Animal in Ottoman Egypt*: 54–5. See also al-Rahim, *al-Rif al-misri* and 'The Egyptian Rural Society at the End of the Eighteenth Century', in R. Mantran, ed., *L'Égypte au XIXe siècle*, Paris: Centre National de la Recherche Scientifique, 1982: 197–210. On the rise of tax farming, see Baber Johansen, *The Islamic Law on Land Tax and Rent: The Peasants' Loss of Property Rights as Interpreted in the Hanafite Legal Literature of the Mamluk and Ottoman Periods*, London: Routledge, 2016; on the colonial consolidation of the system of private property rights, including the powers of dispossession that differentiated this system from earlier property regimes, see my

essay 'Principles True in Every Country', in Mitchell, *Rule of Experts*: 54–79.

83. Irfan Habib suggests that spread of the *saqiya* eastwards into the Indus basin of northwest India in the pre-Mughal period, by connecting animal power with irrigation on a new scale, enabled a major agrarian transformation. 'Technological Changes and Society, 13th and 14th Centuries', *Proceedings of the Indian History Congress*, 31, 1969: 149–55. See also David Gilmartin, *Blood and Water: The Indus River Basin in Modern History*, Berkeley, CA: University of California Press, 2015: 16–18.
84. Abul-Magd, *Imagined Empires*: 91.
85. Willcocks, *Egyptian Irrigation*, 1st edition: 246.
86. Amr Khairy Ahmed, 'Egypt Ignited: How Steam Power Arrived on the Nile and Integrated Egypt into Industrial Capitalism (1820s–76)', PhD thesis, Lund University, 2023: 197–226.
87. Egypt Today Staff, 'Egypt Installs Solar-Powered Pumps at 85 Water Wells in New Valley', *Egypt Today*, 26 June 2022.
88. On the capturing of subaltern voices, see Ranajit Guha, 'The Prose of Counterinsurgency', in Ranajit Guha and Gayatri Chakravorty Spivak, eds, *Selected Subaltern Studies*, New York: Oxford University Press, 1988: 45–86; Gayatri Chakravorty Spivak, 'Can the Subaltern Speak?', in C. Nelson and L. Grossberg, eds, *Marxism and the Interpretation of Culture*, Basingstoke, UK: Macmillan, 1988: 271–313; and Timothy Mitchell, 'Can the Mosquito Speak?'.
89. For example, Qasim 'Abduh Qasim, *al-Nil wa-al-mujtama' al-misri fi 'asr salatin al-mamalik*, Cairo: Dar al-Ma'arif, 1978: 22–39.
90. On the significance of the register, see Alan Mikhail, *Under Osman's Tree: The Ottoman Empire, Egypt, and Environmental History*, Chicago, IL: University of Chicago Press, 2017: 19–33. The register distinguished between *al-jusur al-sultaniyya*, the channels whose upkeep was the responsibility of the Ottoman (Sultanic) authorities, and *al-jusur al-baladiyya*, local embankments, for which the community was responsible. On this division of responsibility, see also Sato, *State and Rural Society in Medieval Islam*: 225–33.
91. 'Ali Mubarak, *al-Khitat*.
92. This is the conclusion one can draw from Alan Mikhail's account of the animal in Ottoman Egypt. See the case of the dispute over an ox whose purchaser harnessed it to a waterwheel and found it dead in the morning, which was a dispute over the purchase of an animal, not the working of the irrigation system: Mikhail, *The Animal in Ottoman Egypt*: 19–21. Another case is discussed on pp. 30–1 of the same.
93. I am grateful to Ibrahim El-Houdaiby for this reference, which is from al-Bujayrami, *Tuhfat al-habib 'ala sharh al-khatib*. On such legal manuals and their place in constructing a political order, see Brinkley Messick, *The Calligraphic State: Textual Domination and History in a Muslim Society*, Berkeley, CA: University of California Press, 1993.
94. Roger Owen, *Cotton and the Egyptian Economy*: 8, citing Jean Mazuel, *L'Oeuvre géographique de Linant de Bellefonds*, Cairo: La Société Royale de Géographie d'Egypte, 1937; Maurice Linant de Bellefonds, *Mémoires sur les principaux travaux d'utilité publique exécutés en Égypte depuis la plus haute*

antiquité jusqu'à nos jours, Paris: Arthus Betrand, 1872–3: 27–8.

95. ʻAli Mubarak, *Kitab nukhbat al-fikr fi tadbir nil Misr*, Cairo: Matbaʻat Wadi al-Nil al-ʻArabiyya wa-l-Ifranjiyya, 1879/80: 187–91, 201–4. On Mubarak, see Ghislaine Alleaume, 'L'Ecole Polytechnique du Caire et ses éleves: la formation d'une élite technique dans l'Égypte du XIXeme siècle', PhD thesis, Université de Lyon II, 1993: 200–10; and Darrell Dykstra, 'A Biographical Study in Egyptian Modernization: 'Ali Mubarak', PhD thesis, University of Michigan, 1977.
96. Willcocks, *Egyptian Irrigation*, 1st edition: 247–51.
97. Ibid.: 27. Repeated in Willcocks and Craig, *Egyptian Irrigation*, 3rd edition: 32. All of the southern Delta, above a line from Tanta to Zagazig, lay above the eight-metre contour. See map, ibid.: 370.
98. Henry Villiers Stuart, an Irish landowner and Member of Parliament who conducted a tour of inspection on behalf of the British government in 1882 reported that 'much might even be done with water-wheels . . . because one water-wheel, worked by two or three pairs of bullocks, can irrigate four acres of cane, five or six acres of cotton, or thirteen acres of cereal crops . . . An important advantage of the water-wheel is that it can be worked at long distances from the Nile by sinking wells below the Nile level. A perpetual supply of water exists far inland at that depth, because the soil is porous.' Henry Villiers Stuart, *Egypt After the War, Being the Narrative of a Tour of Inspection*, London: John Murray, 1883: 242. Barois, *Les irrigations*: 266, 268, gave figures of twelve to fifteen acres (five to six hectares) per waterwheel, and noted that sometimes up to four wheels could operate one well.
99. Willcocks, *Egyptian Irrigation*, 1st edition: 300–1. Another British engineer, N. E. Verschoyle, Inspector of Irrigation for Buhaira Province, complained in 1898 that to improve land close to the new canals, 'still larger areas of unfavourably situated land have been sacrificed'. He concluded that the flood basin system 'was better suited to the country and had more permanency than the system of perennial irrigation whose mushroom growth has had the whole strength of the Government on its side but has had Nature against it', cited in Willcocks and Craig, *Egyptian Irrigation*, 3rd edition: 399.
100. Willcocks, *Egyptian Irrigation*, 1st edition: iii.
101. Ibid.: xi. In fact, even with the labour shortage, the third-season sorghum crop covered about 25 per cent of the cultivated area. William Willcocks, *Report on Perennial Irrigation and Flood Protection for Egypt*, Cairo: National Printing Office, 1894: 61. On other engineers being always promoted over him, see Willcocks, *Sixty Years in the East*, 117.
102. According to Willcocks and Craig, there were 750 pumps on wells in the basins, rated at 11,740 horsepower, compared with 40,631 *saqiyas*: *Egyptian Irrigation*, 3rd edition: 109.
103. Willcocks, *Egyptian Irrigation*, 1st edition: 247.
104. Justin C. Ross, 'Irrigation and Agriculture in Egypt', *The Scottish Geographical Magazine*, 9: 4, 1893: 177.
105. Willcocks changed his mind about the damage caused by perennial irrigation after a study of irrigation schemes in the Po Valley in Italy: Willcocks,

Report on Perennial Irrigation: 60.

106. George P. Foaden and F. Fletcher, *Text-Book of Egyptian Agriculture*, Cairo: Ministry of Education, 1908: 152.
107. Writing about the Delta immediately after the British occupation, Colin Scott-Moncrieff, whom the British appointed as Director of Irrigation in May 1883, remarked that, in addition to about 2,000 steam pumps, 'there are innumerable water-wheels. One wheel waters about five acres, and requires three bullocks and two men to work it. With the recent cattle disease, this is a heavy burden.' He added that, with the building of barrages across the river, 'I hope that two-thirds to three-fourths of them will disappear': Colin Scott-Moncrieff, *Note on the Irrigation Works of Egypt and the Improvements to be Made to Them*, Cairo: Ministry of Irrigation, 1884: 9.
108. On the role of Ottoman imperial expansion in the history of plague, and Ottoman measures to combat the disease, see Nükhet Varlik, *Plague and Empire in the Early Modern Mediterranean World: The Ottoman Experience, 1347–1600*, New York: Cambridge University Press, 2016; Nükhet Varlik, ed., *Plague and Contagion in the Islamic Mediterranean*, Leeds: Arc Humanities Press, 2017. On the impact of waves of human and animal disease in this period, see Mikhail, *The Animal in Ottoman Egypt*: 42–6, 49–53.
109. LaVerne Kuhnke, *Lives at Risk: Public Health in Nineteenth-Century Egypt*, Berkeley, CA: University of California Press, 1990; Yosra Hussein, 'Vectors of Disease: Medicine and Public Health in Egypt's Cholera Epidemic (1883)', MA paper, Middle Eastern, South Asian, and African Studies Department, Columbia University, 2023.
110. Nükhet Varlik, '"Oriental Plague" or Epidemiological Orientalism? Revisiting The Plague Episteme of The Early Modern Mediterranean', in Varlik, ed., *Plague and Contagion*: 57–87.
111. Clive A. Spinage, *Cattle Plague: A History*, New York: Kluwer Academic/Plenum Publishers, 2003: 5.
112. Ahmad al-Hitta, *Tā'rikh Misr al-iqtisadi fi al-qarn al-tasi' 'ashar*, Alexandria: Al-Matba'a al-Misriyya, 1967: 55; and *Ta'rikh al-zira'a al-misriyya fi 'asr Muhammad 'Ali al-kabir*, Cairo: n.p, 1950: 25. Rinderpest afflicts camels (also used to turn waterwheels) as well as cattle, but water buffalo suffer only a benign form of the disease. Spinage, *Cattle Plague*: 31, 34.
113. Mikhail, *Nature and Empire in Ottoman Egypt*: 219; Mikhail, *Under Osman's Tree*: 138–9.
114. 'Abd al-Hamid Subhi Nasif, *Mashru'at al-rayy wa-atharuha fi al-mujtama' al-misri fi al-qarn al-tasi' 'ashar, 1805–1882*, Cairo: al-Hay'a al-Misriyya al-'Amma li-l-Kitab, 2018: 237.
115. Order from Muhammad Ali issued 18 Dhu al-Qi'da 1241 (24 June 1826), cited 'Abd al-'Aẓim Muḥammad Sa'udi, *Tarikh tatawwur al-rayy fi Misr, 1882–1914*, Cairo: Al-Hay'a al-Misriyya al-'Amma li-l-Kitab, 2001: 43.
116. Abul-Magd, *Imagined Empires*: 95–6.
117. Spinage, *Cattle Plague*: 497; Rivlin, *Agricultural Policy*: 156.
118. Lady Duff Gordon, *Letters from Egypt, 1862–1869*, New York: Praeger, 1969: 108–9, 158, 161, 166. On the role of cattle plague in spurring the switch to steam power, see Ahmed, 'Egypt Ignited'.

119. Evidence on the source of the outbreak is inconclusive, but it is probable that the large movement of pack animals associated with the Italian invasion contributed to its rapid spread. Clive A. Spinage, 'The Italian Occupation of Massawa and the Supposed Origin of the African Rinderpest Panzootic', *African Journal of Ecology*, 55: 4, 2017.
120. Richard Grove and George Adamson, *El Niño in World History*, London: Palgrave Macmillan, 2018: 99–102.
121. The information on causes of deforestation is sparse, but sources tend to agree that the highlands had been largely deforested by the end of the century. The link between loss of cattle dung and increased use of firewood, and between deforestation and increased runoff, are my own inferences from this literature. Richard Pankhurst, 'The History of Deforestation and Afforestation in Ethiopia Prior to World War I', *Northeast African Studies*, 2: 1, 1995: 119–33; James C. McCann, 'The Plow and the Forest: Narratives of Deforestation in Ethiopia, 1840–1992', *Environmental History*, 2: 2, 1997: 138–59; Jan Nyssen et al., 'Environmental Conditions and Human Drivers for Changes to North Ethiopian Mountain Landscapes Over 145 Years', *Science of The Total Environment*, 485–6, 2014: 164–79.
122. When Willcocks, the leading British hydraulic engineer, later mentioned abandoned *saqiya* pits, it was partly to note their use for measuring the rising water table and associated salinity, a problem created by the extended system of canal-based irrigation the British had built. In switching from the basin system to perennial irrigation, they had transformed groundwater from a method of storage into a problem of water logging.
123. 'Ali Mubarak, *Kitab nukhbat al-fikr*: 187–91, 201–4.
124. Cookson-Hills, 'Engineering the Nile': 255–65, 306–21. The scheme required both an inlet canal, to fill the 670-square-kilometre depression, and an outlet canal to take water from the surface of the reservoir back to the Nile, below Cairo. Willcocks, *Report on Perennial Irrigation*: 30–1.
125. Alan Mikhail demonstrates how in the case of the eighteenth-century Fayyum, Ottoman authorities had to defer decisions about irrigation to local communities, whose labour and superior knowledge gave them the upper hand in the relationship: 'An Irrigated Empire: The View From Ottoman Fayyum', *International Journal of Middle East Studies*, 42: 4, 2010: 569–90.
126. Willcocks, *Sixty Years in the East*: 98–100.
127. Ross, 'Irrigation and Agriculture in Egypt': 191, cited in Tvedt, *The River Nile*: 26.
128. For an illuminating example of the colonial use of narratives of environmental precarity and decline in North Africa, see Diana K. Davis, *Resurrecting the Granary of Rome: Environmental History and French Colonial Expansion in North Africa*, Athens, OH: Ohio University Press, 2007.
129. Karl W. Butzer, *Early Hydraulic Civilization in Egypt: A Study in Cultural Ecology*, Chicago, IL: University of Chicago Press, 1976: 41; Mikhail, *Nature and Empire in Ottoman Egypt*: 38–81.
130. Carl Petry, *Protectors or Praetorians? The Last Mamluk Sultans and Egypt's Waning As a Great Power*, Albany, NY: State University of New York Press, 1994: 104–5.

131. Girard, 'Memoire sur l'agriculture': 499.
132. Boaz Shoshan, 'Grain Riots and the "Moral Economy": Cairo, 1350–1517', *The Journal of Interdisciplinary History*, 10: 3, 1980: 459–78.
133. Carl Petry, *The Mamluk Sultanate: A History*, Cambridge: Cambridge University Press, 2002: 182.
134. Gladys Franz-Murphy, 'A New Interpretation of the Economic History of Medieval Egypt: The Role of the Textile Industry 254–567/868–1171', in Michael G. Morony, *Manufacturing and Labour*, London: Routledge, 2003; Nelly Hanna, *Making Big Money in 1600: The Life and Times of Isma'il Abu Taqiyya, Egyptian Merchant*, Syracuse, NY: Syracuse University Press, 1998: 85.
135. Girard, 'Memoire sur l'agriculture': 540.
136. Franz-Murphy, 'A New Interpretation'.
137. Lorenzo Bondioli, personal communication, 1 December 2022; and 'Peasants, Merchants, and Caliphs: Capital and Empire in Fatimid Egypt', PhD thesis, History Department, Princeton University, 2021:, 156.
138. Amina Elbendary, 'The Worst of Times: Crisis Management and *Al-Shidda Al-'Uzma*', in Nelly Hanna, ed., *Money, Land and Trade: An Economic History of the Muslim Mediterranean*, London: I.B. Tauris, 2002, 67–83; Michael Chamberlain, 'The Crusader Era and the Ayyubid Dynasty', *The Cambridge History of Egypt*, vol. 1, 640–1571, ed. Carl F. Petry: 221; Mounira Chapoutot-Remadi, 'Une grande crise à la fin du XIIIe siècle en Égypte', *Journal of the Economic and Social History of the Orient*, 26: 3, 1983: 217–45.
139. Petry, *Protectors*; Shoshan, 'Grain Riots'; Aḥmad ibn 'Ali al-Maqrizi, *Kitab al-suluk li-ma'rifat duwal al-muluk*, ed. Muḥammad Mustafa Ziyada, 4, Cairo: Lajnat al-Ta'lif wa-l-Tarjama wa-l-Nashr: 318, 330, 334–5.
140. Aḥmad ibn 'Ali al-Maqrizi and Adel Allouche, *Mamluk Economics: A Study and Translation of Al-Maqrizi's Ighathah*, Salt Lake City, UT: University of Utah Press, 1994, 41. Maqrizi's source for this dubious information on extreme famine was 'Abd al-Latif al-Baghdadi, *Kitab al-ifadah wa-l-i'tibar fi l-umur al-mushahada wa-l-ḥawadith al-mu'ayana bi-ard Misr*, written in 1204, ed. and trans. as *A Physician on the Nile: A Description of Egypt and Journal of the Famine Years*, New York: New York University Press, 2022: 89–114. 'The affluent and moderately well-to-do classes got into the habit [of eating one another],' al-Baghdadi reports, 'some out of necessity, others for the sake of the novel culinary experience' (143). Note that the word 'famine' does not occur in al-Baghdadi's title, while the text itself refers to the episode as *al-ju'*, hunger.
141. Maqrizi and Allouche, *Mamluk Economics*. Stuart J. Borsch proposes that high Nile floods in Maqrizi's time, and across 150 years from 1350, can be explained by a decay of the basin system in Upper Egypt caused by the Black Death (1347–9). With less of the flood absorbed (and stored) in the south, higher volumes of water reached Cairo. 'Nile Floods and the Irrigation System in Fifteenth-Century Egypt', *Mamluk Studies Review*, 4, 2000: 131–45. On Maqrizi's experience as *muhtasib* and its influence on his writings, see Nasser Rabbat, *Writing Egypt: Al-Maqrizi and his Historical Project*,

Edinburgh: Edinburgh University Press, 2023: 80–9.

142. Gaston Wiet, 'Le Traité des Famines de Maqrīzī', *Journal of the Economic and Social History of the Orient*, 5: 1, 1962. The French title was first used by Silvestre de Sacy when referring to it in the translation of another work by Maqrizi: *Traité des Monnoies Musulmanes, Traduit de l'arabe de Makrizi*, published in *Magasin encyclopédique*, 1796, reprinted in Georges Salmon, ed., *Bibliothèque des arabisants français, 1, Silvestre de Sacy 1758–1838*, Cairo: Institut Français d'Archéologie Orientale, 1905: 9–66.

143. D'Souza, *Drowned and Dammed*: 6.

144. Ghislaine Alleaume, 'An Industrial Revolution in Agriculture', in Alan K. Bowman and Eugene Rogan, eds, *Agriculture in Egypt, From Pharaonic to Modern Times*, Oxford: Oxford University Press 1999: 331–45. See also Maha Ghalwash, *State, Peasants, and Land in Mid-Nineteenth-Century Egypt*, Cairo: American University in Cairo Press, 2023.

145. Ahmed, 'Egypt Ignited': 141–226; on the important role of Levantine capitalists, see also Kristen Alff, 'Levantine Joint-Stock Companies, Trans-Mediterranean Partnerships, and Nineteenth-Century Capitalist Development', *Comparative Studies in Society and History*, 60: 1, 2018: 150–77.

146. Henry Day, 'Modern Progress in Egypt', *Scientific American*, 28, 14 June 1873: 372, cited in Ahmed, 'Egypt Ignited': 313.

147. Ahmed, 'Egypt Ignited': 169.

148. Willcocks and Craig, *Egyptian Irrigation*, 3rd edition: 115.

149. Harvested cane loses moisture quickly and must be milled within twenty-four hours, so the mills that extract the juice and boil it to create raw sugar must be built close to the fields; but refined sugar cannot tolerate long sea voyages, as the humidity causes the crystals to coalesce; so for export, the final processing – refining the sugar into crystals – had to take place in the region of consumption, which in the case of Europe also meant a region with more abundant fuel for the refineries. Galloway, *The Sugar Cane Industry*: 17.

150. The exception is wheat, which in Egypt occupied the ground for six months, or about twice the length of time of other grains and legumes. Like cotton and sugar cane, it was grown mostly for export rather than local consumption, reflecting its value as the cereal whose storage protein is uniquely suited to making leavened bread. See Colin Wrigley, 'Wheat: An Overview of the Grain That Provides "Our Daily Bread"', *Encyclopedia of Food Grains*, 2nd edition, vol. 1, Oxford: Academic Press, 2016: 105–16. Unlike cane and cotton, however, wheat was a winter crop, so could be cultivated on a large scale without interfering with the flooding of the field basins.

151. For the physiology of cotton, see Abdul Rehman and Muhammad Farooq, 'Morphology, Physiology and Ecology of Cotton', in Khawar Jabran and Bhagirath Singh Chauhan, eds, *Cotton Production*, Chichester, UK: John Wiley, 2020: 23–46.

152. The indigo plant, which yielded several cuttings, occupied the ground for about eight months, from June. Like cotton and sugar, it could not be grown on flood basins. Girard, 'Memoire sur l'agriculture': 545–6.

153. Ibid.: 543.

154. Tsugitaka Sato, *Sugar in the Social Life of Medieval Islam*, Leiden: Brill, 2015: 26–30; Hanna, *Making Big Money in 1600*; Goitein, *A Mediterranean Society*: 252.
155. Hanna, *Making Big Money in 1600*: 81–4, 98–9; Andre Raymond, *Artisans et Commerçants au Caire au XVIIIe siècle*, Beirut: Presses de l'Ifpo, 2014, vol. 1: 203–41; vol. 2: 373–415.
156. Using what were known as *salaam* contracts: Hanna, *Making Big Money*: 84. See also Bondioli, 'Peasants, Merchants, and Caliphs': 122–4.
157. Girard, 'Memoire sur l'agriculture': 547.
158. Crouchley, *Economic Development*: 149.
159. Brown, 'Who Abolished Corvee Labour'; Esmeir, *Juridical Humanity*.
160. De Bellefonds, *Mémoires*: 4–5; Stuart J. Borsch, *The Black Death in Egypt and England: A Comparative Study*, Austin, TX: University of Texas Press, 2005: 37–9.
161. Alleaume, 'An Industrial Revolution in Agriculture'.
162. Jamie C. Woodward et al., 'The Nile: Evolution, Quaternary River Environments and Material Fluxes', in Avijit Gupta, ed., *Large Rivers: Geomorphology and Management*, London: John Wiley and Sons, 2007: 261–92, 281.
163. Fenn, *The Khedive's Country*.
164. David R. Montgomery, *Dirt: The Erosion of Civilizations*, Berkeley, CA: University of California Press, 2012: 203–6.
165. The *'izba*, from a word meaning distant or remote, originally referred to the temporary housing built on the floodplain, to support the summer well-irrigation season and livestock grazing. It became the term for permanent worker housing on the new estates, and eventually for the estates themselves. Mitchell, *Rule of Experts*: 54–79, and Mona Abaza, *The Cotton Plantation Remembered: An Egyptian Family Story*, Cairo: American University in Cairo Press, 2013.
166. Crouchley, *Economic Development*: 152–3.
167. Richard Elliot Benedick, 'The High Dam and the Transformation of the Nile', *Middle East Journal*, 33: 2, 1979: 135.
168. William Adams, *Nubia: Corridor to Africa*, Princeton, NJ: Princeton University Press, 1977: 19.
169. Sefelnasr and Sherif, 'Impacts of Seawater Rise'.
170. See 'Programme of Land Reclamation by M. Nourisson Bey', in Willcocks and Craig, *Egyptian Irrigation*, 3rd edition: 488–91.
171. Vladimir Borisovich Lutsky, *Modern History of the Arab Countries*, Moscow: Progress Publishers, 1969: 238.
172. See Rana Baker, 'Engineering Profit: Egyptian Railroads and the Unmaking of Prosperity, 1847–1907', PhD thesis, Middle Eastern, South Asian, and African Studies Department, Columbia University, 2022. Tunçer, *Sovereign Debt*: 31–51.

3 Capitalism as a Detour

1. For the US figures, see Board of Governors of the Federal Reserve System, 'Credit and Liquidity Programs and the Balance Sheet', Recent Balance Sheet Trends, federalreserve.gov; for global figures, see Bank for International Settlements, 'Central Bank Total Assets', data.bis.org.
2. For an overview of the issues, in particular the failure of mainstream economic models to account for money creation by banks, see Geoffrey Ingham, *The Nature of Money*, Cambridge: Polity Press, 2004; Servaas Storm, 'Cordon of Conformity: Why DSGE Models Are Not the Future of Macroeconomics', *International Journal of Political Economy*, 50: 2, 2021: 77–98. On the history of monetary theory in the United States, see Perry Mehrling, *The Money Interest and the Public Interest: American Monetary Thought, 1920–1970*, Cambridge, MA: Harvard University Press, 1997. On the longer work of economics in obscuring the nature of money, see Christine Desan, *Making Money: Coin, Currency, and the Coming of Capitalism*, Oxford: Oxford University Press, 2014.
3. Donald MacKenzie, *An Engine, Not a Camera: How Financial Models Shape Markets*, Cambridge, MA: MIT Press, 2006.
4. On these effects of abstraction, the technical methods of producing them, and how such methods are overlooked, see chapters 1 and 2 in Timothy Mitchell, *Colonising Egypt*, Cambridge: Cambridge University Press, 1988; and chapters 2 and 3 in Mitchell, *Rule of Experts: Egypt, Techno-Politics, Modernity*, Berkeley, CA: University of California Press, 2002.
5. Adam Tooze, *Crashed: How a Decade of Financial Crises Changed the World*, London: Penguin Books, 2018.
6. Ingham, *The Nature of Money*: 69–85; Stefano Sgambati, 'Rethinking Banking: Debt Discounting and the Making of Modern Money as Liquidity', *New Political Economy*, 21: 3, 2016: 274–90.
7. Fabian Muniesa and Liliana Doganova, 'The Time That Money Requires: Use of the Future and Critique of the Present in Financial Valuation', *Finance and Society*, 6: 2, 2020: 95–113; Jens Beckert and Richard Bronk, eds, *Uncertain Futures: Imaginaries, Narratives, and Calculation in the Economy*, Oxford: Oxford University Press, 2018.
8. Richard Werner, 'A Lost Century in Economics: Three Theories of Banking and the Conclusive Evidence', *International Review of Financial Analysis*, 46, 2016: 361–79.
9. Alfred Mitchell-Innes, 'What Is Money?', *Banking Law Journal*, 1913.
10. Mitchell-Innes became an ally of the emergent nationalist movement, serving briefly in 1907 as founding president of the national sporting club al-Nadi al-Ahly, organised by a group of leading political figures and becoming the home of the storied Egyptian football club al-Ahly. On Mitchell-Innes and his role in Cairo, see Casey Primel, 'Calculating Futures: Debt, Markets, and the Science of Prices in Colonial Egypt', PhD thesis, Middle Eastern, South Asian, and African Studies Department, Columbia University, 2016. See also Wilson Jacob, *Working Out Egypt: Effendi Masculinity and Subject Formation in Colonial Modernity, 1870–1940*, Durham, NC: Duke University Press, 2011: 85–7.

11. From Washington he sent to London a report on the American system of banking. Foreign Office. Report by [A.] Mitchell-Innes, Councilor, H.M. Embassy, Washington, DC, on the banking and currency system in the United States. 1910. The National Archives of the UK (TNA): T 1/11211/11528.
12. Mitchell-Innes, 'The Credit Theory of Money'.
13. John Maynard Keynes, '*What is Money?* By A. Mitchell Innes', *Economic Journal*, 24: 95, 1914: 421, cited in L. Randall Wray, ed., *Credit and State Theories of Money: The Contributions of A. Mitchell Innes*, Cheltenham, UK: Edward Elgar, 2004: 2.
14. Joseph Schumpeter, *The Theory of Economic Development: An Inquiry into Profits, Capital, Credit, Interest, and the Business Cycle*, Cambridge, MA: Harvard University Press, 1934.
15. Desan, *Making Money*; Tim Di Muzio and Richard Robbins, *Debt as Power*, Manchester: Manchester University Press, 2016.
16. While Smith explains the origin of money in this way in book 1 of *The Wealth of Nations*, New York: Penguin Books, 1986, in chapter 2 of book 2 (p. 395) he offers an account of money based on the creation of credit money by merchants and bankers.
17. Mitchell-Innes appears to be drawing here on Georg Friedrich Knapp, *Staatliche Theorie des Geldes*, 1905. See the abridged edition and translation: Georg Friedrich Knapp, *The State Theory of Money*, trans. from the German by H.M. Lucas and James Bonar, London: Macmillan and Co., 1924.
18. See Herman van der Wee, 'Monetary, Credit and Banking Systems', in D. C. Coleman, P. Mathias, and M. M. Postan, eds, *The Cambridge Economic History of Europe*, vol. 5, Cambridge: Cambridge University Press, 1977.
19. The agriculturalist would fulfil the contract by supplying the crop at its full value, enabling the merchant to profit from the discounted advance sales. For an extensive discussion from the Fatimid period, see Lorenzo Bondioli, 'Peasants, Merchants, and Caliphs: Capital and Empire in Fatimid Egypt', PhD thesis, Department of History, Princeton University, 2021. As Bondioli points out, the practice continued to be widespread in the Ottoman era, as shown by Nelly Hanna, *Making Big Money in 1600: The Life and Times of Isma'il Abu Taqiyya, Egyptian Merchant*, Syracuse, NY: Syracuse University Press, 1998: 83–7; Kenneth M. Cuno, *The Pasha's Peasants: Land, Society, and Economy in Lower Egypt, 1740–1858*, Cambridge and New York: Cambridge University Press, 1992: 56–7; and Beshara Doumani, *Rediscovering Palestine: Merchants and Peasants in Jabal Nablus, 1700–1900*, Berkeley, CA, and London: University of California Press, 1995: 134–48. See also Lorenzo Bondioli, 'Islam, Merchants, and Capitalism: Fifty-Five Years in the Socioeconomic History of the Medieval Islamic World', *Capitalism: A Journal of History and Economics*, 4: 2, 2023: 258–307.
20. S. D. Goitein, *A Mediterranean Society: The Jewish Communities of the Arab World as Portrayed in the Documents of the Cairo Geniza*, vol. 1, Berkeley,

CA: University of California Press, 1988: 197–200; Abraham L. Udovitch, *Partnership and Profit in Medieval Islam,* Princeton, NJ: Princeton University Press, 1970: 77–86; and 'Credit as a Means of Investment in Medieval Islamic Trade', *Journal of the American Oriental Society*, 87: 3, 1967: 260–4; Janet Abu-Lughod, *Before European Hegemony: The World System A.D. 1250–1350*, New York: Oxford University Press, 1989. Lapidus describes the role of *karimi* spice merchants in providing credit in Mamluk Syria: Ira M. Lapidus, *Muslim Cities in the Later Middle Ages,* Cambridge: Cambridge University Press, 2008: 121. On the developing uses of credit in later medieval/early modern England, see Marjorie McIntosh, 'Money Lending on the Periphery of London, 1300–1600', *Albion: A Quarterly Journal Concerned with British Studies*, 20: 4, 1988: 557–71; on the inconvenience of cash and preference for credit throughout medieval Europe, see Meir Kohn, 'Payments and the Development of Finance in Pre-Industrial Europe', *Department of Economics Dartmouth College,* Working Paper, 2001, wpmucdn.com (accessed 25 January 2021).

21. Muhammad ibn Ahmad Sarakhsi, *Kitab al-mabsut*, vol. 22, Cairo: Matba'at al-Sa'ada, 1906–13: 38, cited in Abraham L. Udovitch, *Partnership and Profit*: 79. Udovitch further explains: 'On the most basic commercial level, sales, both wholesale and retail were normally conducted on a credit basis, i.e. on the basis of deferred payment for a fixed period. The documents make clear that the buyer paid a premium for the deferment of payment. This was expressed either by granting a discount from the stated price in return for an immediate cash payment (discounts varied from 2%–4%), or by the quotation of two prices: one for a cash transaction, and a slightly higher one for a credit transaction': Abraham L. Udovitch, 'Reflections on the Institutions of Credits and Banking in the Medieval Islamic Near East', *Studia Islamica*, 41, 1975: 13.
22. On credit networks of the Indian Ocean world, see Najaf Haidar, 'The Network of Monetary Exchange in the Indian Ocean Trade, 1200–1700', in Himanshu Prabha Ray and Edward A. Alpers, eds, *Cross Currents and Community Networks: The History of the Indian Ocean World*, New Delhi: Oxford University Press, 2007: 198–200; for the Mediterranean world and the use of the *suftaja*, or bill of exchange, see S. D. Goitein, *A Mediterranean Society*: 240–2; for the Islamic world more generally, Subhi Y. Labib, 'Capitalism in Medieval Islam', *The Journal of Economic History*, 29: 1, 1969: 88–93; credit and bills of exchange also continued to be widely used in the Ottoman world. For details, see Bruce Masters, *The Origins of Western Economic Dominance in the Middle East: Mercantilism and the Islamic Economy in Aleppo, 1600–1750*, New York: New York University Press, 1988: 147–53.
23. Or a multiple of that standard period (the date due was termed 'usance'). Or in other cases, the period of the time until the next fair, when bills were settled. See chapter 4 of M. T. Boyer-Xambeau, Lucien Gillard, and Ghislain Deleplace, *Private Money and Public Currencies: The Sixteenth Century Challenge*, Armonk, NY: M. E. Sharpe, 1994.
24. But note the argument that 'evil is the root of all money' – that it is lack of

trust that gives rise to money – by John Moore: Nobuhiro Kiyotaki and John Moore, 'Evil Is the Root of All Money', *The American Economic Review*, 92: 2, 2002: 62–6.

25. Goitein, *A Mediterranean Society*: 259–60; Hanna, *Making Big Money in 1600*.
26. Daniel Lord Smail, *Legal Plunder: Households and Debt Collection in Late Medieval Europe*, Cambridge, MA: Harvard University Press, 2016.
27. On the development of the negotiability of bills, see John H. Munro, 'The Medieval Origins of the Financial Revolution: Usury, *Rentes*, and Negotiability', *The International History Review*, 25: 3, 2003: 505–62.
28. Henry Thornton, *An Inquiry into the Nature and Effect of the Paper Credit of Great Britain*, London, 1802: 22–6.
29. David Graeber, *Debt: The First 5,000 Years*, Brooklyn, NY: Melville House, 2014; Julie Claustre, 'Vivre à crédit dans une ville sans banque (Paris: XIVe-XVe siècle)', 109: 3/4, 2013: 567–96.
30. Craig Muldrew, *The Economy of Obligation: The Culture of Credit and Social Relations in Early Modern England*, New York and Basingstoke: St Martin's Press, 1998: 95–8; Smail, *Legal Plunder*: 105–8.
31. Udovitch, 'Reflections on the Institutions of Credits and Banking': 14; Ronald C. Jennings, 'Loans and Credit in Early 17th Century Ottoman Judicial Records: The Sharia Court of Anatolian Kayseri', *Journal of the Economic and Social History of the Orient*, 16: 2/3, 1973: 168–216.
32. Van der Wee, 'Monetary, Credit and Banking Systems': 301.
33. For Europe in the later Middle Ages, see Smail, *Legal Plunder*: 90–1. For early modern England, see Muldrew, *The Economy of Obligation*, 95–119; for the nineteenth century, see Henry Dunning Macleod, *The Theory of Credit*, vol. 2, London, New York: Longmans, Green, and Co, 1891: 339. For New England in the eighteenth and early nineteenth century, see Naomi R. Lamoreaux, 'Rethinking the Transition to Capitalism in the Early American Northeast', *The Journal of American History* 90: 2, 2003: 441–2.
34. Muldrew, *The Economy of Obligation*, 95–119.
35. Van der Wee, 'Monetary, Credit and Banking Systems': 301; Sidney Pollard, 'Capital Accounting in the Industrial Revolution', *Bulletin of Economic Research*, 15: 2, 1963: 78.
36. Mitchell-Innes, 'What Is Money?': 397. On the Tanta fair, see also Catherine Mayeur-Jaouen, *The Mulid of al-Sayyid al-Badawi of Tanta: Egypt's Legendary Sufi Festival*, Cairo: American University in Cairo Press, 2019.
37. Janet Abu-Lughod, *Before European Hegemony*, on the Champagne fairs and long-distance trade of the thirteenth century. See also Charles Kindleberger, *A Financial History of Western Europe*, Abingdon: Routledge, 1984: 36–9.
38. Fernand Braudel, *The Mediterranean and the Mediterranean World in the Age of Philip II*, vol. 1, Berkeley, CA: University of California Press, 1995: 379–80. Braudel adds: 'Only paper changed hands, and not a penny of currency, reports a Venetian observer with only a little exaggeration.' The classic account of the fairs is Henri Pirenne, *Economic and Social History of Medieval Europe*, New York: Harcourt, Brace, and Company, 1937: 96–139.
39. Boyer-Xambeau et al., *Private Money and Public Currencies*: 160–95.

40. Violet Babour, *Capitalism in Amsterdam in the Seventeenth Century*, Baltimore, MD: Johns Hopkins Press, 1950; Immanuel Wallerstein, *The Modern World System*, vol. 2, New York: Academic Press, 1974: 58–60.
41. T. S. Ashton, *The Industrial Revolution, 1760–1830*, London: Oxford University Press, 1948: 103.
42. Desan, *Making Money*: 295–329.
43. Ashton, *Industrial Revolution*: 99–100; Viviana Zelizer, *The Social Meaning of Money: Pin Money, Paychecks, Poor Relief, and Other Currencies*, Princeton, NJ: Princeton University Press, 2017.
44. Keynes, *Indian Currency and Finance*: 17.
45. Ashton, *Industrial Revolution*: 101; Smail, *Legal Plunder*.
46. Liliana Doganova, *Discounting the Future: The Ascendancy of a Political Technology*, Brooklyn, NY: Zone Books, 2024; Jonathan Nitzan and Shimshon Bichler, *Capital as Power: A Study of Order and Creorder*, London: Routledge, 2009: 155–8.
47. Before the rise of discounting, another source of gain provided the main profit from issuing bills: exchange rate difference between the money of account in different cities: Munro, 'The Medieval Origins of the Financial Revolution'.
48. Pirenne, *Economic and Social History of Medieval Europe*: 215–17.
49. Nitzan and Bichler, *Capital as Power*; Jonathan Levy, 'Capital as Process and the History of Capitalism', *The Business History Review*, 91: 3, 2017: 483–510; Michael Perelman, 'Fictitious Capital and Crisis Theory', in *Marx's Crises Theory: Scarcity, Labour, and Finance,* New York: Praeger, 1987: 170–217.
50. See Reinhart Koselleck, *Futures Past: On the Semantics of Historical Time*, New York: Columbia University Press, 2004: 21–2.
51. Major contributions to these debates include: T. H. Aston and C. H. E. Philpin, eds, *The Brenner Debate: Agrarian Class Structure and Economic Development in Pre-industrial Europe*, Cambridge: Cambridge University Press, 1987; Robert C. Allen, *The British Industrial Revolution in Global Perspective*, Cambridge: Cambridge University Press, 2009; E. A. Wrigley, *Continuity, Chance and Change: The Character of the Industrial Revolution in England*, Cambridge: Cambridge University Press, 1990; Jan de Vries, *The Industrious Revolution*, Cambridge: Cambridge University Press, 2008; Jane Humphries, 'Childhood and Child Labour in the British Industrial Revolution', *The Economic History Review*, 66: 2, 2013: 395–418; Katrina Honeyman, *Child Workers in England, 1780–1820: Parish Apprentices and the Making of the Early Industrial Labour Force*, Burlington, VT: Ashgate, 2007; Silvia Frederici, *Caliban and the Witch: Women, the Body, and Primitive Accumulation*, New York: Autonomedia, 2004; Kenneth Pomeranz, *The Great Divergence: China, Europe, and the Making of the Modern World Economy*, Princeton, NJ: Princeton University Press, 2000; Prasannan Parthasarathi, *Why Europe Grew Rich and Asia Did Not: Global Economic Divergence, 1600–1850*, Cambridge: Cambridge University Press, 2011; Jack Goldstone, *Why Europe? The Rise of the West in World History, 1500–1850*, New York: McGraw-Hill Education, 2009; Sven Beckert, *Empire of Cotton: A Global History*, New York: Vintage Books, 2015.

52. See 'Cheshire & Greater Manchester: Quarry Bank', National Trust, n.d., nationaltrust.org.uk (accessed 23 July 2025).
53. *The Mill*, drama series created by John Fay, first broadcast in the UK by Channel 4, July 2013. On the use of the archive, see 'The Mill on Channel 4', Quarry Bank Revealed (blog), 18 July 2013, quarrybankmill.wordpress.com (accessed 11 February 2021).
54. Douglas A. Farnie, 'The Role of Merchants as Prime Movers in the Expansion of the Cotton Industry, 1760–1900', in Douglas A. Farnie and David J. Jeremy, eds, *The Fibre That Changed the World*, Oxford: Oxford University Press, 2004: 15–55, 27.
55. Douglas A. Farnie, *The English Cotton Industry and the World Market, 1815–1896*, Oxford: Oxford University Press, 1979: 209.
56. Maxine Berg, *The Machinery Question and the Making of Political Economy 1815–1848*, Cambridge: Cambridge University Press, 1980: 27–31; Barry Supple, 'The Nature of Enterprise', in D. C. Coleman et al., eds, *The Cambridge Economic History of Europe*, vol. 5, Cambridge: Cambridge University Press, 1977: 427; S. R. H. Jones, 'Technology, Transaction Costs, and the Transition to Factory Production in the British Silk Industry, 1700–1870', *Journal of Economic History*, 47: 1, 1987: 71–96. Keith Tribe, 'Industrialisation as a Historical Category', in *Genealogies of Capitalism*, London: Macmillan, 1981: 101–20.
57. In 1833, raw cotton was imported mostly from the United States (also Brazil, India, the West Indies, and India; briefly from Egypt in 1820s); of this, more than 10 per cent was lost in spinning, and another part in the manufacturing of cloth; of the remainder 72 per cent was exported, about half as yarn and thread and half as manufactured goods, leaving 28 per cent for home consumption and stock. Figures from table in Edward Baines, *History of the Cotton Manufacture in Great Britain*, London: H. Fisher, R. Fisher, and P. Jackson, 1835: 367. Note that domestic consumption was of higher-quality textiles, and that both domestic and exported cloth and yarn appear to have declined rapidly in quality through the first half of the nineteenth century: (Lars G. Sandberg, 'Movements in the Quality of British Cotton Textile Exports, 1815–1913', *The Journal of Economic History*, 28: 1, 1968: 19).
58. The references to 'flimsy' and 'freak' are from William C. Jones Ltd, 'Cotton Waste: A Study of a Great Lancashire Industry', in Philip Sykas, ed., *Pathways in the Nineteenth-Century British Textile Industry*, London: Routledge, 2022: 261. Originally published as Charles Hobson, *Cotton Waste: A Study of a Great Lancashire Industry*, Manchester: William C. Jones Ltd, 1920: 8.
59. Beckert, *Empire of Cotton*, xvii.
60. Subhasis Das and V. K. Kothari, 'Moisture Vapour Transmission Behaviour of Cotton Fabrics', *Indian Journal of Fibre & Textile Research*, 37: 2, 2012: 151–6.
61. See, for example, Gregory Clark, 'What Made Britannia Great? How Much of the Rise of Britain to World Dominance by 1850 Does the Industrial Revolution Explain?', in Tim Hatton, Kevin O'Rourke, and Alan Taylor, eds, *Comparative Economic History: Essays in Honor of Jeffrey Williamson*, Cambridge, MA: MIT Press, 2007: 33–57.

62. Jones, 'Technology, Transaction Costs'; Andreas Malm, *Fossil Capital: The Rise of Steam Power and the Roots of Global Warming*, London: Verso, 2016.
63. Pomeranz, *The Great Divergence*. See also Andre Gunder Frank, *ReORIENT: Global Economy in the Asian Age*, Berkeley, CA: University of California Press, 1998.
64. William Sewell, 'The Temporalities of Capitalism', *Socio-Economic Review*, 6: 3, 2008: 517–37.
65. Alexander Anievas and Kerem Nisancioglu, *How the West Came to Rule: The Geopolitical Origins of Capitalism*, London: Pluto Press, 2015.
66. For a recent example of this reading of capitalism, see Søren Mau, *Mute Compulsion: A Marxist Theory of the Economic Power of Capital*, London: Verso, 2023: 277–8.
67. H. V. Bowen, *War and British Society 1688–1815*, Cambridge: Cambridge University Press, 1998: 18–19.
68. John Charles Herries, *A Review of the Controversy Respecting the High Price of Bullion, and the State of our Currency*, London: J. Budd, 1811: 44–5, cited in Larry Neal, *The Rise of Financial Capitalism: International Capital Markets in the Age of Reason*, Cambridge: Cambridge University Press, 1990: 191.
69. On smuggling, see Niall Ferguson, *House of Rothschild: Money's Prophets, 1798–1848*, New York: Viking, 1998: 58–62. The blockade also spurred the development of cotton manufacturing on the continent: Beckert, *Empire of Cotton*: 157.
70. Neal, *The Rise of Financial Capitalism*: 181. Neal shows the limits of older views, such as those of Williamson, based on a misunderstanding of the nature of capital, that spending on warfare retarded Britain's industrial development by 'crowding out' investment. See Jeffrey G. Williamson, 'Why Was British Growth so Slow During the Industrial Revolution', *Journal of Economic History*, 44: 3, 1984: 687–712.
71. Ferguson, *House of Rothschild*.
72. S. D. Chapman, 'The International Houses: The Continental Contribution to British Commerce, 1800–1860', *Journal of European Economic History*, 6: 1, 1977: 5–48.
73. Ferguson, *House of Rothschild*: 55–6.
74. Farnie, 'The Role of Merchants as Prime Movers': 30–1.
75. Tiago Mata and Robert Van Horn, 'Capitalist Threads: Engels the Businessman and Marx's *Capital*', *History of Political Economy*, 49: 2, 2017: 207–32.
76. Pandelis Michalis Glavanis, 'Aspects of The Economic and Social History of the Greek Community in Alexandria during the Nineteenth Century', PhD thesis, Department of Economic and Social History, University of Hull, 1989. 145–7. The words 'at a slight angle' are from E. M. Forster, *Pharos and Pharillon*, New York: Alfred A. Knopf, 1923: 110.
77. Chapman, 'The International Houses': 35–42. See also Farnie, 'The Role of Merchants'; Fred Halliday, 'The Millet of Manchester: Arab Merchants and Cotton Trade', *British Journal of Middle Eastern Studies*, 19: 2, 1992: 159–76, who also notes the presence of a sizeable community of Arab merchants from Morocco, mostly from Fez.

78. Farnie, 'The Role of Merchants': 33, citing an 1885 report.
79. On motives for the invasion of Egypt and its role as a potential replacement for Caribbean plantations, see François Charles-Roux, *Les Origines de l'expédition d'Egypt*, Paris: Plon-Nourrit et cie, 1910: 303–4. On the significance of the Haitian revolution, see Cedric Robinson, *Black Marxism: The Making of the Black Radical Tradition*, 3rd edition, Chapel Hill, NC: University of North Carolina Press, 2021; Alan Forrest, *The Death of the French Atlantic: Trade, War and Slavery in the Age of Revolution*, Oxford: Oxford University Press, 2020: 3–21; and Adom Getachew, 'Universalism After the Post-Colonial Turn: Interpreting the Haitian Revolution', *Political Theory*, 44: 6, 2016: 821–45.
80. Elena Frangakis-Syrett, *The Commerce of Smyrna in the Eighteenth Century, 1700–1820*, Athens: Centre for Asia Minor Studies, 1992. Also, Daniel Crecelius, 'Egypt in the Eighteenth Century', in M. W. Daly, ed., *The Cambridge History of Egypt*, Cambridge: Cambridge University Press, 1998, on the rise of Levantine Christian merchant communities.
81. Jane Hathaway, *The Arab Lands under Ottoman Rule, 1516–1800*, 2nd edition, London: Routledge, 2019: 81–101; Daniel Crecelius, *The Roots of Modern Egypt: A Study of the Regimes of 'Ali Bey al-Kabir and Muhammad Bey Abu al-Dhahab, 1760–1775*, Minneapolis, MN: Bibliotheca Islamica, 1981; Nelly Hanna, *Artisan Entrepreneurs in Cairo and Early-Modern Capitalism (1600–1800)*, Syracuse, NY: Syracuse University Press, 2011; Alan Mikhail, *Under Osman's Tree: The Ottoman Empire, Egypt, and Environmental History*, Chicago, IL: University of Chicago Press, 2017, 73–92.
82. 'Abd al-Rahman Jabarti, *al-Ta'rikh al-Musamma 'Aja'ib al-athar fi al-tarajim wa-l-akhbar*, vol. 1, Bulaq, Egypt: Dar al-Ṭiba'a, 1880: 382.
83. Madawi al-Rashid, *A History of Saudi Arabia*, Cambridge: Cambridge University Press, 2012: 13–22. On the cloth fair, K. N. Chaudhuri, *Trade and Civilisation in the Indian Ocean: An Economic History from the Rise of Islam to 1750*, Cambridge: Cambridge University Press, 1985: 46; on the significance of the Hajj to transregional commerce, Rishad Choudhury, *Hajj across Empires: Pilgrimage and Political Culture after the Mughals, 1739–1857*, Cambridge: Cambridge University Press, 2024: 25–107.
84. John Bowring, *Report on Egypt and Candia: Addressed to the Right Hon. Lord Viscount Palmerston*, London: W. Clowes and Sons for Her Majesty's Stationery Office, 1840: 64 (quoting a letter from Briggs's partner Robert Thurburn). On the building of the army, see Khaled Fahmy, *All the Pasha's Men: Mehmed Ali, his Army, and the Making of Modern Egypt*, Cambridge: Cambridge University Press, 1997.
85. A. E. Crouchley, *The Economic Development of Modern Egypt*, London: Longmans, Green and Co, 1938: 105. He adds: 'The merchants who engaged in this work had to have considerable funds at their disposal – to be in fact real merchant bankers – but they made big profits from thus combining money-lending and trading.'
86. Frederick Stanley Rodkey, 'The Attempts of Briggs and Company to Guide

British Policy in the Levant in the Interest of Mehemet Ali Pasha, 1821–41', *Journal of Modern History*, 5: 3, 1933: 324–51.

87. G. Dardaud, 'Un ingénieur français au service de Mohamed Ali. Louis Alexis Jumel (1785–1823). D'après les documents inédits des archives du consulat de France du Caire', *Bulletin de l'Institut Égyptien*, 22: 1, 1939: 49–97.
88. On orchards irrigated by waterwheels placed over wells, see Pierre-Simon Girard, 'Memoire sur l'Agriculture, l'Industrie et le commerce de l'Égypte', *Description de l'Égypte: Etat Moderne*, vol. 2, Paris: Imprimerie Impériale: 502.
89. Ibid.: 542–4.
90. George Waddington, *Journal of a Visit to Some Parts of Ethiopia*, London: J. Murray, 1822: 43; John Lewis Burckhardt, *Travels in Nubia*, London: John Murray, 1822: 33, 240.
91. Burckhardt, *Travels in Nubia*: 276.
92. Anders J. Bjørkelo, *Prelude to the Mahdiyya: Peasants and Traders in the Shendi Region, 1821–1885*, Cambridge: Cambridge University Press, 1989: 13, 36–7, 72, 86–8; El Haj Abdalla Bilal Omer, *The Danagla Traders of Northern Sudan: Rural Capitalism and Agricultural Development*, London: Ithaca Press, 1985: 26–36, 57–67; W. R. G. Bond, 'Some Curious Methods of Cultivation in Dongola Province', *Sudan Notes and Records*, 8, 1925: 96–103.
93. On the resistance in Upper Egypt, see Zeinab Abul-Magd, *Imagined Empires, A History of Revolt in Egypt*, Berkeley, CA: University of California Press, 2013: 80–104.
94. 'Abd al-Hamid Subhi Nasif, *Mashru'at al-rayy wa-atharuha fi al-mujtama' al-misri fi al-qarn al-tasi'* 'ashr *1805–1882*, Cairo: al-Hay'a al-Misriyya al-'Amma li-l-Kitab, 2018: 239–42.
95. Ibid.: 235–37; Dardaud, 'Un ingénieur français': 64, where the brothers are identified as Paul and Antoine Gibbara. On al-Jibara, see also Jan Goldberg, 'On the Origins of Majalis al-Tujjar in Mid-Nineteenth Century Egypt', *Islamic Law and Society*, 6: 2, 1999: 209–10.
96. Dardaud, 'Un ingénieur français': 84. The variety became known as 'Maho' cotton in Europe, except in France, where it was traded as 'Jumel' cotton, after the French engineer hired by Muhammad Ali initially to develop woollen manufacturing, who had helped promote the expansion of long-staple cotton during his brief period in Egypt, from 1818 until his death at age thirty-eight in 1823 (Dardaud, 'Un ingénieur français': 66).
97. Rodkey, 'The Attempts of Briggs & Company': 326–7.
98. Christopher Bayly, *Indian Society and the Making of the British Empire*, vol. 2, Cambridge: Cambridge University Press, 1988: 45–78.
99. Ninth report of the select committee of the House of Commons on the Affairs of India, 25 June 1783, reprinted in Edmund Burke, *The Writings and Speeches of Edmund Burke*, vol. 8, Boston, MA: Little Brown, 1901: 43.
100. H. V. Bowen, *The Business of Empire: The East India Company and Imperial Britain, 1756-1833*, Cambridge: Cambridge University Press, 2006: 246–52; Priya Satia, *Empire of Guns: The Violent Making of the Industrial Revolution*, New York: Penguin Books, 2008: 111–12; Chaudhuri, *Trade and Civilisation*

in the Indian Ocean, 105. It was reported that 'The goods which are exported from Europe to India consist chiefly of military and naval stores, of clothing for troops, and of other objects for the consumption of the Europeans residing there; and, excepting some lead, copper utensils and sheet copper, woollen cloth, and other commodities of little comparative value, no sort of merchandise is sent from England that is in demand for the wants or desires of the native inhabitants.' 'Ninth report of the select committee of the House of Commons on the Affairs of India', 25 June 1783, reprinted in Burke, *Writings and Speeches*: 48–9.

101. 'Ninth report' in Burke, *Writings and Speeches*: 72.
102. Anthony Webster, *The Twilight of the East India Company: The Evolution of Anglo-Asian Commerce and Politics, 1790–1860*, Rochester, NY: Boydell Press, 2009; Amales Tripathi, *Trade and Finance in the Bengal Presidency 1793–1833*, Calcutta: Oxford University Press, 1979: 77–119.
103. Robinson, *Black Marxism*; Vincent Brown, *Tacky's Revolt: The Story of an Atlantic Slave War*, Cambridge, MA: Belknap Press, 2022; Michael Craton, *Testing the Chains: Resistance to Slavery in the British West Indies*, Ithaca, NY: Cornell University Press, 1982; Christopher Taylor, *The Black Carib Wars: Freedom, Survival, and the Making of the Garifuna*, Jackson, MI: University Press of Mississippi, 2012. On the crises in West India finance, see S. G. Checkland, 'Finance for the West Indies, 1780–1815', *The Economic History Review*, 10: 3, 1958: 461–9.
104. Jonathan Eacott, *Selling Empire: India and the Making of Britain and America, 1600–1830*, Chapel Hill, NC: University of North Carolina Press, 2016: 207; Ralph Hidy, *The House of Baring in American Trade and Finance: English Merchant Bankers at Work, 1763–1861*, Cambridge, MA: Harvard University Press, 1949.
105. Sir Charles Fawcett, 'The Striped Flag of the East India Company and its Connection with the American Stars and Stripes', *The Mariner's Mirror*, 23: 4, 1937.
106. Caitlin Rosenthal, *Accounting for Slavery: Masters and Management*, Cambridge, MA: Harvard University Press, 2018; Bonnie Martin, 'Slavery's Invisible Engine: Mortgaging Human Property', *Journal of Southern History*, 76: 4, 2010: 817; Maxine Berg and Pat Hudson, *Slavery, Capitalism and the Industrial Revolution*, Cambridge: Polity Press, 2023: 165–85.
107. Joseph Inikori, *Africans and the Industrial Revolution in England: A Study in International Trade and Development*, Cambridge: Cambridge University Press, 2002: 330.
108. Joseph Inikori, 'The Credit Needs of the African Trade and the Development of the Credit Economy in England', *Explorations in Economic History*, 27: 2, 1990: 197–231. See also Berg and Hudson, *Slavery, Capitalism and the Industrial Revolution*: 165–85.
109. Carl Wennerlind, *Casualties of Credit: The English Financial Revolution, 1620–1720*, Cambridge, MA: Harvard University Press, 2011: 199.
110. In 1777, the war reduced the Virginia tobacco trade 'to a trickle'. Glasgow merchants, who had lived off the import and re-export of tobacco, switched

to the importing and processing of cotton. Seymour Shapiro, *Capital and the Cotton Industry in the Industrial Revolution*, Ithaca, NY: Cornell University Press, 1967: 173–5. The American war also interrupted the trade of coarse linens, which flowed in the other direction.

111. S. D. Chapman, 'British Marketing Enterprise: The Changing Roles of Merchants, Manufacturers, and Financiers, 1700–1860', *The Business History Review*, 53: 2, 1979: 218.
112. See details in Michael Janes, *From Smuggling to Cotton Kings: The Greg Story*, Gloucestershire: Memoir Books, 2010; Mary B. Rose, *The Gregs of Quarry Bank Mill: The Rise and Decline of a Family Firm, 1750–1914*, Cambridge: Cambridge University Press, 1986; and Beckert, *Empire of Cotton*, 56–63.
113. Shapiro, *Capital and the Cotton Industry*: 105.
114. Dale Tomich, 'The Second Slavery and World Capitalism: A Perspective for Historical Inquiry', *International Review of Social History*, 63: 3, 2018: 477–501; Robin Blackburn, *The Reckoning: From the Second Slavery to Abolition, 1776–1888*, London: Verso, 2024; Walter Johnson, *River of Dark Dreams: Slavery and Empire in the Mississippi Valley's Cotton Kingdom*, Cambridge, MA: Belknap Press, 2013.
115. Farnie, *The English Cotton Industry*: 15.
116. Pomeranz, *The Great Divergence.*
117. Patrick O'Brien, 'Was the British Industrial Revolution a Conjuncture in Global Economic History?', *Journal of Global History*, 17: 1, 2022: 128–50. See also Emma Rothschild, 'Where Is Capital?', *Capitalism: A Journal of History and Economics*, 2: 2, 2021: 291–371.
118. E. A. Wrigley, *Energy and the English Industrial Revolution*, Cambridge: Cambridge University Press, 2010, explained the industrial revolution as a tipping point. See also Simon Sharpe and Timothy Lenton, 'Upward-Scaling Tipping Cascades to Meet Climate Goals: Plausible Grounds for Hope', *Climate Policy*, 21: 4, 2021: 421–33.
119. Moishe Postone, *Time, Labor, and Social Domination: A Reinterpretation of Marx's Critical Theory*, Cambridge: Cambridge University Press, 1993. On the influence of this understanding, see, for example, Patrick Murray, *The Mismeasure of Wealth: Essays on Marx and Social Form*, Leiden: Brill, 2016, and Sewell, 'The Temporalities of Capitalism'. For a powerful reinterpretation, with parallels to the one presented here, see Carolyn Hardin, *Capturing Finance: Arbitrage and Social Domination*, Durham, NC: Duke University Press, 2021. However, in contrast to Hardin, as I explain in chapter 4, I suggest a different approach to 'abstraction'.
120. Augusto Graziani makes this argument from the perspective of Marx's theory of money: 'The Marxist Theory of Money', *International Journal of Political Economy*, 27: 2, 1997: 26–50.
121. Nicholas Mayhew, *Sterling: the Rise and Fall of a Currency*, London: A. Lane, 1999: 57.
122. See Ashton, *The Industrial Revolution, 1760–1830*, who notes that 'The dearth of coin of small denomination was a serious matter for manufacturers with wages to pay. Many of them spent days riding from place to place in search of shillings' (p. 99). For a corrective, see Smail, *Legal*

Plunder: 91–2, who argues that the lack of small coins and the wide use of credit for everyday purchases was not a sign of economic weakness.

123. Kiyotaki and Moore, 'Evil Is the Root of All Money'. I was fortunate to meet John Moore in the summer of 1976, in a small oasis settlement in southern Algeria, from where I was hoping to cross the Sahara to West Africa, a route that had no roads in those days. He returned north to Cambridge, England (where, as it happens, we were both students), and went on to become one of Britain's leading economists. I continued south across the Sahara. Since then, our paths have never crossed, until this note.
124. As Robert Meister points out, the development of complex credit relations pre-dates the rise of wage labour and was a means to the elimination of feudal relations: *Justice is an Option: A Democratic Theory of Finance for the Twenty-First Century*, Chicago, IL: University of Chicago Press, 2021: 17–22. See also Desan, *Making Money*, and Jairus Banaji, *A Brief History of Commercial Capitalism*, Chicago, IL: Haymarket Books, 2020.
125. The term 'apparatus of capture' comes from Deleuze and Guattari, who draw on the work of the French economist Bernard Schmitt. What follows here is developed partly from Schmitt's work. Gilles Deleuze and Félix Guattari, *A Thousand Plateaus: Capitalism and Schizophrenia*, Minneapolis, MN: University of Minnesota Press, 1987: 437–48; Bernard Schmitt, *Monnaie, Salaires, et Profits*, Paris: Presses Universitaires de France, 1966; Alvaro Cencini, Jean-Luc Bailly, and Sergio Rossi, eds, *Quantum Macroeconomics: The Legacy of Bernard Schmitt*, London: Routledge, 2017; Christian Kerslake, 'Marxism and Money in Deleuze and Guattari's *Capitalism and Schizophrenia*: On the Conflict Between the Theories of Suzanne De Brunhoff and Bernard Schmitt', *Parrhesia*, 22, 2015: 38–78.
126. Pat Hudson, *The Genesis of Industrial Capital: A Study of West Riding Wool Textile Industry, c.1750–1850*, Cambridge: Cambridge University Press, 1986: 151 (which explains in detail how an employer profited from the creation of credit); G. W. Hilton, 'The Truck Act of 1831', *The Economic History Review*, 10: 3, 1958: 470–9. On payment by shop notes, and discounting of these by local shopkeepers, see also Shapiro, *Capital and the Cotton Industry*: 138–9.
127. Samuel Hibbert, *Remarks on the Facility of Obtaining Commercial Credit; or, An Exposure of the Various Deceptions by Which Credit Is Procured*, Manchester: W. Cowdroy, 1806: 33. He adds: 'Any manufacturing house, that disperses notes abroad, in payment for the weekly wages of its workmen or weavers—whenever a house has failed in which are united by the issuing of such paper, the business of a banker and trader, ruin has spread like an infection amongst the habitations of the poor' (34). He then goes on to insist on the importance of imprisonment for debt, to prevent the excessive creation of credit (44–5), stating, 'There is no town I believe in the United Kingdom, where credit is so freely given as in Manchester' (51). Pamphlet published anonymously, addressed to 'the inhabitants of Manchester' by 'a fellow townsman'.
128. Joseph Schumpeter, *Theory of Economic Development*, was an early statement of this principle, although as we saw in chapter 1, his different and more

fleeting experience of the sugar industry led him to draw a different conclusion about the overcoming of technical difficulties.

129. Hudson, *The Genesis of Industrial Capital:* 217–18, on the role of merchant-manufacturers as bankers.
130. Albert Feavearyear, *The Pound Sterling: A History of English Money*, Oxford: Clarendon Press, 1963: 234. But not in Lancashire, where the prominence of Liverpool and Manchester as centres of overseas trade meant that bills of exchange were always available as credit money, and no use was made of coins or local banknotes, the latter being found mostly in other manufacturing regions including Yorkshire and the West Country. Lancashire switched directly to the use of Bank of England banknotes, when the central bank established branch banks in Manchester (1826) and Liverpool (1827), and then to the cheque system. Sidney Pollard, *Peaceful Conquest: The Industrialization of Europe 1760–1970*, Oxford: Oxford University Press, 1981: 38.
131. Schmitt, *Monnaie, Salaires, et Profits*; Alvaro Cencini, *Bernard Schmitt's Quantum Macroeconomic Analysis*, New York: Routledge, 2023.
132. Although, as Viviana Zelizer shows in *The Social Meaning of Money*, other forms of money persisted long after.
133. See chapter 1.
134. Herbert Heaton, 'Financing the Industrial Revolution', in François Crouzet, ed., *Capital Formation in the Industrial Revolution*, London: Methuen, 1972: 90.
135. Mary B. Rose, 'The Role of the Family in Providing Capital and Managerial Talent in Samuel Greg & Company 1784–1840', *Business History*, 19: 1, 1977: 49.
136. Immanuel Wallerstein, 'The Bourgeois(ie) as Concept and Reality', *New Left Review*, 167, 1988.
137. Rose, 'The Role of the Family in Providing Capital'.
138. See Eric Hobsbawm, *The Age of Empire: 1875–1914*, London: Weidenfeld and Nicolson, 1987; Michael Geyer and Charles Bright, 'World History in a Global Age', *American Historical Review*, 100: 4, 1995: 1034–60.
139. Hobsbawm, *Age of Empire*: 60. The term imperialism emerged in the 1870s, but was popularised first by J. A. Hobson, *Imperialism: A Study*, London: J. Nisbet, 1902, and then by Lenin in his 1917 booklet: Vladimir I. Lenin, *Imperialism: The Highest Stage of Capitalism*, New York: International Publishers, 1937.
140. Hobson, *Imperialism*: 81, 86, 89.
141. Caroline Fohlin, 'A Brief History of Investment Banking from Medieval Times to the Present', in Youssef Cassis, Catherine R. Schenk, and Richard S. Grossman, eds, *The Oxford Handbook of Banking and Financial History*, Oxford: Oxford University Press, 2016: 133–62.
142. David Harvey, *Paris, Capital of Modernity*, New York: Routledge, 2006; Carl E. Schorske, *Fin-de-siècle Vienna: Politics and Culture*, New York: Vintage Books, 1981; Janet Abu-Lughod, *Rabat: Urban Apartheid in Morocco*, Princeton, NJ: Princeton University Press, 1980.
143. The term is from Cas Gilbert, 'The Financial Importance of Rapid Building',

Engineering Record, 41, 30 June 1900: 624. See Sharon Irish, 'A "Machine That Makes the Land Pay": The West Street Building in New York', *Technology and Culture*, 30: 2, 1989: 376–97.

144. Henry George, *Progress and Poverty*, 5th edition, London: Kegan Paul, Trench & Co., 1883.
145. Patrick Geddes, *Cities in Evolution: An Introduction to the Town Planning Movement and to the Study of Civics*, London: Williams & Norgate, 1915: 70, emphasis added. On the wider influence of Geddes, and his involvement in colonial ventures, see John Scott and Ray Bromley, *Envisioning Sociology: Victor Branford, Patrick Geddes, and the Quest for Social Reconstruction*, Albany, NY: State University of New York Press, 2013.
146. Ebenezer Howard, *To-morrow: A Peaceful Path to Real Reform*, London: Swan Sonnenschein, 1898: 13.
147. Ibid.
148. Eve Blau, *The Architecture of Red Vienna, 1919–34*, Cambridge, MA: MIT Press, 1999.
149. Mark Le Vine, *Overthrowing Geography: Jaffa, Tel Aviv, and the Struggle for Palestine, 1880–1948*, Berkeley, CA: University of California Press, 2005: 158–72. Geddes was also involved in planning the new Jewish suburbs around Jerusalem: Rana Barakat, 'Urban Planning, Colonialism, and the Pro-Jerusalem Society', *Jerusalem Quarterly*, 65, 2016: 22–34.
150. Henry Dunning Macleod, *The Theory and Practice of Banking*, 6th edition, London: Longman, Greens and Co, 1855–6, 2 vols. On his influence, including his role in revising and codifying the law of bills of exchange, see Henry Dunning Macleod, 'Supplementary Statement and Testimonials of Henry Dunning Macleod, a candidate for the chair of commercial and political economy and mercantile law at the University of Edinburgh', London 1880. See also Macleod, *The Theory of Credit*, vol. 2.
151. Macleod, *The Theory of Credit*, vol. 2: 364.
152. Kiyotaki and Moore, 'Evil Is the Root of All Money': 3.
153. John Gurley and Edward Stone Shaw, *Money in the Theory of Finance*, Washington, DC: Brookings Institution, 1960.
154. Michael McLeay, Amar Radia, and Ryland Thomas, 'Money Creation in the Modern Economy', *Bank of England Quarterly Bulletin* Q1, 2014, bank ofengland.co.uk (accessed 22 January 2021); Richard Werner, 'How Do Banks Create Money, and Why Can Other Firms Not Do the Same? An Explanation for the Coexistence of Lending and Deposit-Taking', *International Review of Financial Analysis*, 36, 2014: 71–7; Richard Werner, 'Can Banks Individually Create Money Out of Nothing? The Theories and the Empirical Evidence', *International Review of Financial Analysis*, 36, 2014: 1–19; Werner, 'A Lost Century'.
155. Sgambati, 'Rethinking Banking', which is an important corrective to the account in Werner, 'How Do Banks Create Money?'
156. Ingham, *The Nature of Money*: 134–51.

4 Reading the Book of the Future

1. See Timothy Mitchell, 'Origins and Limits of the Modern Idea of the Economy', Advanced Study Center, University of Michigan, *Working Papers Series*, 12, 1995; 'Fixing the Economy', *Cultural Studies*, 12: 1, 1998: 82–101; and further references in chapter 5, below.
2. Moses I. Finley, *The Ancient Economy*, London: Chatto and Windus, 1973: 17–34, explores how the ancient Greco-Roman world was not organised in terms of the conceptual elements that constitute what we call 'the economy'.
3. On the performance of economic categories, see Fabian Muniesa, *The Provoked Economy: Economic Reality and the Performative Turn*, London: Routledge, 2014; Koray Çalişkan and Michel Callon, 'Economization, Part 1: Shifting Attention from the Economy Towards Processes of Economization', *Economy and Society*, 38: 3, 2009: 369–98; 'Economization, Part 2: A Research Programme for the Study of Markets', *Economy and Society*, 39: 1, 2010: 1–32; Donald MacKenzie, Fabian Muniesa, and Siu Leung-Sea, eds, *Do Economists Make Markets?: On the Performativity of Economics*, Princeton, NJ: Princeton University Press, 2008; see also Bruno Latour, *Science in Action: How to Follow Scientists and Engineers through Society*, Cambridge, MA: Harvard University Press, 1987. See also Judith Butler, 'Performative Agency', *Journal of Cultural Economy*, 3: 2, 2010: 147–61.
4. Joseph Vogl, *The Specter of Capital*, Stanford, CA: Stanford University Press, 2015. The Greek term *oikonomia* combines *oikos* (dwelling-place, household, household goods) and *nomos* (pasture, sphere of command (νομός), but also habitual practice, convention, law (νόμος)). So it refers to the proper administration of the household, especially its women and slaves; but also to a just or ethical disposition of those affairs. See Keith Tribe, *The Economy of the Word*, New York: Oxford University Press, 2015: 23–6. For a detailed examination of the meanings of *oikonomia* in Hellenistic thought, see Dotan Leshem, 'The Ancient Art of Economics', *European Journal of the History of Economic Thought*, 21: 2, 2014: 201–29.
5. Talal Asad, *Genealogies of Religion: Discipline and Reasons of Power in Christianity and Islam*, Baltimore, MD: Johns Hopkins University Press, 1993; Gil Anidjar, 'Secularism', *Critical Inquiry*, 33: 1, 2006: 52–77.
6. Fatih Ermiş, *A History of Ottoman Economic Thought: Developments Before the Nineteenth Century*, London: Routledge, 2014: 81. See also Subhi Labib, 'Capitalism in Medieval Islam', *The Journal of Economic History*, 29: 1, 1969: 94–6.
7. Tribe, *The Economy of the Word*; the analysis that follows here is indebted to Tribe's pioneering work. See also Istvan Hont, *Jealousy of Trade: International Competition and the Nation-State in Historical Perspective*, Cambridge, MA: Belknap Press, 2005; and Ute Tellmann, *Life and Money: The Genealogy of the Liberal Economy and the Displacement of Politics*, New York: Columbia University Press, 2017.
8. Giorgio Agamben, *The Kingdom and the Glory: For a Theological Genealogy of Economy and Government*, Stanford, CA: Stanford University Press, 2011: 280. See also Alberto Toscano, 'Divine Management: Critical Remarks on

Giorgio Agamben's *The Kingdom and the Glory*', *Journal of the Theoretical Humanities*, 16, 2011: 125–36.

9. Adam Smith, *The Wealth of Nations, Books IV–V*, Penguin Classics, 1982: 260. Smith makes almost no use of the term economy outside his criticism in Book IV of 'systems' (or, as we would say today, theories) of political economy. The handful of occurrences of the term elsewhere in the book use it in the sense of frugality, discussed below.
10. 'Drummond, Henry', *Oxford Dictionary of National Biography*.
11. Richard Whately, *Introductory Lectures on Political-Economy*, London: B. Fellowes, 1831: 2. Whately derived the new name from the Greek verb *katallasso* (καταλλάσσω), to exchange or reconcile.
12. Keith Tribe, *Governing Economy: The Reformation of German Economic Discourse, 1750–1840*, Cambridge: Cambridge University Press, 1988.
13. John Stuart Mill, 'On the Definition of Political Economy; and on the Method of Investigation Proper to It', in *Essays on Some Unsettled Questions of Political Economy*, London: John W. Parker, 1844: 135–7.
14. John Stuart Mill, *Principles of Political Economy, with some of their Applications to Social Philosophy*, London: John W. Parker, 1848: 81; Tribe, *The Economy of the Word*.
15. Agamben, *The Kingdom and the Glory*; Vogl, *The Specter of Capital*; Tellmann, *Life and Money*.
16. Talal Asad, *Formations of the Secular*, Stanford, CA: Stanford University Press, 2003; Anidjar, 'Secularism'; Tomoko Masuzawa, *The Invention of World Religions*, Chicago, IL: University of Chicago Press, 2005.
17. Anidjar, 'Secularism'.
18. William Martin Leake, 'Memoir on the Life and Travels of John Lewis Burckhardt', in John Lewis Burckhardt, *Travels in Nubia*, London: John Murray, 1819: iii–xcii. Burckhardt published under anglicised versions of his given names.
19. 'Drummond, Henry', *Oxford Dictionary of National Biography*.
20. Walter Bagehot, 'The Postulates of English Political Economy, I', *The Fortnightly Review*, 25, 1876: 216. A dinner held that year at the Political Economy Club in London to mark the centenary of *The Wealth of Nations* felt like a funeral ceremony for the field of knowledge. William Stanley Jevons, 'The Future of Political Economy', *The Fortnightly Review*, 26, 1876: 619.
21. Frederic Harrison, 'The Limits of Political Economy', *The Fortnightly Review*, 1, 1865: 362.
22. Bagehot, 'The Postulates of English Political Economy, I': 218.
23. Harrison, 'The Limits of Political Economy': 375.
24. Ibid.: 374–5.
25. The term 'open market' was commonly used, for example by Marshall to describe a 'machinery of exchange' where prices were produced between strangers. 'If it be given that a bottle of wine and a pound of tea can be disposed of for the same price in the same open market at a given period, the gratifications of the purchasers in this market at this time due to the bottle of wine and the pound of tea, have this price as their common exchange

measure; and the machinery of exchange is not concerned with any other of their properties': Alfred Marshall, 'On Mr. Mill's Theory of Value', *The Fortnightly Review*, 25, 1876: 596.

26. Robert J. Steinfeld, *The Invention of Free Labor: The Employment Relation in English and American Law and Culture, 1350–1870*, Chapel Hill, NC: University of North Carolina Press, 1991; Marc W. Steinberg, *England's Great Transformation: Law, Labor, and the Industrial Revolution*, Chicago, IL: University of Chicago Press, 2016.
27. Michel Callon, *Markets in the Making: Rethinking Competition, Goods, and Innovation*, Brooklyn, NY: Zone Books, 2021.
28. Bagehot, 'The Postulates of English Political Economy, I': 220.
29. Timothy Mitchell, 'The Work of Economics: How a Discipline Makes Its World', *European Journal of Sociology*, 46: 2, 2005: 297–320.
30. Adam Smith, *The Wealth of Nations Books I–III*, London: Penguin Classics, 1982: 109–10; Jean-Louis Peaucelle and Cameron Guthrie, 'How Adam Smith Found Inspiration in French Texts on Pin Making in The Eighteenth Century', *History of Economic Ideas*, 19: 3, 2011: 41–67.
31. Brian Bonnyman, *The Third Duke of Buccleuch and Adam Smith: Estate Management and Improvement in Enlightenment Scotland*, Edinburgh: Edinburgh University Press, 2014: 75.
32. David Ricardo, 'An Essay on the Influence of a Low Price of Corn on the Profits of Stock', in Piero Sraffa, ed., *The Works and Correspondence of David Ricardo*, vol. 4, Cambridge: Cambridge University Press, 1951–73: 1–41. Mary S. Morgan, *The World in the Model: How Economists Work and Think*, Cambridge: Cambridge University Press, 2012: 44–90. Whatever his interest in experimental agriculture, as Gauthier Lanot and Keith Tribe note, Ricardo was also drawing on the wider discourse on pauperization triggered by the arguments over the Corn Laws: 'Before Political Economy: Debate over Grain Markets, Dearth and Pauperism in England, 1794–96', *History of European Ideas*, 51: 2, 2024: 226–56.
33. Charles Babbage, *On the Economy of Machinery and Manufactures*, Philadelphia, PA: Carey & Lea, 1832: 27. On Marx and machinery, including his reading of Ure and Babbage, see Keith Tribe, 'De l'atelier au procès de travail: Marx, les machines et la technologie', in François Jarrige, ed., *Dompter Prométhée: Technologies et socialismes à l'âge romantique (1820–1870)*, Besançon: Presses universitaires de Franche-Comté, 2016: 229–50. On Marx and Babbage, see Simon Schaffer, 'Babbage's Intelligence: Calculating Engines and the Factory System', *Critical Inquiry*, 21: 1, 1994: 203–27; and Matteo Pasquinelli, 'On the Origins of Marx's General Intellect', *Radical Philosophy*, 206, 2019: 43–56. For the concept of domination through abstraction, see Moishe Postone, *Time, Labor, and Social Domination: A Reinterpretation of Marx's Critical Theory*, Cambridge: Cambridge University Press, 1993.
34. William Reddy, *The Rise of Market Culture: The Textile Trade and French Society, 1750–1900*, Cambridge: Cambridge University Press, 1987; Marc W. Steinberg, *England's Great Transformation: Law, Labor, and the Industrial Revolution*, Chicago, IL: University of Chicago Press, 2016.

35. For a broader account of the 'marginalist revolution' in the context of nineteenth-century science, see Philip Mirowski, *More Heat than Light: Economics as Social Physics, Physics as Nature's Economics*, Cambridge: Cambridge University Press, 1989.
36. Charles Kindleberger, 'The Panic of 1873', in *Historical Economics: Art or Science?*, New York: Harvester Wheatsheaf, 1990: 310–25.
37. Bagehot, 'The Postulates of English Political Economy, I': 221.
38. Ibid.
39. Bagehot added a third example, the creation of credit by European banks on an unprecedented scale to fund the payment of French reparations of 5 billion francs (£200 million) to Germany after the 1870–1 war: 'The Postulates of English Political Economy, I': 222. This was the largest sum ever paid under a single contract 'since the system of payments began'. Walter Bagehot, 'The Postulates of English Political Economy, II', *The Fortnightly Review*, 25, 1876: 739.
40. William Stanley Jevons, *Investigations in Currency and Finance*, London: Macmillan and Co, 1884: 216.
41. Ibid.: 233.
42. Karuna Mantena, *Alibis of Empire: Henry Maine and the Ends of Liberal Imperialism*, Princeton, NJ: Princeton University Press, 2010.
43. 'Colonial materialism' is an alternative to the term colonial 'economism' proposed in the important study by Aaron Jakes, *Egypt's Occupation: Colonial Economism and the Crises of Capitalism*, Stanford, CA: Stanford University Press, 2020. The term 'economism' at that time had a different meaning. Lenin used it as a label to discredit his rivals for the leadership of Russian revolutionary socialism, by associating them with the demand of the Social Democrats for economic legislation (a shorter working day, improved wages), rather than the struggle for political rights. Lars T. Lih, *Lenin Rediscovered:* What Is to Be Done? *In Context*, Leiden: Brill, 2005.
44. The Earl of Cromer (Evelyn Baring), *Modern Egypt*, vol. 2, New York: The Macmillan Company, 1909: 527–8.
45. Karl Marx, *Capital: A Critique of Political Economy*, vol. 2, Chicago, IL: Charles H. Kerr and Co., 1909: 278–9. In the original German, Marx wrote, 'erfordert ... zu einer regelmäßigen Wirthschaft [sic], einen größren Flächenraum als die Getreidekultur'. German second edition of 1888: 226. The Penguin Classics translation removes the term 'economy', writing that forest cultivation 'requires a greater surface area than the cultivation of grain, if it is to be conducted on a regular commercial basis'. Karl Marx, *Capital: A Critique of Political Economy*, vol. 2, New York: Penguin Books, 1990–2: 195. The other exception in Marx is the use of 'economy' to distinguish 'money economy' from 'credit economy', and both of these from barter, or 'natural economy' (*Geldwirthschaft*, *Kreditwirthschaft*, and *Naturalwirthschaft*). In the 1909 English edition (cited above), these are rendered as 'natural economy, the money-system, and the credit-system', 132–3. The Progress Publishers version in the 1950s and the Penguin Classics edition cope differently, the former adding the definite article

('*the* money economy', '*the* credit economy') and the latter oscillating between the two formulations. Karl Marx, *Capital: A Critique of Political Economy*, vol. 2, Moscow: Progress Publishers, 1956. Karl Marx, *Capital*, Penguin Classics edition: 2: 195–6.

46. Karl Marx, *Marx's Economic Manuscript of 1864–1865*, Fred Moseley, ed., Leiden: Brill, 2015: 173. Italics in the original. This is the only use of the term 'economy' in *Capital*, vol. 3 (the text is a new edition of that work).
47. J. A. Hobson, *Imperialism: A Study*, London: James Nisbet & Co., 1902: 192, 195.
48. See the parallel use of 'world economy' by German writers in the same period, in a similar sense. Quinn Slobodian, 'How to See the World Economy: Statistics, Maps, and Schumpeter's Camera in the First Age of Globalization', *Journal of Global History*, 10: 2, 2015: 307–32.
49. This confusion is often found in discussions of the pioneering work of Friedrich List, *National System of Political Economy*, Philadelphia, PA: J.B. Lippincott & Co., 1856: 185–6. See Timothy Mitchell, *Carbon Democracy: Political Power in the Age of Oil*, London: Verso, 2011: 126.
50. See chapters 5 and 8.
51. See Jairus Banaji, *A Brief History of Commercial Capitalism*, Chicago, IL: Haymarket Books, 2020; Lorenzo Bondioli, 'Peasants, Merchants, and Caliphs: Capital and Empire in Fatimid Egypt', PhD thesis, History Department, Princeton University, 2021.
52. Ellen Meiksins Wood, 'Separation of the Economic and Political Under Capitalism', *New Left Review*, 127, 1981; and *The Origin of Capitalism: A Longer View*, London: Verso, 2002. Such arguments often reference the famous passage from the Preface to Marx's *Contribution to the Critique of Political Economy*, 1859, where he writes 'The totality of these relations of production constitutes the economic structure of society [*die ökonomische Struktur der Gesellschaft*], the real foundation [*die reale Basis*], on which arises a legal and political superstructure.' However, to distinguish the structure of a building from its foundation is not the same as imagining society to consist of separate 'realms' of the economic and the political; and in Marx, this distinction is a characteristic of all societies. Karl Marx, *A Contribution to the Critique of Political Economy*, Chicago, IL: C.H. Kerr, 1911: 11; and Karl Marx, *Zur Kritik der Politischen Ökonomie*, Stuttgart: J.H.W. Dietz Nachf., 1909: lv.
53. Georg Lukács, *History and Class Consciousness: Studies in Marxist Dialectics*, trans. Rodney Livingstone, Cambridge, MA: MIT Press, 1971: 87.
54. On the different approach to abstraction developed here, see Timothy Mitchell, *Colonising Egypt*, Cambridge: Cambridge University Press, 1988.
55. See Postone, *Time, Labor, and Social Domination*: 24–34, 123–85. Postone (p. 16) faulted Lukács and the Frankfurt School for retaining a transhistorical concept of labour, even as they followed Marx in describing the process of its abstraction.
56. Carolyn Hardin, *Capturing Finance: Arbitrage and Social Domination*, Durham, NC: Duke University Press, 2021; Maurizio Lazzarato, *Governing by Debt*, Cambridge, MA: MIT Press, 2015; Jonathan Levy, *Freaks of Fortune:*

The Emerging World of Capitalism and Risk in America, Cambridge, MA: Harvard University Press, 2012.

57. See Timothy Mitchell, 'Culture and Economy', in Tony Bennett and John Frow, eds, *The Sage Handbook of Cultural Analysis*, London: SAGE, 2008: 447–66.
58. Callon, *Markets in the Making.*
59. Muniesa, *The Provoked Economy*; Çalişkan and Callon, 'Economization, Part 1' and 'Economization, Part 2'; Michelle Murphy, *The Economization of Life*, Durham, NC: Duke University Press, 2017.
60. Mill, 'On the Definition of Political Economy': 138.
61. Ibid.: 135.
62. See Mitchell, *Colonising Egypt.*
63. Karl Marx, 'Speech on Free Trade', *Northern Star*, 9 October 1847, in Karl Marx and Friedrich Engels, *Historisch-Kritische Gesamtausgabe*, vol. 1, Berlin: Marx-Engels-Verlag, 1927–35: 429–30, cited in Tribe, *The Economy of the Word*: 171.
64. Volume two of *Capital* appeared in 1885 and volume three in 1894, decades after Marx had abandoned their drafting. A fourth volume was assembled by Karl Kautsky from earlier notebooks and published in German in four volumes between 1905 and 1910 as *Theorien über den Mehrwert*, a selection from which was published in one volume in English as Karl Marx, *Theories of Surplus Value*, London: Lawrence and Wishart, 1951.
65. See Gareth Steadman Jones, *Karl Marx: Greatness and Illusion*, and 'Karl Marx's changing picture of the end of capitalism', *Journal of the British Academy*, 6, 2018: 187–206; Keith Tribe, 'Karl Marx's "Critique of Political Economy": A Critique', in *The Economy of the Word*, 171–254; and Kohei Saito, *Marx in the Anthropocene: Towards the Idea of Degrowth Communism*, Cambridge: Cambridge University Press, 2022.
66. Saito, *Marx in the Anthropocene*: 186–210. See also Harry Harootunian, *Marx After Marx: History and Time in the Expansion of Capitalism*, New York: Columbia University Press, 2015.
67. 'Hence the Ten Hours' Bill was not only a great practical success', wrote Marx in 1864; 'it was the victory of a principle; it was the first time that in broad daylight the political economy of the middle class succumbed to the political economy of the working class. But there was in store a still greater victory of the political economy of labour over the political economy of property. We speak of the co-operative movement, especially the co-operative factories raised by the unassisted efforts of a few bold "hands". The value of these great social experiments cannot be over-rated': Karl Marx, 'Inaugural Address of the International Working Men's Association', [28 September 1864], in Marx and Engels, *Collected Works*, vol. 20, London: Lawrence & Wishart, 1975: 11.
68. Karl Marx, *Capital*, vol. 3, in Karl Marx and Frederick Engels, *Collected Works*, vol. 37, London: Lawrence & Wishart, 1975: 434–8. On Marx's attempt to make sense of fictitious capital, and an effort to reconstruct how his fragmentary thoughts on the subject might have been developed, see

Michael Perelman, *Marx's Crises Theory: Scarcity, Labor, and Finance*, New York: Praeger, 1987: 170–217.

69. Perelman, *Marx's Crises Theory*: 173–82. Engels wrote (in his supplement at the end of *Capital*, vol. 3), 'Since 1865, when the book was written, a change has taken place which today assigns a considerably increased and constantly growing role to the stock exchange, and which, as it develops, tends to concentrate all production, industrial as well as agricultural, and all commerce, the means of communication as well as the functions of exchange, in the hands of stock exchange operators, so that the stock exchange becomes the most prominent representative of capitalist productions itself'. Engels notes how different areas of industry were converted to joint stock, then banks, then agriculture, and ends: 'Then colonisation. Today this is purely a subsidiary of the stock exchange, in whose interests the European powers divided Africa a few years ago, and the French conquered Tunis and Tonkin. Africa leased directly to companies (Niger, South Africa, German South-West and German East Africa), and Mashonaland and Natal seized by Rhodes for the stock exchange': Karl Marx, *Capital: A Critique of Political Economy*, vol. 3, New York: International Publishers, 1967: 644–5.
70. David Landes, *Bankers and Pashas: International Finance and Economic Imperialism in Egypt*, Cambridge, MA: Harvard University Press, 1979: 47–53.
71. Helen M. Davies, *Emile and Isaac Pereire: Bankers, Socialists and Sephardic Jews in Nineteenth-Century France*, Manchester: Manchester University Press, 2015.
72. Davies, *Emile and Isaac Pereire*: 210–13; on the failure of Marseilles to benefit from the Suez Canal, see D. A. Farnie, *East and West of Suez: The Suez Canal in History, 1854–1956*, Oxford: Clarendon Press, 1969: 143–5.
73. Marx and Engels, *Collected Works*, vol. 37: 439.
74. Leo Marx, 'Technology: The Emergence of a Hazardous Concept', *Social Research*, 64: 3, 1997: 965–88.
75. Simon J. Cook, *The Intellectual Foundations of Alfred Marshall's Economic Science: A Rounded Globe of Knowledge*, Cambridge: Cambridge University Press, 2009.
76. Preface to the fifth edition, cited in C. W. Gillebaud, 'The Evolution of Marshall's *Principles of Economics*', *The Economic Journal*, 52: 208, 1942: 330–49. Marshall eventually published a separate work, one on industry and another on credit. These largely descriptive accounts were never reconciled with the analysis of the *Principles*.
77. J. A. Hobson, *The Industrial System: An Inquiry into Earned and Unearned Income*, New York: C. Scribner's Sons, 1910: v.
78. Alfred Marshall, *Principles of Economics*, 5th edition, London: Macmillan and Co., 1907: 762.
79. Thorstein Veblen, *The Theory of Business Enterprise*, New York: C. Scribner's Sons, 1905.
80. Wesley Clair Mitchell, *Business Cycles*, Berkeley, CA: University of California Press, 1913: 23–7.

81. Henry V. Poor, *Manual of the Railroads of the United States for 1868–69*, New York: H.V. and H.W. Poor, 1868: 30–1.
82. Poor wrote economic analysis columns for the *New York Times*. He 'wrote on other fields as well, notably economics: *The Money Question: A Handbook for our Times*, New York: H.V. & H.W. Poor, 1898 includes an interesting discussion of the differences between "metallic money" and "symbolic money", or moneys of exchange and moneys of account, showing how the latter were coming to be of more importance in the United States as the economy continued to grow. It is his work on information and management, however, that stands out from other writings on business of the time': Morgen Witzel, ed., *The Encyclopedia of the History of American Management*, Bristol: Thoemmes Continuum, 2005.
83. Timothy J. Sinclair, *The New Masters of Capital: American Bond Rating Agencies and the Politics of Creditworthiness*, Ithaca, NY: Cornell University Press, 2005: 23–6. For the parallel story of Charles Dow, founder of the Wall Street Journal, and the computation of the Dow Jones Industrial Average, see Marieke de Goede, *Virtue, Fortune, and Faith: A Genealogy of Finance*, Minneapolis, MN: University of Minnesota Press, 2005: 87–120.
84. Alfred Chandler, *Henry Varnum Poor: Business Editor, Analyst, and Reformer*, Cambridge, MA: Harvard University Press, 1956; Thomas K. McCraw, 'The Challenge of Alfred D. Chandler, Jr.: Retrospect and Prospect', *Reviews in American History*, 15: 1, 1987: 165.
85. McCraw, 'The Challenge of Alfred D. Chandler, Jr.'
86. Ibid., 166.
87. Paul Uselding, 'Business History and the History of Technology', *The Business History Review*, 54: 4, 1980: 443–52.
88. Alfred D. Chandler, *The Visible Hand: The Managerial Revolution in American Business*, Cambridge, MA: Belknap Press, 1977.
89. Richard Swedberg, *Max Weber and the Idea of Economic Sociology*, Princeton, NJ: Princeton University Press, 1998.
90. Adam Tooze, *Statistics and the German State, 1900–1945: The Making of Modern Economic Knowledge*, Cambridge: Cambridge University Press, 2001; Quinn Slobodian, *Globalists: The End of Empire and the Birth of Neoliberalism*, Cambridge, MA: Harvard University Press, 2018. Keith Tribe notes that, while theories of the management of the economy are attributed to Keynes and other writers of the mid-twentieth century, it was in the late nineteenth and early twentieth century that 'the modern language of economic management and politico-economic strategy was forged': *Governing Economy*: 6. For uses on the term economy in colonial contexts, especially India, see Manu Goswami, *Producing India: From Colonial Economy to National Space*, Chicago, IL: University of Chicago Press, 2010; and U. Kalpagam, *Rule by Numbers: Governmentality in Colonial India*, London: Lexington Books, 2014.
91. Max Weber, *Economy and Society: A New Translation*, ed. and trans. from the German by Keith Tribe, Cambridge, MA: Harvard University Press, 2019. Tribe's meticulous version (which also restores the original typographic layout, essential to the structuring of its argument), together with his

introduction to the text and an appendix discussing the translation of key terms, reveals what earlier translations have concealed. Note: this and the following paragraphs are based upon Timothy Mitchell, 'Economy Shall (No Longer) Mean Economisation', *Journal of Cultural Economy*, 15: 6, 2022: 838–42.

92. The term is rendered by Tribe with the more elegant English word 'sociation'.
93. Weber, *Economy and Society*: 146.
94. Ibid.: 144.
95. Eugen von Böhm-Bawerk, *The Positive Theory of Capital*, trans. from the German by William Smart, London: Macmillan and Co., 1888: 80.
96. Çalişkan and Callon, 'Economization, Part 1'; Callon, *Markets in the Making*: 413–18. Tribe solves the difficulty in a different way. He translates *Wirtschaften* as 'economic action', opting for a more readable phrase and one that better fits Weber's concern for the role of economic agents in the process of economising. Weber, *Economy and Society*: 485–6.
97. William Callison, 'The Politics of Rationality in Early Neoliberalism: Max Weber, Ludwig von Mises, and the Socialist Calculation Debate', *Journal of the History of Ideas*, 83: 2, 2022: 269–91.
98. Mitchell, 'Fixing the Economy': 82–101.
99. Weber, *Economy and Society*: 216. Although English writers like Hobson, in a passage cited above, were able to use the term 'national economy' to refer to a process, equivalent to the German term *Volkswirtschaft*, by the interwar period it had become more difficult to render *Volkswirtschaft* into English. Joseph Schumpeter, in *The Theory of Economic Development* (1912, translated into English in 1934), laid out a picture of the 'circular flow of economic life' formed by every household and firm producing or consuming according to its individual needs. He added: 'The total of all commodities produced and marketed in a community [*Volkswirtschaft*] within an economic period may be called the social product.' Here, the English translator rendered the German 'national-economy' as 'community'. Schumpeter added that the social product 'does not exist as such . . . But it is a useful abstraction'. Joseph Schumpeter, *The Theory of Economic Development: An Inquiry into Profits, Capital, Credit, Interest, and the Business Cycle*, trans. from the German by Redvers Opie, Cambridge, MA: Harvard University Press, 1934: 9. The German text reads in full: 'Die Summe alles dessen, was in einer Volkswirtschaft in einer Wirtschaftsperiode produziert und auf den Markt gebracht wird, kann man das Sozialprodukt derselben nennen. Es ist für unsern Zweck nicht nötig, näher auf die Bedeutung dieses Begriffes einzugehen. Das Sozialprodukt existiert nicht als solches. Es ist als solches ebensowenig ein von irgend jemand bewußt angestrebtes Resultat planvoller Tätigkeit, als die Volkswirtschaft als solche eine nach einem einheitlichen Plane arbeitende "Wirtschaft" ist': Joseph Schumpeter, *Theorie der wirtschaftlichen Entwicklung*, Leipzig: Verlag von Duncker & Humblot, 1912: 9.
100. Max Weber, *Wirtschaft und Gesellschaft*, trans. from the German by Paul

Siebeck, Tübingen: J. C. B. Mohr, 1922: 31; cf. Weber, *Economy and Society*: 143.

101. See the appendix to Tribe's translation.
102. Tribe prefers to stress this continuousness of operation, so translates *Wirtschaftsbetrieb* as 'the pursuit of economic activity'. The phrase mirrors his rendering of economisation as 'economic action', although 'pursuit', while suggesting continuousness, barely captures the repeating, expanding process suggested by Weber's gloss of the term as 'continuously ordered'. Once again, with the help of Tribe's discussion of terms in the appendix, one might read 'activity' in the broader sense of economisation.
103. Weber, *Economy and Society*: 191.
104. Ibid.: 182.
105. Ibid.: 207.
106. Callison, 'The Politics of Rationality'; Slobodian, *Globalists*.
107. See chapter 5.
108. Ludwig von Mises, 'Die Wirtschaftsrechnung im sozialistischen Gemeinwesen', *Archiv für Sozialwissenschaft und Sozialpolitik*, 47, 1920–1: 100. English translation: 'Economic Calculation in the Socialist Commonwealth', in F. A. Hayek, ed., *Collectivist Economic Planning*, London: Routledge & Kegan Paul 1935: 105. Mises later adopted the new use of 'the economy' but referred to it only using the term 'the market economy', dismissing the possibility of a planned economy. 'The paradox of "planning" is that it cannot plan, because of the absence of economic calculation. What is called a planned economy is no economy at all. It is just a system of groping about in the dark': Ludwig von Mises, *Human Action: A Treatise on Economics*, Auburn, AL: Ludwig von Mises Institute, 1998: 696.
109. F. A. Hayek, 'The Nature and History of the Problem', in Hayek, ed., *Collectivist Economic Planning*: 9, 29, 34, 37.
110. Hayek spelled out his objections to the concept of the economy in *Law, Legislation and Liberty*, vol. 2, *The Mirage of Social Justice*, Chicago, IL: University of Chicago Press, 1976, reprinted in one volume, London: Routledge and Kegan Paul 1982: 2: 107–9, arguing that 'whenever we speak of the economy of a country, or of the world, we are employing a term which suggests that these systems ought to be run on socialist lines.' Ludwig von Mises used 'catallactics' in *Human Action: A Treatise on Economics*, New Haven, CT: Yale University Press, 1949, taking it from Whately's *Introductory Lectures*: 2.
111. Ibid.: 2, 4, 21, 34.
112. Tribe, 'Introduction', in Weber, *Economy and Society*.
113. Max Weber, *The Theory of Social and Economic Organization*, Talcott Parsons, ed. and trans. from the German by A. M. Henderson and Talcott Parsons, Glencoe, IL: Free Press & Falcon's Wing Press, 1947, emphasis added. See Timothy Mitchell, 'Economists and the Economy in the Twentieth Century', in George Steinmetz, ed., *The Politics of Method in the Human Sciences: Positivism and Its Epistemological Others*, Durham, NC: Duke University Press, 2005: 126–41.
114. Max Weber, *Economy and Society*: 63.

115. Michel Callon, 'Introduction: The Embeddedness of Markets in Economics', *Sociological Review Monographs*, 46, 1998; and Çalişkan and Callon, 'Economization, Part 1'.

5 Economentality: How the Future Entered Government

1. This chapter is a revised version of a paper written for a conference on Around 1948: Realignments in Politics and Culture, University of Chicago, April 2011, and published as Timothy Mitchell, 'Economentality: How the Future Entered Government', *Critical Inquiry*, 40: 4, 2014: 479–507.
2. Hector Prud'homme, 'Nile Project – visit of Mr. Daninos', memo., 27 January 1953, Documents relating to the Report on the Agricultural Aspects of the Sudd el Aali Project (High Aswan Dam Project), WB 1589793, World Bank Group Archives. The memo spells Daninos's first name as Adrien.
3. Ahmad Shokr, 'Hydropolitics, Economy, and the Aswan High Dam in Mid-Century Egypt', *Arab Studies Journal*, 17, 2009: 9–31, which brings to light the origins and scope of Daninos's plan and its political and engineering context. Other studies of the dam include John Waterbury, *Hydropolitics of the Nile Valley*, Syracuse, NY: Syracuse University Press, 1979; and Tom Little, *High Dam at Aswan: The Subjugation of the Nile*, New York: John Day Co, 1965. A correspondent for the London *Times* and *The Economist* and head of the Arab News Agency (ANA), Little was part of an MI6 spy ring in Cairo. When the Egyptian secret police broke up the ring in 1956, arresting members of the ANA staff, they left Tom Little in place and supplied him with disinformation which he fed back to London; see Stephen Dorril, *MI6: Inside the Covert World of Her Majesty's Secret Intelligence Service*, New York: Free Press, 2002: 631–2.
4. The company's profits had increased more than threefold between 1947 and 1950; Caroline Piquet, 'An International Company in Egypt: Suez, 1858–1956', paper presented at European Business History Association, 11th Annual Conference, Geneva, 13–15 September 2007: 12.
5. F. D. Gregh, 'EGYPT – Visit of Mr. J. Georges-Picot to Mr. Garner and Myself', memo., 26 May 1954, Documents relating to the Report on the Agricultural Aspects of the Sudd el Aali Project (High Aswan Dam Project), WB 1589793, World Bank Group Archives. Jacques Georges-Picot was the General Manager of the Suez Canal Co.
6. Amy L. S. Staples and Amy L. Sayward, *The Birth of Development: How the World Bank, Food and Agriculture Organization, and World Health Organization Changed the World, 1945–1965*, Kent, OH: Kent State University Press, 2006; Michele Alacevich, *The Political Economy of the World Bank: The Early Years*, Stanford, CA: Stanford University Press, 2009; and Devesh Kapur, John P. Lewis, and Richard Webb, *The World Bank: Its First Half Century*, vols 1–2, Washington, DC: Brookings Institution, 1997.
7. 'Memo on meeting of Mr Betts with Mackenzie', in 'EGYPT: Development Schemes – High Aswan Dam – Sterling Releases', memo., 16 September

1955, National Archives of the UK, Public Record Office, BT, 213/44, 1955; IBRD, 'External Aid for Egypt', in 'EGYPT: Development Schemes – High Aswan Dam – Sterling Releases', memo., 18 November 1955, BT, 213/44, 1955.

8. On the interwar period in the US, see Onur Özgöde, 'Institutionalism in Action: Balancing the Substantive Imbalances of "the Economy" through the Veil of Money', *History of Political Economy*, 52: 2, 2020: 307–39; and Timothy Shenk, "Inventing the American Economy," in Romain Huret, Nelson Lichtenstein, and Jean-Christian Vinel, eds, *Capitalism Contested: The New Deal and Its Legacies* (Philadelphia, PA: University of Pennsylvania Press, 2020). The work of Wesley Clair Mitchell, *Business Cycles*, Berkeley, CA: University of California Press, 1913, is sometimes referenced as a pioneering account of 'the economy' some two to three decades before the general emergence of the idea (cf. Daniel Breslau, 'Economics Invents the Economy: Mathematics, Statistics, and Models in the Work of Irving Fisher and Wesley Mitchell', *Theory and Society*, 32: 3, 2020: 79–411). However, *Business Cycles* refers to 'money economy' (and occasionally 'the money economy') not as a subsidiary part of 'the economy' (a phrase that Mitchell does not use), but as a form of social life in which pecuniary gain replaces the satisfaction of material wants. On the term 'money economy', see chapter 4.
9. Eliot Janeway, 'Japan's Economy Is Strained; Protests Go to Him', *New York Times*, 9 January 1938: E4, and Anne O'Hare McCormick, 'Europe; Germany Pushes Experiment in Controlled Economy', *New York Times*, 29 January 1938: 14.
10. Gerhard Colm and Fritz Lehman, *Economic Consequences of Recent American Tax Policy*, New York: New School for Social Research, 1938.
11. Will Lissner, 'New Deal Policies Blamed for Slump', *New York Times*, 23 January 1938: 4.
12. Ibid. The number of individuals in the US filing tax returns more than doubled, from 3.7 million in 1933 to 7.7 million in 1939; see US Treasury Department, Bureau of Internal Revenue, *Statistics of Income for 1933*, Washington, DC: 1935 and *Statistics of Income for 1939: Part 1*, Washington, DC: 1942, irs.gov (accessed 3 August 2025).
13. The bank partner was Paul Mazur; see ibid. On Colm's relationship to Keynesianism, see Luca Fiorito and Matias Vernengo, 'Gerhard Colm on John Maurice Clark's *Economics of Planning Public Works*: An Unpublished Letter', *Research in the History of Economic Thought and Methodology*, 29A, 2011: 83–93.
14. Albert Gailord Hart, 'Review of *Economic Consequences of Recent American Tax Policy* by Gerhard Colm and Fritz Lehmann', *Journal of Political Economy*, 47, 1939: 438–9. As stated earlier in the review, 'The authors recognize that such a forecast is very insecure, particularly in view of the likelihood of lower interest rates. But their argument is garnished with a display of statistics, which in view of the shakiness of the individual items is more likely to lull the unwary into false security than to reinforce other arguments in the judgment of the critical' (p. 437).

15. Robert M. Collins, *The Business Response to Keynes, 1929–1964*, New York: Columbia University Press, 1981: 12.
16. J. Tinbergen and J. B. D. Derksen, 'Recent Experiments in Social Accounting: Flexible and Dynamic Budgets', *Econometrica*, 17: S, 1949: 195–204; Carol S. Carson, 'The History of the United States National Income and Product Accounts: The Development of an Analytic Tool', *Review of Income and Wealth*, 21, 1975: 153–81.
17. On the violent suppression of the revolt, see Matthew Hughes, 'The Banality of Brutality: British Armed Forces and the Repression of the Arab Revolt in Palestine, 1936–39', *The English Historical Review*, 124: 507, 2009, 313–54. On the wider history, Rashid Khalidi, *The Hundred Years' War on Palestine: A History of Settler Colonialism and Resistance*, New York: Metropolitan Books, 2020: 56–64.
18. Michel Foucault, *Security, Territory, Population: Lectures at the Collège de France 1977–1978*, trans. from the French by Graham Burchell, Basingstoke, UK: Palgrave Macmillan, 2007: 109.
19. Koray Çalişkan and Michel Callon, 'Economization, Part 1: Shifting Attention from the Economy towards Processes of Economization', *Economy and Society*, 38: 3, 2009: 369–98; 'Economization, Part 2: A Research Programme for the Study of Markets', *Economy and Society*, 39: 1, 2010: 1–32; and Michel Callon, 'Introduction: The Embeddedness of Economic Markets in Economics', *Sociological Review*, 46: 1, 1998: 1–57.
20. Timothy Mitchell, 'The Work of Economics: How a Discipline Makes its World', *European Journal of Sociology*, 46, 2005: 297–320.
21. Coal miners were not 'protected' by the Taft–Hartley Act because the miners' union refused to sign its oath that leaders were not communists.
22. Robert J. Donovan, *Conflict and Crisis: The Presidency of Harry S. Truman, 1945–1948*, Columbia, MO: University of Missouri Press, 1996: 208–11.
23. Louis Stark, 'Men Are Informed; Word Is Sent to Stay on Job as Leaders Bow at Last Minute Delay in Some Notices', *New York Times*, 11 May 1948: 4.
24. In 1948 the United States accounted for 59 per cent of total world production but consumed just over 60 per cent. In other words, the net imports supplied barely 1 per cent of consumption; see 'Table Db190–197, Selected Mineral Fuels – Imports and Exports: 1867–2001', in Susan B. Carter et al., eds, *Historical Statistics of the United States*, New York: Cambridge University Press, 2006; and *Twentieth Century Petroleum Statistics*, Dallas, TX: DeGolyer and MacNaughton, 2009.
25. Kenneth Campbell, 'Marshall-Aid Ship Cheered in France: First "Official" Vessel Brings 8,800 Tons of U.S. Wheat as Premier Praises ERP', *New York Times*, 11 May 1948: 20.
26. Timothy Mitchell, *Carbon Democracy: Political Power in the Age of Oil*, London: Verso, 2011, provides an extended version of the argument summarised in this and the following paragraph.
27. The Iraq Petroleum Company had five owners. The entrepreneur who assembled the Company, Calouste Gulbenkian, controlled 5 per cent. The remainder was owned in four equal parts by the Anglo-Iranian Oil Company (today BP), Shell, a consortium of US oil companies led by Standard Oil, and a

consortium of French oil interests, the last of these partly owned by the first three: ibid., 97.

28. Hanna Batatu, *The Old Social Classes and the Revolutionary Movements of Iraq*, London: Saqi Books, 2004: 622.
29. Ibid., 627.
30. On oil worker protests in Iran, see Katayoun Shafiee, *Machineries of Oil: An Infrastructural History of BP in Iran*, Cambridge, MA: MIT Press, 2018: 121–55; on Saudi Arabia, see Robert Vitalis, *America's Kingdom: Mythmaking on the Saudi Oil Frontier*, Stanford, CA: Stanford University Press, 2007; on Bahrain, see Fuad I. Khuri, *Tribe and State in Bahrain: The Transformation of Social and Political Authority in an Arab State*, Chicago, IL: University of Chicago Press, 1980: 194–217; on Egypt, see Joel Beinin and Zachary Lockman, *Workers on the Nile: Nationalism, Communism, Islam, and the Egyptian Working Class, 1882–1954*, Princeton, NJ: Princeton University Press, 1987.
31. On the controversial nature of the name 'Israel', which previously referred to the Jewish people as a whole, see Simon Rawidowicz, 'Israel, the People, the State', in *State of Israel, Diaspora, and Jewish Continuity: Essays on the 'Ever-Dying People'*, Hanover, NH: University Press of New England, 1998: 182–93.
32. On the slight variation in the naming of the two localities, see below, note 60.
33. Council of Economic Advisers, *The Economic Report of the President*, Washington, DC: United States Government Printing Office, 1953: 155.
34. Michael A. Bernstein, *A Perilous Progress: Economists and Public Purpose in Twentieth-Century America*, Princeton, NJ: Princeton University Press, 2004: 108.
35. Robert M. Collins, *More: The Politics of Economic Growth*, New York: Oxford University Press, 2000.
36. William J. Barber, *Designs within Disorder: Franklin D. Roosevelt, the Economists, and the Shaping of American Economic Policy, 1933–1945*, Cambridge: Cambridge University Press, 1996; Roger B. Porter, 'The Council of Economic Advisers', in Colin Campbell et al., eds, *Executive Leadership in Anglo-American Systems*, Pittsburgh, PA: University of Pittsburgh Press, 1991: 189.
37. Charles Maier, 'The Politics of Productivity: Foundations of American International Economic Policy after World War II', *International Organization*, 31: 4, 1977: 607–33.
38. Council of Economic Advisers, *The Economic Report of the President*, Washington, DC: United States Government Printing Office, 1948: 77–8.
39. John Maynard Keynes, *The General Theory of Employment, Interest and Money*, New York: Harcourt Brace Jovanovich, 1964: 161, 149–50, 155.
40. Philip Mirowski, *More Heat Than Light: Economics as Social Physics, Physics as Nature's Economics*, Cambridge: Cambridge University Press, 1991.
41. Reinhart Koselleck, *Futures Past: On the Semantics of Historical Time*, trans. from the German by Keith Tribe, New York: Columbia University Press, 2004: 22.

42. Ibid.: 267.
43. John Dewey, *The Public and Its Problems: An Essay in Political Inquiry: An Essay in Political Inquiry*, Melvin L. Rogers, ed., University Park, PA: Penn State University Press, 2012: 117.
44. Norman Wengert, 'Antecedents of TVA: The Legislative History of Muscle Shoals', *Agricultural History*, 26: 4, 1952: 141–7.
45. Dewey, *The Public and Its Problems*: 117.
46. Georg Simmel, *Philosophy of Money*, London: Routledge, 1978: 416, 485.
47. Quoted in Stephen Kotkin, *Magnetic Mountain: Stalinism as a Civilization*, Berkeley, CA: University of California Press, 1997: 30.
48. Ibid.: 32. See also Alan M. Ball, *Imagining America: Influences and Images in Twentieth-Century Russia*, Lanham, MD: Rowman and Littlefield, 2003: 119–43.
49. See, for example, John Maynard Keynes, 'Economic Possibilities for Our Grandchildren (1930)', in *Essays in Persuasion*, New York: W.W. Norton & Co., 1963: 358–73.
50. Collins, *More*: 142–3.
51. Quoted in ibid., 5.
52. Mitchell, *Carbon Democracy*.
53. Daniel Speich, 'Travelling with the GDP Through Early Development Economics' History', *The Nature of Evidence: How Well Do 'Facts' Travel?* Working Paper No. 33:08, London School of Economics, 2008; Mary S. Morgan, '"On a Mission" with Mutable Mobiles', *The Nature of Evidence: How Well Do 'Facts' Travel?* Working Paper No. 34:08, London School of Economics, 2008.
54. Arturo Escobar, *Encountering Development: The Making and Unmaking of the Third World*, Princeton, NJ: Princeton University Press, 1995.
55. Louis J. Halle, 'On Teaching International Relations', *Virginia Quarterly Review*, 40, 1964: 16, 17, 12–13. See also Gilbert Rist, *The History of Development: From Western Origins to Global Faith*, 3rd edition, trans. from the French by Patrick Camiller, New York: Zed Books, 2008: 70.
56. Harry S. Truman, 'Truman's Inaugural Address', 20 January 1949, truman library.org (accessed 3 August 2025).
57. On visions of modernisation and development in the postwar Middle East, including those of Arab and Turkish modernisers, see among others Nathan Citino, *Envisioning the Arab Future: Modernization in US-Arab Relations, 1945–1967*, Cambridge: Cambridge University Press, 2017, and Begüm Adalet, *Hotels and Highways: The Construction of Modernization Theory in Cold War Turkey*, Stanford, CA: Stanford University Press, 2018.
58. Daniel Drache, 'The Short but Significant Life of the International Trade Organization: Lessons for Our Time', Centre for the Study of Globalisation and Regionalisation, Working Paper No. 62, University of Warwick, 2000.
59. Truman, 'Inaugural Address'.
60. Walid Khalidi, *All That Remains: The Palestinian Villages Occupied and Depopulated by Israel in 1948*, Washington, DC: Institute for Palestine Studies, 1992: 95. Although named after a figure from the Iraqi town of Fallūja (الفلّوجة), the Palestinian village is spelled Fālūja (الفالوجة).

61. For details, see Rashid Khalidi, *The Iron Cage: The Story of the Palestinian Struggle for Statehood*, Boston, MA: Beacon Press, 2006; Nur Masalha, *Expulsion of the Palestinians: The Concept of 'Transfer' in Zionist Political Thought, 1882–1948*, Washington, DC: Institute for Palestine Studies, 1992; Ilan Pappe, *The Ethnic Cleansing of Palestine*, Oxford: Oneworld Publications, 2006; Norman Finkelstein, *Image and Reality of the Israel-Palestine Conflict*, New York: Verso, 2003; and Ahmad H. Sa'di and Lila Abu-Lughod, eds, *Nakba: Palestine, 1948, and the Claims of Memory*, New York: Columbia University Press, 2007.
62. Benny Morris, *The Birth of the Palestinian Refugee Problem Revisited*, Cambridge: Cambridge University Press, 2004: 521–4.
63. Beinin and Lockman, *Workers on the Nile*: 355–62.
64. Ibid., and Joel Gordon, *Nasser's Blessed Movement: Egypt's Free Officers and the July Revolution*, New York: Oxford University Press, 1992: 127–74, 192.
65. Staples and Sayward, *The Birth of Development*; Harold N. Graves, Jr., 'The Bank as International Mediator: Three Episodes', in Edward S. Mason and Robert E. Asher, eds, *The World Bank Since Bretton Woods*, Washington, DC: Brookings Institution, 1973: 595–646.
66. Letter, F. Dorsey Stephens to Joseph Rudinsky, Beirut, 17 February 1956, enclosing memorandum by Leslie Carver recording a meeting with Eugene Black, Cairo, 3–4 February 1956. Documents relating to the Report on the Agricultural Aspects of the Sudd el Aali Project (High Aswan Dam Project), WB 1589793, World Bank Group Archives.
67. W. Scott Lucas and Alistair Morey, 'The Hidden "Alliance": The CIA and MI6 Before and After Suez', in David Stafford and Rhodri Jeffreys-Jones, eds, *American-British-Canadian Intelligence Relations, 1939–2000*, Portland, OR: Frank Cass, 2000: 95–120. See also chapter 8 in Peter Wright and Paul Greengrass, *Spycatcher*, New York: Viking Press, 1987, and Keith Kyle, *Suez: Britain's End of Empire in the Middle East*, London: I. B. Tauris, 2002.
68. On the surprise to the Bank, see 'Transcript of Oral History Interview with Sir William Iliff Held on August 12 and 16, 1961', *World Bank Group Archives Oral History Program*, Washington, DC: World Bank Group: 32–3, documents.worldbank.org (accessed 3 August 2025).
69. Quoted in Lucas and Morey, 'The Hidden "Alliance"': 107.
70. Nathan J. Citino, *From Arab Nationalism to OPEC Eisenhower, King Sa'ūd, and the Making of U.S.-Saudi Relations*, Bloomington, IN: Indiana University Press, 2002, 87–111.
71. United Kingdom, Board of Trade, UK High Commission, Bonn (R. W. Jackling) to S Mackenzie, BoT, 15 April 1955. National Archives of the UK: Public Record Office: BT, 213/44, 1955, 'EGYPT : Development Schemes – High Aswan Dam – Sterling Releases'.
72. Gregh, 'EGYPT – Visit of Mr. J. Georges-Picot'.
73. On Rosenstein-Rodan's disagreements with Bank policy, see 'Transcript of Oral History Interview with Paul Rosenstein-Rodan held on August 14, 1961', *World Bank Group Archives Oral History Program*, Washington, DC: World Bank Group, documents.worldbank.org (accessed 3 August 2025). Rosenstein-Rodan worked at the Bank from 1947 to 1953.

74. P. N. Rosenstein-Rodan, 'Problems of Industrialisation of Eastern and South-Eastern Europe', *The Economic Journal*, 53: 210/211, 1943: 203.
75. See Alacevich, *The Political Economy of the World Bank*: 144–5.

6 The Properties of Markets

1. This chapter is an expanded and revised version of a paper published as 'The Properties of Markets', in Donald MacKenzie, Fabian Muniesa, and Siu Leung-Sea, eds, *Do Economists Make Markets? On the Performativity of Economics*, Princeton, NJ: Princeton University Press, 2008.
2. Wendy Brown, *Undoing the Demos: Neoliberalism's Stealth Revolution*, New York: Zone Books, 2017; *In the Ruins of Neoliberalism: The Rise of Antidemocratic Politics in the West*, New York: Columbia University Press, 2019. Studies of the history of neoliberalism include Daniel Steadman Jones, *Masters of the Universe: Hayek, Friedman, and the Birth of Neoliberal Politics*, Princeton, NJ: Princeton University Press, 2012; Dieter Plehwe, Bernhard J. A. Walpen, and Gisela Neunhöffer, *Neoliberal Hegemony: A Global Critique*, London: Routledge, 2006; Quinn Slobodian, *Globalists: The End of Empire and the Birth of Neoliberalism*, Cambridge, MA: Harvard University Press, 2018; Philip Mirowski and Dieter Plehwe, eds, *The Road from Mont Pelerin: The Making of the Neoliberal Thought Collective*, Cambridge, MA: Harvard University Press, 2009; and Angus Burgin, *The Great Persuasion: Reinventing Free Markets Since the Depression*, Cambridge, MA: Harvard University Press, 2012.
3. Friedrich Hayek, 'Economics and Knowledge', *Economica*, 4: 13, 1937: 33–54; 'The Use of Knowledge in Society', *American Economic Review*, 35: 4, 1945.
4. Koray Çalişkan and Michel Callon, 'Economization, Part 1: Shifting Attention from the Economy towards Processes of Economization', *Economy and Society*, 38: 3, 2009: 369–98.
5. Donald MacKenzie, *An Engine, Not a Camera: How Financial Models Shape Markets*, Cambridge, MA: MIT Press, 2006. See also Mary S. Morgan, 'Economics', in Theodore M. Porter and Dorothy Ross, eds, *The Cambridge History of Science, Volume 7, The Modern Social Sciences*, Cambridge: Cambridge University Press, 2003: 275–305.
6. Michel Callon, *Laws of the Market*, London: Wiley, 1998; *Markets in the Making: Rethinking Competition, Goods, and Innovation*, Brooklyn, NY: Zone Books, 2021.
7. Katharina Pistor, *The Code of Capital: How the Law Creates Wealth and Inequality*, Princeton, NJ: Princeton University Press, 2016.
8. Kristin Asdal and Tone Huse, *Nature-Made Economy: Cod, Capital, and the Great Economization of the Ocean*, Cambridge, MA: MIT Press, 2023. The other examples in this paragraph are from MacKenzie, Muniesa, and Leung-Sea, eds, *Do Economists Make Markets?*
9. Timothy Mitchell, *Rule of Experts: Egypt, Techno-Politics, Modernity*, Berkeley, CA: University of California Press, 2002: 54–79.

10. The car parking example is discussed, along with several others, in Asef Bayat, 'Uncivil Society: The Politics of the "Informal People"', *Third World Quarterly*, 18: 1, 1997: 53–72.
11. Partha Chatterjee, *The Politics of the Governed: Reflections on Popular Politics in Most of the World*, New York: Columbia University Press, 2004; Julia Elyachar, *Markets of Dispossession: NGOs, Economic Development, and the State in Cairo*, Durham, NC: Duke University Press, 2005; Lawrence Liang, 'Porous Legalities and Avenues of Participation', in Monica Narula et al., eds, *Sarai Reader 05: Bare Acts*, New Delhi: The Sarai Program, Centre for the Study of Developing Societies, 2005: 6–17.
12. 'The Good Think Tank Guide', *The Economist*, 21 October 1991.
13. Hernando de Soto, *The Mystery of Capital: Why Capitalism Triumphs in the West and Fails Everywhere Else*, New York: Basic Books, 2000: 5–6.
14. Hernando de Soto, interview on Damien Carrick, 'The Law Report', ABC News, 11 December 2001, abc.net.au (accessed 4 August 2025).
15. De Soto, *The Mystery of Capital*: 5–6.
16. The Institute for Liberty and Democracy created a pilot property registry programme in 1992–4 and ran it on behalf of the government. In 1996, the government replaced it with the Commission for the Formalization of Informal Property, which was run by the government and staffed by existing and former ILD personnel, with funding from the World Bank. World Bank, 'Implementation Completion Report (SCL-43840) on a Loan in the Amount of US $36.12 Million to the Republic of Peru for an Urban Property Rights Project', Washington, DC: World Bank, 2004.
17. De Soto, *The Mystery of Capital*: 5.
18. Hernando De Soto, *The Other Path: The Invisible Revolution in the Third World*, New York: Harper Collins, 1989; retitled second edition: *The Other Path: The Economic Answer to Terrorism*, 2nd edition, New York: Basic Books, 2002.
19. Dieter Plehwe et al., *Neoliberal Hegemony*. See note 2 for more references.
20. Friedrich Hayek, 'Letter to Antony Fisher', 1 January 1980, atlasusa.org (accessed 4 August 2025).
21. Ray Bromley, 'A New Path to Development? The Significance and Impact of Hernando De Soto's Ideas on Underdevelopment, Production, and Reproduction', *Economic Geography*, 66: 4, 1990: 328–48.
22. Timothy Mitchell, 'The Work of Economics: How a Discipline Makes its World', *European Journal of Sociology*, 47: 2, 2005: 297–320.
23. Ibid.
24. These endorsements are carried inside the front cover of the paperback edition of Hernando de Soto's *The Mystery of Capital*, alongside favourable quotations from reviews in more than a dozen leading US and British journals.
25. For a list of awards, see Institute for Liberty and Democracy, 'ILD Awards', ild.org.pe, archived at web.archive.org (accessed 16 August 2025).
26. The review actually stated that de Soto's book 'has already led the cognoscenti to put him in the pantheon of great progressive intellectuals of our age'.

Mark Leonard, 'Liberty, Equality, Property', *New Statesman*, 4 September 2000: 9–10; Institute for Liberty and Democracy, 'Press and Academic Reviews', ild.org.pe, archived at web.archive.org (accessed 16 August 2025). Nevertheless, Leonard's review reported great enthusiasm for de Soto among advisors to the British Prime Minister Tony Blair. Several other commentators from the centre-left of British and American politics were equally enthusiastic.

27. Quoted inside the front cover of the paperback edition of de Soto, *The Mystery of Capital.*
28. Mitchell, 'The Work of Economics'.
29. World Bank, 'Implementation Completion Report (SCL-43840)'.
30. See Helge Onsrud, 'The High Level Commission for Legal Empowerment of the Poor', Centre for Property Rights and Development, Norwegian Mapping and Cadastre Authority. The commission was supported and funded by the governments of Canada, Denmark, Finland, Iceland, Norway, Sweden, and the United Kingdom.
31. 'Norway and Hernando de Soto. Bankrolling the Presidents' Man', *Development Today*, 10–11, 18 July 2005.
32. Ibid.
33. ILD worked with a local partner, the Egyptian Center for Economic Studies. See 'Formalization Action Plan', Egyptian Center for Economic Studies (ECES), eces.org.eg (accessed 5 August 2025).
34. Steve Forbes, 'Mideast Miracle?', CATO Institute, 16 February 2004, cato.org (accessed 4 August 2025).
35. Mitchell, *Rule of Experts:* 59–74; 'Ali Barakat, *Tatawwur al-milkiyya al-zira'iyya fi Misr wa-atharuhu 'ala al-haraka al-siyasiyya*, Cairo: Dar al-Thaqafa al-Jadida, 1977.
36. David S. Landes, *Bankers and Pashas: International Finance and Economic Imperialism in Egypt*, Cambridge, MA: Harvard University Press, 1979.
37. Mitchell, *Rule of Experts*: 73. The quotation is from Henry Villiers Stuart, *Egypt After the War, Being the Narrative of a Tour of Inspection*, London: Macmillan, 1883: 157. The effect of debt on the erecting of waterwheels is mentioned on p. 242.
38. Mitchell, *Rule of Experts*: 73. On the broader social origins of the 'Urabi movement, as the popular movement was known, after its leader Ahmad 'Urabi, see Juan Cole, *Colonialism and Revolution in the Middle East: Social and Cultural Origins of Egypt's 'Urabi Movement*, Princeton, NJ: Princeton University Press, 1993.
39. Mitchell, *Rule of Experts*: 84–93.
40. The English acre and Egyptian feddan are roughly equivalent measures and are used interchangeably in this chapter. One acre is 4,047 square metres (0.4047 hectares). One feddan is 4,200 square metres (0.42 ha) or 1.037 acres. On the standardization of the Egyptian feddan, see H. G. Lyons, *The Cadastral Survey of Egypt 1892-1907*, Cairo, National Print. Department, 1908.
41. Ibid.: 71.
42. See especially Ranajit Guha, *A Rule of Property for Bengal: An Essay on the*

Idea of Permanent Settlement, 2nd edition, New Delhi: Orient Longman, 1981.

43. Mike Davis, *Late Victorian Holocausts: El Niño Famines and the Making of the Third World*, London: Verso, 2001.
44. De Soto, *Mystery of Capital*: 62. As noted in the case of Egypt above, the protections to which de Soto objects often derive not from Roman or early colonial law, but from struggles of the twentieth century.
45. 'Breathing Life into Dead Capital', *The Economist*, 17 January 2004.
46. In the decade 1980–9 alone, the number of farms in the United States declined by 11 per cent. Unlike the attrition pattern of previous decades, which was concentrated in small farms, the collapse in the 1980s spread to medium-sized operations. In the 1990s the collapse continued. 'Agricultural Outlook (October 1989)', *United States Department of Agriculture*, Washington, DC: Economic Research Service: 20; Nicholas D. Kristof, 'As Life for Family Farmers Worsens, the Toughest Wither', *New York Times*, 2 April 2000, Section 1: 18.
47. The term 'inclusive exclusion' is borrowed from the discussion of the topologies (and powers of property) implicit in the logic of sovereignty in Giorgio Agamben, *Homo Sacer: Sovereign Power and Bare Life*, Stanford, CA: Stanford University Press, 1998, although my use of the term here owes more to Derrida than Agamben.
48. Christopher Woodruff, 'Review of de Soto's *The Mystery of Capital*', *Journal of Economic Literature*, 39: 4, 2001: 1215–23.
49. Figures from a 1998 survey by the United States Small Business Administration show that among businesses in the United States with sales under $25,000 a year, 40 per cent borrow no funds at all. Only about a quarter of businesses take traditional loans, either from banks and other financial institutions, or from family and friends. Twice as many businesses, almost 50 per cent, use nontraditional credit. This consists largely of funds borrowed on credit cards, in some cases business credit cards but in most cases personal cards. See the appendix of 'Financing Patterns of Small Firms: Findings from the 1998 Survey of Small Business Finance', United States Small Business Administration, Office of Advocacy, 2003.
50. UK Office on National Statistics, 'Dwelling Stock by Tenure', ons.gov.uk; United Nations Economic Commission for Europe, *Annual Bulletin of Housing and Building Statistics for Europe and North America 2000*, table 2, unece.org (accessed 4 August 2025); Soula Proxenos, 'Home Ownership Rates: A Global Perspective', *Housing Finance International*, 17, 2002: 3–7.
51. Ranald C. Michie, *The City of London: Continuity and Change, 1850–1990*, Basingstoke, UK: Macmillan, 1992: 102.
52. Proxenos, 'Home Ownership Rates'.
53. Hassan Fathy, *Gurna: A Tale of Two Villages*, Cairo: Ministry of Culture, 1969, reprinted as *Architecture for the Poor*, Chicago, IL: University of Chicago Press, 1973. Other early discussions include J. F. C. Turner, 'The Squatter Settlement: Architecture that Works', *Architectural Design*, 38, 1968: 355–60. For subsequent critiques, see P. M. Ward, *Self-Help Housing: A*

Critique, London: Mansell, 1982, and, for the limits of Fathy's vision, Mitchell, *Rule of Experts*: 184–95.

54. The observations are based on the study of eight house-building projects in 2004 and additional information on numerous other cases gathered over the previous decade. For more details on the village, see Mitchell, *Rule of Experts*: 249–66.
55. Pollen from the male date palm is brought to fertilise the female tree by hand, so only one male is needed for every fifteen or twenty females. The other males can be felled for use in building.
56. Law No. 116 of 1983 established penalties ranging from E£10,000 to E£50,000 for unauthorised building on agricultural land. The Prime Minister's Military Order no. 1 of 1996, passed under emergency regulations promulgated in 1981, gave the government powers to destroy such buildings. Estimates of the area of agricultural land lost to construction in the two decades that followed the 1983 law range from 600,000 to 1.2 million acres, although there are no reliable figures from that period. More recent studies using remote sensing data show a much slower rate of loss. In Cairo and the Egyptian Delta, where the rate is probably greatest, 74,600 ha (184,340 acres) were lost in the twenty-four years from 1992 to 2015. Gamal Essam El-Din, 'Opposition Wants Deeper Reforms', *al-Ahram Weekly*, 9 October 2003; David Sims, *Egypt's Desert Dreams: Development or Disaster?*, Cairo: American University Press in Cairo, 2015; Ahmed M. Soliman, *Urban Informality: Experiences and Urban Sustainability Transitions in Middle East Cities*, Cham, Switzerland: Springer Nature, 2021: 255; Taher M. Radwan et al., 'Dramatic Loss of Agricultural Land Due to Urban Expansion Threatens Food Security in the Nile Delta, Egypt', *Remote Sensing*, 11:3, 2019: 332.
57. 'Parliament approves toughened penalty on agricultural encroachments', *Egypt Today*, 17 October 2022, egypttoday.com (accessed 17 August 2025).
58. The problem of building on agricultural land is more serious around large cities, and in the Nile Delta north of Cairo, where most land is a long way from the desert margin. Radwan et al., 'Dramatic Loss of Agricultural Land'.
59. Yahia Shawkat, 'Sixty Years of Legalizing Informal Buildings in Egypt 1956-2019', 10 October 2019, The Built Environment Observatory/Marsad al-Umran, marsadomran.info (accessed 19 August 2025).
60. One qirat is 1/24 of an Egyptian acre (or feddan), or 175 square metres. E£ indicates Egyptian pounds. In spring 2004, E£1,000 was equal to about US$160 (US$1 = E£6.20).
61. Arab Republic of Egypt, Ministry of Housing and Reconstruction, Advisory Committee for Reconstruction, and United Kingdom, Ministry of Overseas Development, *Ismailiya Demonstration Projects*, Cairo: Ministry of Housing and Reconstruction, 1978.
62. Ahmed M. Soliman, 'The Right to Land: To Whom it Belongs after a Reconciliation Law in Egypt', *Journal of Contemporary Urban Affairs*, 6: 2, 2022: 99–111.
63. Michael Carter and Pedro Olinto, 'Getting Institutions Right for Whom? Credit Constraints and the Impact of Property Rights on the Quantity and

Composition of Investment', *Agricultural and Applied Economics Staff Paper*, 433, University of Wisconsin-Madison, 2000, cited in Woodruff, 'Review of de Soto': 1218. An updated version of the paper was published in the *American Journal of Agricultural Economics*, 85: 1, 2003: 173–86. For other evidence that land titling tends to benefit disproportionately the better off, see Klaus Deininger and Hans Binswanger, 'The Evolution of the World Bank's Land Policy: Principles, Experience, and Future Challenges', *World Bank Research Observer*, 14, 1999: 250.

64. Deininger and Binswanger, 'The Evolution of the World Bank's Land Policy': 253.
65. Ibid.: 253, citing P. D. Bidinger et al., 'Consequences of Mid-1980s Drought: Longitudinal Evidence from Mahbubnagar', *Economic and Political Weekly*, 26, 1991: A105–14.
66. Deininger and Binswanger, 'Evolution of the World Bank's Land Policy': 254, citing Rachel E. Kranton and Anand V. Swamy, 'The Hazards of Piecemeal Reform: British Civil Courts and the Credit Market in Colonial India', World Bank, Development Research Group, Washington, DC, 1997. See Mitchell, *Rule of Experts*: 54–79, for a discussion on the role of distress sales in the concentration of land ownership in nineteenth-century Egypt.
67. Erica Field and Maximo Torero, 'Do Property Titles Increase Credit Access among the Urban Poor? Evidence from a Nationwide Titling Program', Princeton University, September 2002, scholar.harvard.edu (accessed 5 August 2025); see also Erica Field, 'Entitled to Work: Urban Property Rights and Labor Supply in Peru', *Quarterly Journal of Economics*, 112: 4, 2007: 1561–602; Erica Field, 'Urban Property Rights and Household Welfare in Developing Countries', PhD thesis, Princeton University, 2003; J. Calderon Cockburn, 'Regularisation of Urban Land in Peru', *Land Lines*, Lincoln Institute of Land Policy, 1998; Ayako Kagawa, 'Policy Effects and Tenure Security Perceptions of Peruvian Urban Land Tenure Regularisation Policy in the 1990s', Workshop Paper, ESF/N-AERUS International Workshop, Leuven and Brussels, Belgium, 23–6 May 2001; Maximo Torrero, 'Estudio de la Oferta, Demanda y Fuentes de Credito Informal', Documento Suplementario del Estudio: Perfil de la Demanda y Oferta del Credito Formal y Informal, COFOPRI office: Grupo de Analisis para el Desarollo, 1999. All four are cited in Field, 'Entitled to Work'. See also Mitchell, 'The Work of Economics'.
68. Field and Torero, 'Do Property Titles Increase Credit Access'.
69. Gershon Feder et al., *Land Policy and Farm Productivity in Thailand*, Baltimore, MD: Johns Hopkins University Press, 1988. The only effect was that those with title received somewhat larger loans.
70. Deininger and Binswanger, 'Evolution of the World Bank's Land Policy': 260.
71. Ibid.: 266, citing J. W. Bruce, 'A Perspective on Indigenous Land Tenure Systems and Land Concentration', in R. W. Downs and S. P. Reyna, eds, *Land and Society in Contemporary Africa*, Hanover, NH: University Press of New England, 1988; and Jean-Philippe Plateau, 'The Evolutionary Theory of Land Rights as Applied to Sub-Saharan Africa: A Critical Assessment', *Development and Change*, 27, 1996: 29–86.

72. Deininger and Binswanger, 'Evolution of the World Bank's Land Policy'; online conference on 'Land, Real Estate, and the Economy', hosted by the World Bank Group's Land and Real Estate Initiative, November 1999.
73. World Bank, 'Implementation Completion Report (SCL-43840) On a Loan in the Amount of US $36.12 Million to the Republic of Peru for an Urban Property Rights Project', Washington, DC: World Bank, 2004.
74. Lee Alston, Gary Libecap, and Robert Schneider, 'The Determinants and Impact of Property Rights: Land Titles on the Brazilian Frontier', *Journal of Law, Economics, and Organization*, 12: 1, 1996: 1569–614, cited in Woodruff, 'Review of *The Mystery of Capital*': 1221.
75. Woodruff, 'Review of *The Mystery of Capital*': 1221.
76. Deininger and Binswanger, 'Evolution of the World Bank's Land Policy': 252.
77. Field, 'Entitled to Work'; Alan B. Krueger, 'Economic Scene: A Study Looks at Squatters and Land Titles in Peru', *New York Times*, 9 January 2003: C2; Erica Field, 'Urban Property Rights and Household Welfare in Developing Countries', PhD thesis, Princeton University, 2003; J. Bradford DeLong, 'The Future of Economics', archived at web.archive.org.
78. See Mitchell, 'The Work of Economics'.
79. Arvind Subramanian, 'The Egyptian Stabilization Experience: An Analytical Retrospective', Working Papers of the International Monetary Fund, WP/97/105, International Monetary Fund, Middle Eastern Department, 1997. For an account of the reforms and their outcome, see Mitchell, *Rule of Experts*: 272–303.
80. Tamir Moustafa, 'Law Versus the State: The Judicialization of Politics in Egypt', *Law and Social Inquiry*, 28: 4, 2003: 883–931.
81. Manufacturing as a proportion of GDP declined from 17 per cent in 1989 to 16 per cent in 2009 and 14 per cent in 2024; 'Manufacturing, value added (% of GDP)', data.worldbank.org. The labour force participation rate (those in the labour force as a percentage of the working age population) fell from 54 per cent in 1989 to 50 per cent in 2009 and 45 per cent in 2022, 'ILOSTAT: Labour Force Participation Rate', ILO.org.
82. On the history of this collaboration between the ILD and the Egyptian Center for Economic Studies, see 'The Case for Formalization of Business in Egypt', ECES Policy Viewpoint, 17, September 2005, eces.org (accessed 15 August 2025).
83. Law No. 148 of 2001. The regulatory body was the General Authority for Real Estate Finance. Pierre Loza, 'The Not-So-Real Estate Company', *Al-Ahram Weekly*, 4–10 March 2004.
84. Economist Intelligence Unit, 'Egypt Regulations: Mortgage Law to be Amended', *EIU ViewsWire*, 19 January 2005. In 2004 the International Financial Corporation (the arm of the World Bank that lends to the private sector) and the German Investment and Development Company put up funds to launch Egypt's first private mortgage company, the Egyptian Housing Finance Company, with an authorised capital of E£100 million. The other share owners were the Egyptian American Bank, holding 40 per cent of shares, the Bank of Alexandria, and the Indian-owned Housing Development Finance Corporation: Pierre Loza, 'The Not-So-Real Estate Company.'

85. Maha Hasan, 'al-Tamwil al-'aqari: hal yukhaddam mahdudi al-dakhl?', *al-Ahram Weekly*, 28 March 2004: 16; Jeson Ingraham, 'Make or Break Year?', *Business Today Egypt*, February 2004, businesstodayegypt.com (accessed 6 August 2025).
86. De Soto, *Mystery of Capital*: 56.
87. This right was introduced in the 2012 Constitution and retained in the revised 2014 Constitution. Neither constitution explained how the right was to be enforced. But its status as a constitutional right helped reinforce popular sentiment against housing evictions, even in cases where the right was violated. See Tadamun, 'The Right to Adequate Housing in the Egyptian Constitution', tadamun.co (accessed 5 August 2025).
88. De Soto, *Mystery of Capital*: 6–7.
89. Ibid.: 56.
90. Alain Pottage, 'The Measure of Land', *The Modern Law Review*, 57: 3, 1994: 361–84; and 'The Originality of Registration', *Oxford Journal of Legal Studies*, 15: 3, 1995: 371–401.
91. See C. Dent Bostick, 'Land Title Registration: An English Solution to an American Problem', *Indiana Law Journal*, 63: 1, 1987: 55–111. On the history of the US system of property title registry, including its origins in settler colonial concerns to secure European property rights against native nations and enslaved people, see K. Sue Park, 'Property and Sovereignty in America: A History of Title Registries & Jurisdictional Power', *The Yale Law Journal*, 133: 5, 2024: 1487–581.
92. Even in the cities, domestic animals are a significant source of food security for low-income households. In Cairo, it was estimated in 2000 that 16 per cent of households raise domestic animals, usually chickens or other small animals raised on the roof of an apartment building or in a rear yard. See Jörg Gertel and Said Samir, 'Cairo: Urban Agriculture and Visions for a "Modern" City', in RUAF, *Growing Cities, Growing Food*, 19 November 2000: 209–34, ruaf.org (accessed 19 September 2021).
93. Mitchell, *Rule of Experts*: 66–70.
94. Joseph F. Nahas, *Situation économique et sociale du fellah égyptien*, Paris: Arthur Rousseau, 1901: 141.

7 Infrastructures Work on Time

1. Some parts of this chapter appeared previously as 'Infrastructures Work on Time', in New Silk Roads, *e-flux architecture*, January 2020, e-flux.com (accessed 6 August 2025).
2. See, for example, David Harvey's influential analysis of 'space-time compression', in *The Condition of Postmodernity: An Inquiry into the Origins of Cultural Change*, Cambridge: Blackwell Publishers, 1990. But see also Geoffrey Bowker, 'All Together Now: Synchronization, Speed, and the Failure of Narrativity', *History and Theory*, 53: 4, 2014: 563–76.
3. For an excellent sampling and review of recent scholarship on the politics of infrastructure, see Brian Larkin, 'The Politics and Poetics of

Infrastructure', *Annual Review of Anthropology*, 42, 2013: 327–43; Stephen J. Collier, James Christopher Mizes, and Antina von Schnitzler, eds, *Limn 7: Public Infrastructures/Infrastructural Publics*, 2016; Nikhil Anand, Akhil Gupta, and Hannah Appel, eds, *The Promise of Infrastructure: A School for Advanced Research Advanced Seminar*, Durham, NC: Duke University Press, 2018; and Hillary Angelo and Christine Hentschel, 'Interactions with Infrastructure as Windows into Social Worlds: A Method for Critical Urban Studies: Introduction', *City*, 19: 2–3, 2015: 306–12. For an older overview, see Paul Edwards et al., 'Introduction: An Agenda for Infrastructure Studies', *Journal of the Association for Information Systems*, 10: 5, 2009.

4. Nicholas Hildyard and Xavier Sol, *How Infrastructure Is Shaping the World: A Critical Introduction to Infrastructure Mega-Corridors*, Brussels: Counter Balance: Challenging Public Investment Banks, 2017.
5. Michele Ruta et al., *Belt and Road Economics: Opportunities and Risks of Transport Corridors*, Washington, DC: World Bank, 2019, openknowledge.worldbank.org (accessed 6 August 2025).
6. Dave Donaldson, 'Railroads of the Raj: Estimating the Impact of Transportation Infrastructure', *American Economic Review*, 108: 4–5, 2018: 899–934. Dave Donaldson and Richard Hornbeck, 'Railroads and American Economic Growth: A "Market Access" Approach', *Quarterly Journal of Economics*, 101: 7, 2016: 799–858. On the environmental and other destruction brought by US railroad construction, see Richard White, *Railroaded: The Transcontinentals and Making of Modern America*, New York: W.W. Norton & Co., 2011.
7. For the role of Indian railways in causing famine, see Stuart Sweeney, 'Indian Railways and Famine 1875–1914: Magic Wheels and Empty Stomachs', *Essays in Economic & Business History*, 26, 2008: 147–58.
8. For the role of railways in spreading malaria, cholera, and other disease in India, see Ritika Prasad, *Tracks of Change: Railways and Everyday Life in Colonial India*, Cambridge: Cambridge University Press, 2015: 165–99.
9. Michael Williams, *Deforesting the Earth: from Prehistory to Global Crisis – An Abridgement*, Chicago, IL: University of Chicago Press, 2006: 244, 337–8.
10. Kate Galbraith, 'Delivering Unwelcome Species to the Mediterranean', *New York Times*, 4 March 2015.
11. The phenomenon is known as 'induced traffic'. Peter Hills, 'What is Induced Traffic?', *Transportation*, 23, 1996: 5–16.
12. Suez Canal Authority, 'Canal Characteristics', n.d., suezcanal.gov.eg (accessed 6 August 2025).
13. Brendan Greeley, 'The Bank Effect and the Big Boat Blocking the Suez', *Financial Times*, 25 March 2021.
14. Bent Flyvbjerg, 'Over Budget, Over Time, Over and Over Again: Managing Major Projects', in Peter W. G. Morris, Jeffrey K. Pinto, and Jonas Söderlund, eds, *The Oxford Handbook of Project Management*, Oxford: Oxford University Press, 2013: 321–44. On the open-ended 'promise' of infrastructure, see also Penny Harvey, 'Infrastructures in and out of Time', in Anand, Gupta, and Appel, eds, *The Promise of Infrastructure*: 80–101.

15. Wright, *Railroaded.* China's Belt and Road Initiative was promoted as a means to connect oceans and continents, ports and industrial hubs – but it operated as a vast debt programme, by which Chinese banks and construction companies organised repayment from dozens of countries. Many loans were soon underperforming and requiring further 'rescue loans' to allow financial extraction to continue. 'China Reckons with Its First Overseas Debt Crisis', *Financial Times*, 20 July 2022; 'China's Debt Threat: Time to Rein in the Lending Boom', *Financial Times*, 24 July 2018.
16. Timothy Mitchell, *Carbon Democracy: Political Power in the Age of Oil*, London: Verso, 2011.
17. Thorstein Veblen, *Absentee Ownership and Business Enterprise in Recent Times: The Case of America*, London: George Allen and Unwin, 1923.
18. Mitchell, *Carbon Democracy*; on contemporary forms of sabotage, see also Robert Meister, *Justice Is an Option: A Democratic Theory of Finance for the Twenty-First Century*, Chicago, IL: Chicago University Press, 2021.
19. On the 'hubris' of technology, see Sheila Jasanoff, 'Technologies of Humility: Citizen Participation in Governing Science', *Minerva*, 41: 3, 2003: 223–44, reprinted in Sheila Jasanoff, *Science and Public Reason*, Abingdon, UK: Routledge, 2012, 167–84.
20. For examples of the material production of financial instruments, see William Cronin, 'Pricing the Future', in *Nature's Metropolis: Chicago and the Great West*, New York: W.W. Norton & Co., 1992; and, for a contemporary case, Canay Özden-Schilling, 'Economy Electric', *Cultural Anthropology*, 30: 4, 2015: 578–88.
21. On capital as the capitalisation of future revenue streams, see Jonathan Nitzan and Shimshon Bichler, *Capital as Power: A Study of Order and Creorder*, London: Routledge, 2009; Fabian Muniesa et al., *Capitalization: A Cultural Guide*, Paris: Presses des Mines, 2017; and chapter 1 of this book.
22. On money as an accounting system based on transferable credit, see Felix Martin, *Money: The Unauthorized Biography*, New York: Knopf, 2014: 27. On the political stakes that are raised by rival explanations of money, especially for democratic politics, see Stefan Eich, *The Currency of Politics: The Political Theory of Money from Aristotle to Keynes*, Princeton, NJ: Princeton University Press, 2022; on the history of credit money, see also David Graeber, *Debt: The First 5,000 Years*, Brooklyn, NY: Melville House, 2011.
23. Katharina Pistor, *The Code of Capital: How the Law Creates Wealth and Inequality*, Princeton, NJ: Princeton University Press, 2019.
24. Nitzan and Bichler, *Capital as Power*; Tim Di Muzio and Richard Robbins, *Debt as Power*, Manchester: Manchester University Press, 2016.
25. Adam Smith, *An Inquiry into Nature and Causes of the Wealth of Nations*, vol. 1, Edwin Cannan, ed., Chicago, IL: University of Chicago Press, 1976: 321. Thanks to Stefan Eich for this reference. 'Waggon' was spelled this way.
26. Brian Bonnyman, *The Third Duke of Buccleuch and Adam Smith: Estate Management and Improvement in Enlightenment Scotland*, Edinburgh: Edinburgh University Press, 2014; Henry Hamilton, 'The Failure of the Ayr Bank, 1772', *The Economic History Review*, 8: 3, 1956: 405–17; John Rae, *Life of Adam Smith*, London: Macmillan, 1895; Antoin E. Murphy, *The Genesis of*

Macroeconomics: New Ideas from Sir William Petty to Henry Thornton, Oxford: Oxford University Press, 2009: 171–85.

27. The exact present value of $100 discounted annually at 10 per cent for seven years is $51.32.
28. Jonathan Levy, 'Capital as Process and the History of Capitalism', *Business History Review*, 91: 3, 2017: 483–510. Levy's excellent account of capital as process does not, however, pay much attention to the technopolitical work that must be done to construct an apparatus that successfully claims the future.
29. For a survey of the law and practice of credit and discounting in medieval Islamic societies, see Nicholas Dylan Ray, 'The Medieval Islamic System of Credit and Banking: Legal and Historical Considerations', *Arab Law Quarterly*, 12: 1, 1997: 43–90; and A. L. Udovitch, 'Credit as a Means of Investment in Medieval Islamic Trade', *Journal of the American Oriental Society*, 87: 3, 1967: 260–4.
30. Larrie D. Ferreiro, *Ships and Science: The Birth of Naval Architecture in the Scientific Revolution, 1600–1800*, Cambridge, MA: MIT Press, 2017: 24.
31. Ibid.; John Law, 'On the Methods of Long-Distance Control: Vessels, Navigation and the Portuguese Route to India', *The Sociological Review*, 32: 1, 1984.
32. Di Muzio and Robbins, *Debt as Power*; Giovanni Arrighi, *Adam Smith in Beijing: Lineages of the Twenty-First Century*, London: Verso, 2007; for the role of debt collection in state formation, see Daniel Lord Smail, 'Violence and Predation in Late Medieval Mediterranean Europe', *Comparative Studies in Society and History*, 54: 1, 2012: 7–34.
33. In England, for example, the mortgaging of property was not formally established until the Law of Property Act of 1925. From the eighteenth century a variety of techniques was employed to circumvent this limit and use land as a security for debt, with little need for evidence of collateral. This led to frequent default, the growth of multiple encumbrances on land, and land itself becoming less liquid, with static or decreasing sales. B. L. Anderson, 'Provincial Aspects of the Financial Revolution of the Eighteenth Century', *Business History*, 11: 1, 1969: 12–13.
34. See Meister, *Justice Is an Option*, for this reading of industrialisation as a shift in credit relations.
35. Suez Canal Authority, 'Tolls Table', n.d., suezcanal.gov.eg (accessed 6 August 2025). After earning a record $8 billion from Suez Canal revenues in 2022, Egypt increased the fees in 2023 by 15 per cent.
36. See Donald MacKenzie, 'Just How Fast?', *London Review of Books*, 41: 5, 2019: 23–4.
37. See Nikhil Anand, *Hydraulic City: Water and the Infrastructures of Citizenship in Mumbai*, Durham, NC: Duke University Press, 2017.
38. Julia Elyachar, 'Relational Finance: Ottoman Debt, Financialization, and the Problem of the Semi-Civilized', *Journal of Cultural Economy*, 16: 3, 2023: 323–36.
39. Lauren A. Benton, *A Search for Sovereignty: Law and Geography in European Empires, 1400–1900*, New York: Cambridge University Press, 2010.

40. Deborah Cowen, *The Deadly Life of Logistics: Mapping Violence in Global Trade*, Minneapolis, MN: University of Minnesota Press, 2014.
41. Aihwa Ong, *Neoliberalism as Exception: Mutations in Citizenship and Sovereignty*, Durham, NC: Duke University Press, 2006; Quinn Slobodian, *Crack-Up Capitalism: Market Radicals and the Dream of a World Without Democracy*, London: Penguin Books, 2023.
42. Laleh Khalili, *Sinews of War and Trade: Shipping and Capitalism in the Arabian Peninsula*, London: Verso, 2021: 101–10, 154–8.
43. Keller Easterling, *Extrastatecraft: The Power of Infrastructure Space*, London: Verso, 2014; Andrew Barry, *Material Politics: Disputes along the Pipeline*, Chichester: Wiley-Blackwell, 2013.
44. Stefano Sgambati, 'Rethinking Banking: Debt Discounting and the Making of Modern Money as Liquidity', *New Political Economy*, 21: 3, 2016: 274–90.
45. This paragraph draws on Geoffrey Ingham, *The Nature of Money*, Cambridge: Polity Press, 2004: 134–51.
46. Kean Birch and Fabian Muniesa, eds, *Assetization: Turning Things into Assets in Technoscientific Capitalism*, Cambridge, MA: MIT Press, 2020; Brett Christophers, *Rentier Capitalism: Who Owns the Economy, and Who Pays for It?*, London: Verso, 2020. Infrastructure projects have now become an asset class: see Nicholas Hildyard, *Licensed Larceny: Infrastructure, Financial Extraction and the Global South*, Manchester: Manchester University Press, 2016.
47. Pistor, *The Code of Capital*; Paddy Ireland, 'Capitalism Without the Capitalist: The Joint Stock Company Share and the Emergence of the Modern Doctrine of Separate Corporate Personality', *The Journal of Legal History*, 17: 1, 1996.
48. Susan Leigh Star, 'The Ethnography of Infrastructure', *American Behavioral Scientist*, 43: 3, 1999: 377–91.
49. Barry, *Material Politics*; David Edgerton, *The Shock of the Old: Technology and Global History since 1900*, Oxford: Oxford University Press, 2007; Edwards et al., 'Introduction: An Agenda for Infrastructure'; Akhil Gupta, 'The Future in Ruins', in Anand, Gupta, and Appel, eds, *The Promise of Infrastructure*.
50. Bruno Latour, *We Have Never Been Modern*, Cambridge, MA: Harvard University Press, 1993.

8 A Better Tomorrow, Tomorrow? Climate Crisis and the Alibi of Growth

1. Parts of this chapter are republished as 'The Great EconoCon: Climate Crisis and the Misconception of Growth', *Critical Times*, 9: 1–2 (2026).
2. Calculated from Global Carbon Project, 'Global Carbon Budget 2021', Supplemental data of Global Carbon Budget 2021 (Version 1.0) (Data set). On the problem with periodising anthropogenic climate change, see Christophe Bonneuil and Jean-Baptiste Fressoz, *The Shock of the Anthropocene: The Earth, History and Us*, London: Verso, 2016.

3. Daniel Speich, 'Traveling with the GDP Through Early Development Economics' History', *Working Papers on the Nature of Evidence: How Well Do 'Facts' Travel?*, 33: 8, Department of Economic History, London School of Economics and Political Science, 2008; Michele Alacevich, *The Political Economy of the World Bank: The Early Years*, Stanford, CA: Stanford University Press, 2009.
4. Herman Daly, *Beyond Growth: The Economics of Sustainable Development*, Boston, MA: Beacon Press, 1996; J. K. Gibson-Graham, *A Postcapitalist Politics*, Minneapolis, MN: University of Minnesota Press, 2006; Jean Gadrey and Florence Jany-Catrice, *Les nouveaux indicateurs de richesse*, Paris: La Découverte, 2016; Florence Jany-Catrice and Dominique Méda, 'Well-Being and the Wealth of Nations: How Are They to Be Defined?', *Review of Political Economy*, 2013: 444–60; see also Joseph E. Stiglitz, Amartya Sen, and Jean-Paul Fitoussi, *Mismeasuring Our Lives: Why GDP Doesn't Add Up*, New York: The New Press, 2010.
5. Giorgos Kallis et al., *The Case for Degrowth*, Cambridge: Polity Press, 2020; Jason Hickel, *Less Is More: How Degrowth Will Save the World*, London: Penguin Books, 2021; Kohei Saito, *Marx in the Anthropocene: Towards the Idea of Degrowth Communism*, Cambridge: Cambridge University Press, 2022; Matthias Schmelzer, Andrea Vetter, and Aaron Vansintjan, *The Future Is Degrowth: A Guide to a World Beyond Capitalism*, London: Verso, 2022.
6. Philipp Lepenies, *The Power of a Single Number: A Political History of GDP*, New York: Columbia University Press, 2016. These defects in the measurement of the economy and its growth were not discovered after the fact. They pre-date the invention of the economy and the definition of GDP. Simon Kuznets, the US statistician who pioneered in the interwar period the methods of calculating national income, strongly opposed the including of harmful expenditure such as gambling, prostitution, and the manufacture and purchasing of weapons. Criticisms of the way national income is measured have been made ever since, and have led to continuous and largely unsuccessful attempts to redefine the measure, or replace it; from Mahbub ul-Haq and Amartya Sen's alternative indicators of development (UNDP, *Human Development Report 1990*, Oxford: Oxford University Press, 1990); to Daly and Cobb's Index of Sustainable Economic Welfare and its refinement in the Genuine Progress Indicator; to the 2010 book by Stiglitz, Sen, and Fitoussi, *Mismeasuring Our Lives*. The failure of any of these alternatives to displace the role of GDP suggests that the problem of the measurement of growth in terms of GDP is something more than just errors in the way we 'represent' the economy. On the history of GDP, see also Eli Cook, *The Pricing of Progress: Economic Indicators and the Capitalization of American Life*, Cambridge, MA: Harvard University Press; Benjamin Mitra-Khan, 'Redefining the Economy: How the "Economy" Was Invented in 1620, and Has Been Redefined Ever Since', PhD thesis, Department of Economics, City University London, 2011; Jacob Assa, 'Finance, Social Value, and the Rhetoric of GDP', *Finance and Society*, 4: 2, 2018: 144–58.
7. William D. Nordhaus and James Tobin, 'Is Growth Obsolete?', in William D. Nordhaus and James Tobin, eds, *Economic Research: Retrospect and Prospect*, vol. 5, Cambridge, MA: National Bureau of Economic Research, 1972: 1–80.

8. Nancy Fraser, 'Climates of Capital: For a Trans-Environmental Eco-Socialism', *New Left Review*, 127, 2021: 94–127; Jason W. Moore, *Capitalism in the Web of Life: Ecology and the Accumulation of Capital*, London: Verso, 2015.
9. Alyssa Battistoni, *Free Gifts: Capitalism and the Politics of Nature*, Princeton, NJ: Princeton University Press, 2025.
10. Kate Raworth, *Doughnut Economics: Seven Ways to Think Like a 21st-Century Economist*, New York: Random House, 2017.
11. Timothy Mitchell, 'Society, Economy, and the State Effect', in George Steinmetz, ed., *State/Culture: State-formation After the Cultural Turn*, Ithaca, NY: Cornell University Press, 1999: 76–97.
12. Liliana Doganova, *Discounting the Future: The Ascendancy of a Political Technology*, Brooklyn, NY: Zone Books, 2024.
13. Raworth, *Doughnut Economics*.
14. On the mechanism of pooling, see Stefano Sgambati, 'Rethinking Banking: Debt Discounting and the Making of Modern Money as Liquidity', *New Political Economy*, 21: 3, 2016: 274–90, and the discussion in chapter 3. On capital as a relation of power, see Jonathan Nitzan and Shimshon Bichler, *Capital as Power: A Study of Order and Creorder*, London: Routledge, 2009.
15. Donald MacKenzie, *An Engine, Not a Camera: How Financial Models Shape Markets*, Cambridge, MA: MIT Press, 2006. On derivatives and capitalisation, see also Dick Bryan and Michael Rafferty, *Capitalism with Derivatives: A Political Economy of Financial Derivatives, Capital and Class*, Basingstoke, UK: Palgrave Macmillan, 2006; Randy Martin, *Financialization of Daily Life*, Philadelphia, PA: Temple University Press, 2002; Robert Meister, *Justice Is an Option: A Democratic Theory of Finance for the Twenty-first Century*, Chicago, IL: University of Chicago Press, 2021.
16. Meister, *Justice Is an Option*. See also Carolyn Hardin, *Capturing Finance: Arbitrage and Social Domination*, Durham, NC: Duke University Press, 2021. Meister's and Hardin's insightful studies of this engineering of finance and the modes of political domination it secures can be compared with accounts relying on notions of the 'imagination' of 'fictional' futures, such as Jens Beckert, *Imagined Futures: Fictional Expectations and Capitalist Dynamics*, Cambridge, MA: Harvard University Press, 2016. For other recent approaches to the study of futurity, see Sandra Kemp and Jenny Andersson, eds, *Futures*, Oxford: Oxford University Press, 2021.
17. Maurizio Lazzarato, *The Making of the Indebted Man: An Essay on the Neoliberal Condition*, Cambridge, MA: MIT Press, 2012.
18. Lisa Adkins, Melinda Cooper, and Martijn Konings, *The Asset Economy: Property Ownership and the New Logic of Inequality*, Medford, MA: Polity Press, 2020.
19. Brett Christophers, *Our Lives in Their Portfolios: Why Asset Managers Own the World*, London: Verso, 2023.
20. Adkins, Cooper, and Konings, *The Asset Economy*.
21. Comedy Central, *The Colbert Report*, 10 March 2011. I owe the reference to Robert Vitalis, *White World Order, Black Power Politics: The Birth of American International Relations*, Ithaca, NY: Cornell University Press, 2015: 170–1.

22. The Factory Acts, regulating hours of work and other conditions, and eventually mealtimes, holidays, and the rate of the wage itself, began to standardise, make abstract, the payment of a wage – to construct 'abstract' labour – an effect created more through legislation and statistics than the actual messiness of the factory floor. The 'wage' was a product of the organisation of postponed payment.
23. Michel Callon, *Markets in the Making: Rethinking Competition, Goods, and Innovation*, Brooklyn, NY: Zone Books, 2021.
24. Maxine Berg, *Luxury and Pleasure in Eighteenth-Century Britain*, Oxford: Oxford University Press, 2007; Jonathan Eacott, *Selling Empire: India in the Making of Britain and America, 1600–1830*, Chapel Hill, NC: University of North Carolina Press, 2016: 170–80; William Sewell, *Capitalism and the Emergence of Civic Equality in Eighteenth-Century France*, Chicago, IL: University of Chicago Press, 2021.
25. See Yanis Varoufakis, *Technofeudalism: What Killed Capitalism*, London: Melville House, 2024, from which I take the term 'fiefdoms'; and Cédric Durand, *How Silicon Valley Unleashed Techno-feudalism: The Making of the Digital Economy*, trans. David Broder, London: Verso, 2024.
26. This understanding of money was developed by Bernard Schmitt and explored by Gilles Deleuze. Bernard Schmitt, *Monnaie, Salaires, et Profits*, Paris: Presses Universitaires de France, 1966, and Gilles Deleuze and Félix Guattari, *A Thousand Plateaus: Capitalism and Schizophrenia*, Minneapolis, MN: University of Minnesota Press, 1987. See also Deleuze's lecture at Vincennes, 21 December 1971: 'Anti-Œdipe et Mille Plateaux', webdeleuze.com, n.d (accessed 7 August 2025).
27. Meister describes this difference as an 'arbitrage opportunity' created by the organisation of production, and the source of the surplus appropriated by the owner of capital: *Justice Is an Option*: 19–22.
28. D. W. Jones, *War and Economy in the Age of William III and Marlborough*, Oxford: Blackwell, 1988.
29. J. G. A. Pocock draws attention to the relation between terms such as 'national debt' and the ideas of 'state' and 'nation' in European history, in 'Commerce, Credit and Sovereignty: The Nation-State as Historical Critique', in Béla Kapossy et al., eds, *Markets, Morals, Politics: Jealousy of Trade and the History of Political Thought*, Cambridge, MA: Harvard University Press, 2019: 265–84.
30. Geoffrey Ingham, *The Nature of Money*, Cambridge: Polity Press, 2004: 126–31.
31. Albert Feavearyear, *The Pound Sterling: A History of English Money*, Oxford: Clarendon Press, 1963: 99–305.
32. Marcello De Cecco, *Money and Empire: The International Gold Standard, 1890–1914*, Totowa, NJ: Rowman and Littlefield, 1975: 20–61, 76–102.
33. Ingham, *The Nature of Money*: 145.
34. Doganova, *Discounting the Future*.
35. John W. Kendrick, 'Measurement of Real Product', in National Bureau of Economic Research Conference on Research in Income and Wealth, *A Critique of the United States Income and Product Accounts*, Princeton, NJ: Princeton University Press, 1958: 419.

36. Diane Coyle, *GDP: A Brief but Affectionate History*, Princeton, NJ: Princeton University Press, 2014.
37. Callon, *Markets in the Making.*
38. Keeanga-Yamahtta Taylor demonstrates the significance of race to this practice of capture, showing how redlining gave way to the 'predatory inclusion' of African Americans in the housing market, with the capture of surplus relying on the increased opportunities for foreclosure. *Race for Profit: How Banks and the Real Estate Industry Undermined Black Homeownership*, Chapel Hill, NC: University of North Carolina Press, 2019.
39. Brett Christophers, *Banking across Boundaries: Placing Finance in Capitalism*, Chichester, UK: John Wiley & Sons, 2013.
40. Tim Di Muzio and Richard H. Robbins, *Debt as Power*, Manchester: Manchester University Press, 2016. The figures are from 2006.
41. On forced growth, see Pascal van Griethuysen, '*Bona diagnosis, bona curatio*: How Property Economics Clarifies the Degrowth Debate', *Ecological Economics*, 84, 2012: 262–9; and Mathias Binswanger, 'Is there a Growth Imperative in Capitalist Economies', *Journal of Post Keynesian Economics*, 31: 4, 2009: 707; 'The Growth Imperative Revisited: A Rejoinder to Gilányi and Johnson', *Journal of Post Keynesian Economics*, 37: 4, 2015: 648–60 (but note that Binswanger's model considers only bank loans, and only those bank loans used to finance business activities – thus excluding mortgages, which constitute about 60 per cent of all bank loans). See also Tim Jackson, *Prosperity Without Growth: Foundations for the Economy of Tomorrow*, 2nd edition, London: Routledge, 2017; *Post Growth: Life after Capitalism*, Cambridge: Polity Press, 2021; Michael Rowbotham, *The Grip of Death: A Study of Modern Money, Debt Slavery and Destructive Economics*, Charlbury, UK: Jon Carpenter Publishing, 1998.
42. John Robert McNeill and Peter Engelke, *The Great Acceleration: An Environmental History of the Anthropocene since 1945*, Cambridge, MA: Harvard University Press, 2014.
43. Thorfinn Stainforth and Bartosz Brzezinski, 'More than Half of All CO_2 Emissions Since 1751 Emitted in the Last 30 Years', Institute for European Environmental Policy, 29 April 2020, ieep.eu (accessed 7 August 2025).
44. China, the rest of East Asia, Russia, Europe, and other large industrialised regions produce roughly similar quantities of greenhouse gases per capita (about 6 to 8 tonnes of CO_2 equivalent per year). Most countries of the Global South produce from half to a quarter of that quantity per person, and some far less. The United States, along with Canada and Australia, produce carbon emissions at almost 15 tonnes per person per year. The highest per capita emissions are found among a handful of small oil- and gas-producing states – Qatar, Bahrain, and four other Gulf countries, plus Trinidad and Tobago and Brunei. See Global Carbon Project, 'Data: CO_2 Emissions per Capita', in Population Based on Various Sources – with major processing by Our World in Data, 2024, ourworldindata.org (accessed 7 August 2025), based largely on data from J. C. Minx et al., 'A Comprehensive and Synthetic Dataset for Global, Regional, and National Greenhouse Gas Emissions by Sector 1970–2018 with an Extension to 2019', *Earth Syst. Sci. Data*, 13: 11, 2021: 5213–52.

45. For the arguments in this and the following two paragraphs, see Timothy Mitchell, *Carbon Democracy: Political Power in the Age of Oil*, 2nd edition, London: Verso, 2023.
46. On car ownership as a continuing mode of capitalisation, even as demand for vehicles stagnates, see Tom Haines-Doran, 'The Financialisation of Car Consumption', *New Political Economy*, 29: 3, 2024: 337–55.
47. See Adam Hanieh, 'The Commodities Fetish? Financialisation and Finance Capital in the US Oil Industry', *Historical Materialism*, 29: 4, 2021: 70–113.
48. On representing Islam, defending Zionism, and devising the war on terror, see the chapter 'McJihad: Islam in the U.S. Global Order', in Mitchell, *Carbon Democracy*, 200–30. As noted there, this does not imply reducing these developments to the forces of capital. On the contrary, as the term 'McJihad' designates, it registers the dependence of capital on disjunctive forces, including movements such as jihadism or Zionism, as well as militarism.
49. For an earlier exploration of this theme, see the essay 'Can the Mosquito Speak', in Timothy Mitchell, *Rule of Experts: Egypt, Techno-Politics, Modernity*, Berkeley, CA: University of California Press, 2002: 19–53, and its reworking here in the history of the waterwheel in chapter 2.

Index